The Sky is Your Laboratory
Advanced Astronomy Projects for Amateurs

Robert K. Buchheim

The Sky is Your Laboratory

Advanced Astronomy Projects for Amateurs

Published in association with
Praxis Publishing
Chichester, UK

Robert K. Buchheim
Coto de Caza
California
USA

SPRINGER–PRAXIS BOOKS IN POPULAR ASTRONOMY
SUBJECT *ADVISORY EDITOR*: John Mason B.Sc., M.Sc., Ph.D.

ISBN 978-0-387-71822-4 Springer Berlin Heidelberg New York

Springer is part of Springer-Science + Business Media (springer.com)

Library of Congress Control Number: 2007923856

Cover design: Jim Wilkie
Cover image: Wally Pacholka
Project management: Originator Publishing Services Ltd, Gt Yarmouth, Norfolk, UK

Printed on acid-free paper

Astronomy is one of the few sciences where the experienced amateur can conduct original research, and contribute to professional studies. If you are wondering whether you can do something useful, valuable, and permanent with your observational skills, the answer is, "Yes, you can!" You *can* do this: you *are* capable of contributing to astronomical science.

This book will show you how to conduct observational research projects during your nights under the stars. I hope that it will motivate you to try.

Bob Buchheim
Coto de Caza, California, 2007

Contents

APPENDICES

Acknowledgments

I am grateful for the love, support, and encouragement that my wife, Eileen, has given me during our twenty years together. She has joined me on astronomical travels, spent nights alone while I was staring skyward, graciously gave up a portion of her garden for my backyard observatory, and was a constant cheerleader while I prepared this book. She is, in all ways, a wonderful partner.

Russ Sipe, John Hoot, Myke Collins, and Minor White introduced me to the world of amateur astronomical research. I hope that some of their energy and enthusiasm is reflected in this book. The Orange County Astronomers (OCA) dark-sky site and observatory have been important resources for me and for other research-oriented members of the OCA. I am grateful to the Trustees and Volunteers of the OCA for their many activities in support of amateur astronomy in southern California.

This book would not exist without the people who agreed to review and evaluate the rough draft material. Dr. Rick Feinberg's willingness to read the partially-complete first draft provided me with a great incentive to finish this project. I am particularly grateful to Dr. Tim Castellano, Brian Warner, Dr. David Dunham, Bob Gill, Lee Snyder, Bob Stephens, Dale Mais, and Brian Cudnick for the corrections and recommendations that they provided. Their efforts dramatically improved the quality of the final result. Thank you all for the time, energy, and knowledge that you shared with me to make this book a reality!

Bob Buchheim
Coto de Caza, California, 2007

Figures

TABLE

Abbreviations and acronyms

A/D	Analog-to-Digital
AAA	American Automobile Association
AAVSO	American Association of Variable Star Observers
ABG	Anti-Blooming Gate
ADS	Astrophysics Data System
ADU	Analog–Digital Units
AFOEV	*Association Française des Observateurs d'Etoiles Variables*
AGC	Automatic Gain Control
ALPO	Association of Lunar and Planetary Observers
alt	Altitude
AMS	American Meteor Society
ASA	American Standards Association (most often seen in reference to film speed)
ASAS	All-Sky Automated Survey
ASCII	American Standard Code for Information Interchange
ASL	Above Sea Level
AU	Astronomical Unit (the mean distance from Sun to Earth)
BAA	British Astronomical Association
CALL	Collaborative Asteroid Lightcurve Link
CBA	Center for Backyard Astrophysics
CBAT	Central Bureau for Astronomical Telegrams
CCD	Charge Coupled Device
CDS	Coronal Diagnostic Spectrometer
CHU	Call letters of the Canadian National Research Council Radio Station
comp	Comparison (as in "comp star")
DM	Bonner Durchmusterung star catalog
Dob	Dobsonian telescope

EFL	Effective Focal Length
FITS	Flexible Image Transport System (standard format for astronomical data)
FM	Frequency Modulation
FOV	Field Of View
FWHM	Full-Width-at-Half-Maximum
GCVS	*General Catalog of Variable Stars*
GMT	Greenwich Mean Time
GPS	Global Positioning System
GSC	*Guide Star Catalog*
HD	Henry Draper star catalog
HTML	HyperText Markup Language
IAU	International Astronomical Union
IBVS	International Bulletin of Variable Stars
ICRF	International Celestial Reference Frame
ICRS	International Celestial Reference System
ILOC	International Lunar Occultation Center
IM	Instrumental Magnitude
IMO	International Meteor Organization
IOTA	International Occultation Timing Organization
IRAF	Image Reduction and Analysis Facility
IRAS	InfraRed Astronomical Satellite
JD	Julian Date or Julian Day
JDSO	*Journal of Double Star Observations*
JPEG	Joint Photographic Experts Group (casually used as the name of a standard for coding images of photographic quality)
lat	Latitude
LINEAR	Lincoln Near-Earth Asteroid Research
LONEOS	Lowell Observatory Near-Earth Object Search
long	Longitude
LOS	Line Of Sight
MACHO	MAssive Compact Halo Object
mag	Magnitude
MIRA	Microcomputer Image Reduction and Analysis
m	Meter
MPC	Minor Planet Center
MPCORB	Minor Planet Center ORBital database of asteroids
MSFC	Marshall Space Flight Center
NABG	Non Anti-Blooming Gate
NASA	National Aeronautics and Space Administration (USA)
NEA	Near-Earth Asteroid
NEO	Near-Earth Object
NIST	National Institute of Standards and Technology (USA)

NTSC	National Television Standard Committee (television broadcasting system used in North America and Japan)
O – C	Observed minus Calculated
OGLE	Optical Gravitational Lensing Experiment
PAL	Phase Alternate Line (television broadcasting system used in much of Europe)
PC	Personal Computer
PDT	Pacific Daylight Time
PROBLICOM	PROjector BLInk COMparator
PST	Pacific Standard Time
RA	Right Ascension
RANZ	Royal Astronomical Society of New Zealand
RASC	Royal Astronomical Society of Canada
RMS	Root-Mean-Square
RXTE	Rossi X-ray Timing Explorer
SAS	Society for Astronomical Sciences
SBIG	Santa Barbara Instruments Group (manufacturer of astronomical CCD imagers)
SCT	Schmidt–Cassegrain Telescope
sec	Second
SGS	Self-Guided Spectrograph (SBIG product name)
SNR	Signal-to-Noise Ratio
SOHO	Solar and Heliospheric Observatory
TAI	International Atomic Time (in French *Temps Atomique International*), also known as Atomic Clock Time
TASS	The Amateur Sky Survey
TIFF	Tagged Image File Format
topo	Topographic (as in "topo map")
UBVRI	Ultraviolet-Blue-Visual-Red-Infrared
UCAC	*USNO CCD Astrographic Catalog*
USB	Universal Serial Bus
USGS	United States Geological Survey
USNO	United States Naval Observatory
UT	Universal Time
UTC	Coordinated Universal Time
VCR	Video Cassette Recorder
VLBI	Very-Long Baseline Interferometry
VSX	Acronym for AAVSO's Variable Star indeX
WDS	*Washington Double-Star Catalog*
WFOV	Wide Field Of View
WGS	World Geodetic System
WGS84	World Geodetic System 1984

WWV	Call letters of US National Institute of Science and Technology radio station
WWVH	Call letters of US National Institute of Science and Technology radio station
ZHR	Zenith Hourly Rate

Products and trademarks

Product or trademark	Company
Excel	Microsoft Corp.
TheSky	Software Bisque
Deep Sky	David Chandler
SkyMap Pro	Chris Marriott
Magellan	Magellan Corp.
AstroArt	MSB Software
CCDSoft	Santa Barbara Instruments Group and Software Bisque
MPO Canopus	Brian Warner (Bdw Publishing)
MPO PhotoRed	Brian Warner (Bdw Publishing)
IRAF	National Optical Astronomy Observatories
Kiwi OSD	PFD Systems
Macintosh	Apple
Windows, DOS	Microsoft Corp.
NexStar	Celestron
ST-8XE	Santa Barbara Instruments Group
MaximDL	Diffraction-Limited
TimeCube	Tandy Corp.
Peranso	Tonny Vanmunster
Telrad	Telrad, Inc.
Star Spectroscope	Rainbow Optics
Needle Eye	Van Slyke Engineering
Thinkpad	IBM

Introduction

The foundations of astronomy were laid by a cast of dedicated observers, craftsmen, and thinkers, many of whom were, in a sense, "amateurs". Their scientific efforts were driven by a thirst for knowledge and the joy of discovery, and they funded their research from resources provided by their non-astronomical "day jobs". Galileo built his own little telescopes, and saw things that no man had seen (and few had imagined), while scheming for a salary increase for his job as professor of mechanics. Copernicus developed the idea that the Sun was the center of the Solar System during the time he could spare from his duties as Administrator of the Diocese of Allenstein and advisor to the Prussian government on monetary matters. Herschel was, for many years, a musician who followed the stars as his avocation. Isaac Newton was paid to be a lecturer and professor, and spent his nights filling notebooks with insights that would become *Principia Mathematica*, containing the mathematical basis of orbital mechanics. Curiosity and a razor-sharp mind prompted him to invent the telescope that is now known by his name. Ah, those must have been heady times to live through!

Today, when ground-based telescopes are behemoths with mirrors over twenty feet across, their images enhanced by adaptive optics and ten-million-pixel electronic imagers; when spacecraft visit other worlds to touch and analyze their environments; when astronomers are funded by governments and taxpayers; in this world, is there actually a niche for the backyard star-gazer to conduct meaningful research? That's a fair question.

Can the amateur astronomer do real science? The answer is absolutely, yes [1]. I offer as evidence my paraphrase of the testimony of Dr. Rick Feinberg (Editor in Chief of *Sky & Telescope* magazine). After considering the question, Mr. Feinberg concluded [2]:

> Yes, the amateur astronomer can do good science. Experienced amateur astronomers have some very real advantages in this regard. They have intimate knowledge of the night sky, potentially unlimited telescope time, and access to

reasonably priced, very effective telescopes and CCDs. But for an amateur to do real science, he or she must also have the necessary knowledge of operational procedures and statistics—or collaborate with someone who does. For example, a diligent amateur can generate data that will help understand the evolution of a binary-star system, or the physical properties of an asteroid. Real science requires real rigor in the procedures that are used to gather the data. Sensors must be characterized and calibrated, relevant observational parameters (e.g., airmass and spectral band) must be recorded, and their effects incorporated in the data reduction. Statistical assessment of the accuracy and precision of the data is critical to interpreting the significance of the results.

Are you a 21st century amateur astronomer who feels a twinge of jealousy when you read about the 18th and 19th century astronomers who made headline discoveries just by carefully looking into an eyepiece? Do you sometimes wonder if any of your astro-images might have scientific value, in addition to their aesthetic appeal? If so, then this book is written for you.

Are you a student in search of a research-oriented astronomical experience? You may need help in defining a project, making the observations, reducing and analyzing the data, and reporting your results. Or perhaps you are that student's professor, in need of a "laboratory manual" that will bridge the gap between qualitative amateur astronomy, and quantitative professional research methods. This book is written with both of you in mind.

A wondrous thing has happened in the early years of the 21st century: high-quality, sophisticated telescopes are now commercially available, and surprisingly-affordable CCD imagers are popular among astrophotographers. Powerful computers are common household appliances which, with a few hundred dollars worth of software, can perform data analysis that would have made a professional astronomer drool only a decade or so ago. With a little electronic skill, the observer can also press that computer into service as an observatory controller, commanding the telescope and imager to take data all night while the owner/astronomer still gets a full night's sleep, and arrives refreshed at his "day job".

If you share the dedication to science for the sheer pleasure of learning and discovery that characterized the "gentlemen scientists" of the 18th century, you can continue in their tradition, now enhanced by 21st century scientific technology. It is a fabulous time to be an amateur astronomer!

REFERENCES

[1] Offutt, W.B. "Brief case histories of five successful Professional–Amateur Collaborations", in Percy, J.R. and Wilson, J.B. (eds.), *Amateur–Professional Partnerships in Astronomy*, Astronomical Society of the Pacific Conference Series, Vol. 220, San Francisco (2000).
[2] Feinberg, Rick, "Can Amateur Astronomers Do Good Science?" keynote presentation at 2003 IAPPP Symposium on Telescope Science, Big Bear, CA (2003).

1

Meteor studies

1.1 INTRODUCTION

Some of my favorite memories are of dark, crisp nights with the stars shining steadily against the blackness, so close I can almost touch them. Coyotes howl in the distance, tiny creatures are rustling in the bushes nearby, when suddenly out of the celestial darkness the corner of my eye catches a little spark. In an instant a "shooting star" makes a quick streak spanning ten or twenty degrees, flashes, and disappears. A particularly bright meteor might leave a thinly glowing trail, faintly visible after the meteor itself is gone. A real fireball might take a few seconds to streak from one horizon to the other, dropping occasional sparks or splitting into two or more pieces before its terminal flash. During the Leonid meteor shower of 2003, with the constellation Leo near the zenith, I watched a nearly head-on meteor flash and explode. It left a ghostly glowing smoke-ring that remained clearly visible for 5 minutes, as it slowly expanded and dissipated.

Particles ranging in size from dust-motes to sand-grains to pebbles—and the occasional boulder—are scattered throughout the inner Solar System. They are occasionally swept up by the Earth as it moves on its orbital path around the Sun. The Earth's thick atmosphere is our primary protection from these interplanetary visitors. The velocity with which they "hit" the atmosphere is huge—typically tens of miles per second. The resulting frictional heating vaporizes virtually all of the small particles. The rapid deceleration and thermal stresses shatter the larger ones, so that almost all of their pieces are also vaporized. The vaporized particle leaves in its wake a thin "tunnel" of ionized gas. It is the glowing ionized gas that we actually see as a "shooting star" or meteor.

The particles ("meteoroids") are orbiting the Sun, and many are in orbits that are very elliptical, betraying their relationship to comets. As a comet sheds particles, those particles continue to follow the comet's orbital path quite closely. However, since each particle departed from the comet with a slight relative velocity, and these departure velocities are random, the particles gradually spread out along the comet's orbit, and become a broad stream of debris. If this stream happens to intersect the Earth's orbit, then we'll see a "meteor shower". If you consider the vector-sum of the Earth's velocity and the meteoroid's velocity, as shown in Figure 1.1, you'll find that all of the meteoroids will appear to come from the same direction in the sky. This direction is called the "radiant" of the shower. The best-known meteor showers are identified by the constellation in which this radiant resides: the Geminids (December 12–13) appear to come from the constellation Gemini, the Perseids (August 12) appear to come from the constellation Perseus, etc. Since the Earth circles the Sun once per year, we return to the point where Earth passes through the meteor stream on the same day each year. Hence, a meteor shower is normally an annual event. ("Normally" rather than "always" because some meteor streams are dense—and the shower strongly active—only when the parent comet passes near Earth.)

In addition to shower meteors, there is a background of meteors that do not appear to be associated with a known, active shower. These are called "sporadic" meteors. The rate of background sporadic meteors is about a half-dozen per hour, more or less, depending on the time of night and the time of year. The rate of sporadics is generally lower between November through June, and a bit higher in July through October. The identification of a meteor as "sporadic" is a bit ambiguous, because it can depend on the observer and the project. If you are making a count of the Geminid meteor shower, for example, then any meteor that isn't clearly a Geminid is, for you, a "sporadic". However, on the same night, your friend might be

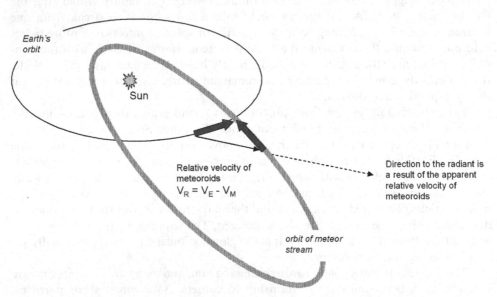

Figure 1.1. The direction of the "radiant" of a meteor shower is defined by the Earth's and the meteoroid stream's velocity vectors.

monitoring the Leo Minorid shower (a less-active shower, of mostly fainter meteors, that peaks at the same time as the Geminids). Some of your "sporadics" may very well be Leo Minorid "shower meteors" for your friend. A more experienced meteor observer might spend the night keeping separate counts of the Geminids and the Leo Minorids; in which case only true "loners" would be counted as sporadics. So, the identification of a meteor as a "sporadic" is an observational definition, not a fundamental property of the meteoroid itself.

Why is it important to monitor meteor activity? First, of course, is the pure scientist's curiosity about the nature of the things in the universe. Careful observers can characterize the density of the debris stream that comprises a meteor shower, and may discover previously unknown shower streams. Second is the practical benefit that this curiosity can have. We are dependent on spacecraft for a variety of important purposes (weather forecasting, telecommunications, resource monitoring), and there have been quite a few events in which satellites were damaged or destroyed by meteor impacts. Meteor activity must be understood so that (a) satellites can be constructed to survive the expected flux of meteor particles, or (b) spacecraft can be launched on schedules and trajectories that will avoid the most dense streams of meteor particles. Third, there are a variety of scientific objectives (meteor stream orbit determination, meteor composition determination, and association of meteor streams with their parent comets) that depend on accurate knowledge of the meteor activity so that detailed observations can be scheduled for a time and location that maximizes their probability of success. Amateur astronomers have historically been important practitioners of meteor observation. We have the time, the knowledge of the sky, and the curiosity to conduct careful observations of "shooting stars" for the benefit of science and technology.

If you've ever watched a meteor shower, you've observed two critical features that may not come across clearly in some written descriptions of these events:

- The meteors appear to "come from" the radiant in a special sense, illustrated in Figure 1.2. The meteor can appear anywhere in the sky, but it will always travel away from the radiant. If you trace the path of the meteor backwards, the extended path will pass through the radiant. If you trace the paths of several meteors backwards, then the location where their extended paths intersect defines the location of the radiant.
- The action is decidedly slow, except in the most unusual meteor storms. For example, the Geminid meteor shower usually displays a peak "zenith hourly rate" of about 50 meteors per hour. At first blush, that sounds pretty spectacular, until you do a little math. Even if you actually experience the Zenith Hourly Rate (which you probably won't, as explained below), 50 meteors per hour is roughly 1 meteor per minute—not exactly a fireworks show! So, a dedicated meteor-shower monitoring expedition will, at times, be almost exciting as watching grass grow.

The "Zenith Hourly Rate" (ZHR) is the theoretical rate of meteors that an individual observer would see if the radiant were at the zenith, the sky completely unobstructed

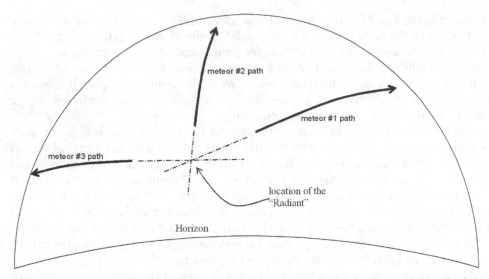

Figure 1.2. The "radiant" of a meteor stream is the point in the sky from which all of the meteors appear to have originated.

and free of light pollution, and the observer has excellent vision (magnitude 6.5 stars visible).

   If the radiant isn't at the zenith, or the sky or observer conditions aren't perfect, then the actual observed hourly rate will be lower. Why? Imagine that the radiant were exactly at the horizon. A meteor that traveled "upward" would be visible to you. But a meteor that started at the horizon and traveled "downward" would never show itself to you. Similarly, if there were scattered clouds, or obstructions such as mountains, trees, or buildings that blocked your view of some parts of the sky, then you'd miss any meteors whose paths were hidden behind them. You know that you can see many more stars from a high, dark mountain top than you can from a city park. The combination of light pollution's veiling glare and atmospheric extinction has the same effect on meteors. If you are observing from a less-than-pristine dark sky site, you'll miss the fainter meteors. Finally, if the observer is impaired by cold or wind, or distracted by additional tasks (such as recording observations), then fewer meteors will be observed.

   All of these effects are accounted for in the definition of ZHR. Suppose that you have observed N meteors in a time period Δt. The defining equation is [1]:

$$\text{ZHR} = k \cdot c_p \cdot (N/\Delta t) \cdot r^{(6.5-Lm)} \cdot [\sin(h_r)]^{-\gamma}$$

where $\Delta t = (T_{\text{end}} - T_{\text{start}} - T_{\text{dead}})$ is the effective duration of observing.
   T_{start} = the starting time of the session.
   T_{end} = the ending time of the session.
   T_{dead} = the accumulated "dead time", during which you were doing some-
        thing other than monitoring the sky (e.g., recording your notes or

resting your eyes). Δt must be greater than 0.4 hours, with dead time less than 20%, to provide a meaningful hourly rate.

K = fraction of the sky that was visible. This factor accounts for obstruction of the sightline (by clouds, hills, etc.).

N = number of shower-member meteors observed in the observing time interval (i.e., sporadics are not counted in N).

Lm = limiting magnitude. A limiting magnitude of 6.5 is assumed to represent "perfect conditions". This factor accounts for less-than-perfect sky conditions (extinction and/or light pollution).

h_r = altitude of the radiant ($h_r = 90$ degrees if the radiant is at the zenith).

c_p = observational bias factor (for a standard observer, not influenced by others, $c_p = 1.0$).

$\gamma = 1$ for visual observations.

r = the "population index" of the meteor shower. For most meteor showers, r is in the range $r = 2.5$ to 3.5. By recording the magnitude of each meteor during the session, the observer can contribute to better determination of the population index for the meteor shower being monitored.

This is the formula that the collecting organizations (e.g., the American Meteor Society and the International Meteor Organization) will use to analyze your meteor count records. With this formula, each observer's data is transformed into a consistent metric of ZHR. It's worthwhile to use this equation to understand what you can expect to see, under realistic conditions:

$$(N/\Delta t) = \frac{\text{ZHR} \cdot \sin(h_r)}{c_p \cdot r^{(6.5 - Lm)}}$$

With the radiant low in the sky, your observed meteor rate will be significantly lower than it would be if the radiant were higher in the sky. For example, when the radiant is 30 degrees above the horizon, the meteor rate is only one-half of what is expected when the radiant is at the zenith [because $\sin(30°) = 0.5$]. Note that the factor $\sin(h_r)$ is only meaningful if the radiant is above the horizon (i.e., $h_r > 0$). Theoretically, you can observe a few shower meteors even when the radiant is slightly below the horizon, but as a practical matter you shouldn't start counting shower meteors until the radiant is at least 10 degrees above the horizon.

The factor involving limiting magnitude has a very potent effect on the observed meteor rate. For example, if the limiting magnitude is $Lm = 6.0$, and $r = 3$, then the observed rate is less than 60% of what it would be under "perfect" ($Lm = 6.5$) conditions. As a practical matter, this correction factor is only useful if the limiting magnitude is better than 5.0. When conditions are worse than that, meteor counts will not be trustworthy.

The amateur scientist can contribute to several important subjects by monitoring meteor activity:

- Determining the characteristics of the "meteor stream" of the known showers.
- Determining the ZHRs and radiants of low-density, poorly-observed meteor streams and sporadic meteors.
- Discovering previously-unknown or only-suspected meteor showers.
- Measuring the altitude of the meteors.
- Monitoring "daytime" meteor activity.

These projects are listed (approximately) in order of increasing difficulty.

1.2   PROJECT A: VISUAL COUNTS OF MAJOR METEOR SHOWERS

This project is the best way to become experienced at meteor observing, by gathering relatively simple, but quite useful, data. It focuses on observing the rate of meteors in the major meteor showers, within a day or two of their predicted peaks. These are: the Quadrantids (January 4th), the March Geminids (March 22nd), the Delta-Aquarids (July 28th), the Orionids (October 20th), and the two most reliable showers of each year: the Perseids (August 12th) and the Geminids (December 14th).

Think about the process of creating a meteor stream—a comet sheds gas, dust, and larger particles along its orbit—and combine it with what you know (and may have observed) about comets. When the comet is far from the Sun, it is pretty much a frozen, inert object. As it comes closer to the Sun, the most volatile components begin to sublime. For example, if there's any methane on board, it will turn from solid to gas when the comet is at roughly the orbit of Jupiter. Activity gets more violent the closer the comet is to the Sun, with the melting or subliming ices carrying or blasting solid bits of sand, rock, and dust away from the comet. All of these are somewhat random processes. A particularly violent jet may be actively blowing material from the comet for a few days, and then "shut down". So the stream of material from a single orbit of the comet is not expected to be a nice, uniform band. Instead, it will be punctuated by dense regions and sparse areas. Since the particles left the comet with some velocity, they don't travel in exactly the same orbit as the comet. Hence, in addition to its initially random density distribution, the cometary stream will widen and disperse as time passes. Each orbit of the comet injects new material into the stream, so the gradual evolution of the comet's orbit under a variety of forces results in a very complex pattern of particle density. That particle density is the prime factor in determining the rate at which we see meteors when the Earth passes through the stream.

Careful observation and recording of the meteor rate can provide a surprising variety of information. The concept is simplicity itself: You select one of the known major meteor showers (e.g., the Perseids, that peak on August 12–13 each year). Beginning a few nights before the predicted peak, and continuing for a few nights past the predicted peak, you spend each night counting the number of Perseid meteors. Throughout the night, you record the total number of Perseids observed in each 10 or 15-minute interval. If all goes well, you can plot your data in a form similar to Figure 1.3, to display the density of the meteor stream over time. If you provide

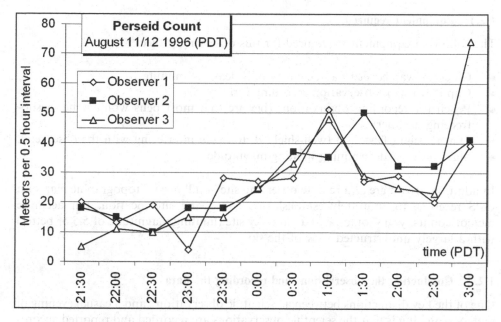

Figure 1.3. Typical nightly meteor-count of a major meteor shower.

your data to one of the central collecting organizations, it can be combined with that of other observers from around the world to map a complete picture of the part of the meteor stream that Earth passed through. Combining many years' worth of world-wide data enables astronomers to map and understand the meteor stream's density pattern. That information may give some indication of the history of the comet that spawned these meteors.

Staying up all night recording meteors can be a trying chore if you're all alone. You'll always wonder if you missed a meteor while you were looking at your note-book to record your observations, and you'll get tired and dry-eyed (and hence become a poorer observer) as the night progresses. For these reasons, a meteor-count is a wonderful group activity. I've done a few with junior-high students, who seemed to enjoy the prospect of staying up all night with their friends, and who managed to learn a bit of astronomy and mathematics along the way. I break the gang up into groups of four or five students. Each group consists of one "recorder" plus three or four "observers", and each group is assigned a one-hour observing period. During their assigned period, the "observers" are arranged in a circle, facing outward. Each observer keeps his or her individual count of meteors observed. The "recorder" keeps track of the time. Every 10 minutes, the recorder polls the observers for their counts of "shower" and "sporadic" meteors during the preceding 10 minutes and records the data. During the final 10-minute interval of their assigned hour, the recorder rouses the next group to prepare for their shift, records his group's final reports, and hands the data sheets off to the next group's recorder.

1.2.1   Equipment required

The following equipment is required for this observing project:

- Clock or watch, set to an accuracy of at least ±1 minute.
- Clipboard with observation-recording forms.
- Pencil for recording observations (beware that most pens won't write in sub-freezing weather).
- Dim red-light flashlight (well-shielded, to avoid interfering with the observers).
- Star chart for determining limiting magnitude.

In addition, if you are at a remote observing site you'll need a topographic map or a GPS receiver, to accurately establish your observing site location. For accurate meteor counts, you should be at a dark-sky site (limiting magnitude of 5.5 or better) with a largely unobstructed view of the sky.

1.2.2   Conducting the observations and recording the data

One of the key distinctions between a "scientific observation" and a casual evening of meteor watching is that the scientific observations are recorded and reported according to standardized methods. In that way, one observer's results can be combined with those of other observers with confidence that all reports have the same meaning, adjustments for special factors (such as different limiting magnitudes) can be properly made, and the sum of everyone's observations comprises a valid statistical database of meteor activity. Two casual meteor watchers might say, "Wow, the meteor shower was really active!" and "Yes, particularly after midnight." The scientist needs to be able to translate his or her observations into standardized, quantitative results, and report that "the rate peaked at 40 per hour between 03:00 and 04:00, after discounting sporadic meteors that weren't members of the shower."

In order for your data to be properly combined with that of other observers, your observations and data record must be made in accordance with certain agreed-upon standard methods. By following these standards, you make observations that will be useful to the scientific community. The key standards defined by the American Meteor Society are as follows:

- Each observer acts independently of the others. This is true even if you observe as a member of a group: each individual's observations are treated in isolation, and reported separately. In a group (such as the student groups I mentioned above), you want to take care that the observers don't influence each other's counts (e.g., by excitedly asking "Hey, did you see that one?"). Sometimes, two observers will see the same meteor. That's fine: each observer adds it to his/her count. (Remember the definition of ZHR considers the number of meteors per hour seen by a *single* observer.)
- Each meteor must be identified as being either (a) a member of the shower under observation, or (b) a sporadic meteor not associated with the shower under

study. The "member" meteors will all appear to originate from the shower's radiant. Any meteor whose path can't be traced back to the radiant is considered to be a "sporadic" meteor.

- The observing conditions must be properly recorded in order for the raw observations to be translated into ZHR. The important conditions include:

  o Date and time of night: Be sure to indicate if this is local time, daylight saving time, or some other such as Universal Time. It's really embarrassing to be going over the data a couple of years later and not be sure what your time-record means! For astronomical projects, recording time in UTC (Coordinated Universal Time) is preferred. See Appendix A if you're not familiar with UTC.

  o Location: This should include both the address (town, state, and country) and the site's latitude/longitude/elevation coordinates. The coordinates are easy and convenient to get from an inexpensive GPS receiver. Be sure to include in your notes whether you're at north or south latitude, and east or west longitude!

  o Sky conditions: The percent of sky blocked by clouds or other obstructions (this may be different for each observer if they're looking in different directions), and the limiting visual magnitude (a parameter that accounts for both sky clarity and light-pollution) should be recorded. If the conditions change during the course of the night, then your observing notes should include the changes.

  o Direction of the center of your "gaze". Human eyes can reliably monitor a limited region of the sky, roughly a 50-degree cone. Therefore, during each hour of the observing session, you will want to select a region of sky to monitor, and record the direction of the center of your gaze (preferably as RA and Declination).

  o "Dead time": Any observer's effectiveness will degrade after an hour or so of continuous monitoring and recording. Therefore, a 5-minute "break" every hour is a wise thing to include in your plan. Obviously, you won't see/record any of the meteors that happen during that time. Similarly, the time spent recording your notes isn't effective "observing time". The sum of breaks, recording, and other interruptions is called "dead time", and should be recorded in your notes so that it can be subtracted from your total observing time during data reduction.

Figure 1.4 shows a format for your data sheet that will ensure that you record all of the information that is required by the American Meteor Society to publish and make use of your observations.

If you have organized a group of observers, then you should use one sheet per observer. Each observer's count is gathered separately; and each observer records the fraction of clear sky and limiting magnitude for the area of sky that he/she is monitoring. The coordinator then organizes and reports all data sheets from a single expedition.

Name: _____
Address:_____   *(your mailing address)*
Date:_____   *(month-day-year;*
                                          use two-day format if observations span midnight)
Type of time used:_____   *(specify EST, EDT, UTC, etc.)*
Observing location: _____   *(longitude, latitude; or town, county, state, country;*
                                          specify N or S latitude and E or W longitude)

observing interval:		total # meteors	# shower meteors	# sporadic meteors	fraction of clear sky	limiting magnitude
beginning	**ending**					

Figure 1.4. Reporting format for simple meteor count of a major/active meteor shower. (Used with the kind permission of the American Meteor Society)

The practice of distinguishing "shower member" vs. "sporadic" meteors is critical to accurate monitoring of meteor shower activity, especially when shower activity is modest. A shower meteor can be identified by evaluating three features:

- Its path, if extended backward, passes through the radiant. This is the most fundamental criterion—any meteor whose trail can't be extended backward to the radiant isn't a shower member.
- Shower members all have nearly the same speed. A meteor whose path is close to that of shower members, but which is markedly slower or faster than shower members seen in the same part of the sky, should be counted as a sporadic. (This "same speed" rule is derived from the fact that all shower members are traveling along essentially the same orbit, hence intersect the Earth with identical velocities. It has to be applied by comparison to obvious shower members *in the same part of the sky*, because the apparent speed of a meteor is affected by your perspective: if it's coming almost directly toward you, then it will appear to move slowly, but if it's traveling past you, it will appear to move more rapidly.)
- The length of a shower-member meteor's path is related to its distance from the radiant. Shower members that appear close to the radiant will have short paths. Members that are seen far from the radiant will have longer paths. The most extreme case is a meteor that is aimed directly at you: if it's a shower member, then it will appear as a brief flash right at the radiant (a "point meteor"). As a very rough rule of thumb, a meteor's path will be roughly half as long as its distance from the radiant.

Near the peak of the major showers, tracing the meteor's path backward will be adequate to determine if it's a "member" or a "sporadic". When the action is slower (i.e., a few days before or after the predicted peak), you'll need to apply all three criteria to distinguish "members" from "sporadics". As a practical matter, after studying a couple of the major showers (e.g., the Perseids or Geminids), you'll find that it isn't too hard to make the distinction.

1.2.3   Reducing, analyzing, and submitting your results

With your campaign of meteor observing complete for a particular shower (which may be a single night, or several nights bracketing the build-up, peak, and wind-down of the shower), your primary responsibility is to double-check that your observing notes and data sheets are correct and complete. You may wish to do some graphing of your data, to get a picture of the hourly rates that you observed, applying the ZHR equation, but that isn't required.

The central collecting organization for meteor observations in the USA is the American Meteor Society. Address your written report to:

> AMS Visual Program Coordinator
> Kim Youmans
> 556 Maurice Drive
> Swainsboro, GA 30401
> E-mail: *meteorsga@bellsouth.net*

and include copies of your filled-in data sheets (per Figure 1.4). I recommend also checking the AMS website, in case the Visual Program Coordinator has changed. Their website is: *http://www.amsmeteors.org/*

European observers should report their results to the International Meteor Organization, in care of:

> IMO Visual Meteor Database
> Rainer Arlt
> Freidenstr. 5
> D-14109 Berlin
> Germany
> E-mail: *visual@imo.net*

The AMS and IMO will calculate the ZHR from your observations, and add your report to those submitted by observers around the world, to characterize the density of the meteor stream.

1.3   PROJECT B: CHARACTERIZATION OF MINOR
METEOR SHOWERS

In the previous project, we assumed that the meteor shower would keep you busy enough that you'd have time for little more than counting shower and sporadic meteors. For the major showers (e.g., the Perseids and Geminids), that is often a good assumption. But what about the sparse showers? Most years, the minor showers show ZHRs of fewer than 15 meteors per hour. That's an average of one meteor every 4 minutes or so, which gives you time to observe and record additional data about each individual meteor. The useful additional data, roughly in order of value, includes:

- The exact time of each meteor.
- The brightness (peak magnitude) of the meteor.
- Unique characteristics such as color, speed, duration (tenths of a second), length (degrees from beginning to end), and existence of a persistent train.

Of particular interest is the time and magnitude of each meteor.

By recording the exact time of each meteor, it may be possible to correlate your observations with those of other observers in the same region who saw the same meteor.

By recording the magnitude of each meteor, you are gathering data that will enable the analysts to validate the "population index" of the shower (or determine it for the first time). Since the meteors in a single shower stream all hit the Earth's atmosphere with the same velocity, and they are assumed to be of similar composition, their brightness is primarily determined by their size. Therefore, the population index contains information about the size distribution of the particles in the meteor stream. The greater the population index, the greater the proportion of tiny particles in the stream. As the meteor stream ages, we expect that the smallest particles are most likely to be scattered out of the main stream (due to a variety of subtle effects that preferentially alter the orbits of smaller particles). Therefore, we expect that a low population index should be a characteristic indicator of an old meteor stream. Estimating the magnitude of a meteor is a tricky skill, but one worth learning. Careful observations by experienced meteor observers are needed to gain understanding of the population index of meteor showers, particularly the less-frequently-observed minor showers.

1.3.1   Equipment needed

The equipment needed for this project is identical to that for meteor counting of major meteor showers, with the addition of a good star chart of the region you'll be facing to help you judge the brightness of meteors by comparing them with nearby stars:

- Clock or watch, set to an accuracy of at least ± 1 minute.
- Pencil and clipboard with observation-recording forms.
- Dim red-light flashlight.
- Star chart for determining your limiting magnitude (Lm) and the magnitude of field stars that you'll use as comparison stars to estimate the meteor's magnitude.

In addition, if you are at a remote observing site you'll need a topographic map or a GPS receiver, to accurately establish your location. You will want to find a dark-sky site with a largely unobstructed view of the sky and limiting magnitude of 5.5 or better.

As part of your preparation, be sure to check a complete listing for known and suspected meteor showers that may be active on the night of your observations and take note of their radiant positions. This will help you make the critical distinction between "shower member" and "sporadic" meteors that you observe. The American Meteor Society website (*http://www.amsmeteors.org*) maintains a useful list of "minor" showers and their characteristics.

1.3.2   Conducting the observations and recording the data

Early in the evening, select a convenient direction for your gaze and spend some time studying the constellations and asterisms, and learning the stellar magnitudes of a range of stars. This will help you to make "on the fly" estimates of the brightness of the meteors. Your goal is to estimate the peak brightness of each meteor to within a half-magnitude, by comparing it with known stars. Get comfortably seated and check the arrangement of your equipment, to be sure it's easily accessible. Also check your site's limiting magnitude, and record both the limiting magnitude and the time of your estimate in your notebook. When you're ready to begin observing, record your starting time in your notebook.

The sequence of observation for each meteor will be as follows:

- When you detect a meteor, follow it for its full duration.
- Take note of its path (is it a member of a known low-rate shower?).
- Compare its peak brightness with convenient, recognizable stars whose magnitude you know, or that are listed on your star chart.
- Pay attention to any special features such as color, a persistent train, or unusual speed.
- Check the time.
- Record the data on this meteor on your observing record.

The AMS-recommended record form for this project is shown in Figure 1.5. The top portion of the form provides the same information that you recorded for the simple meteor count (observer, location, and sky conditions) plus the direction of your primary gaze (to help correlate your results with those of other observers). The bottom portion provides a table for you to record the details of each meteor:

- Time of appearance (to the nearest minute).
- Visual magnitude at brightest (to whole magnitude).
- Color (record only if it was definitely perceivable).
- Type (name of shower, if a member; or "sporadic". This distinction is absolutely essential, even if other data isn't recorded!).
- Speed (slow, medium or fast: this will help distinguish different families of sporadics from shower members).
- Train (note if a persistent train was observed).
- Accuracy (how certain are you about this observation—e.g., if you caught the meteor out of the corner of your eye, it may be uncertain).

VISUAL METEOR OBSERVING FORM

DATE: ____ (year) ____ (month) ____ (day) Begin ____ h ____ m End ____ h ____ m (____)
LOCATION: Long = ____________ W. Lat = ____________ N. Elevation = ____________ m

OBSERVER: ______________________ PLACE: ______________________

LIMITING MAGNITUDE: ____@____:____ ____@____:____ ____@____:____ ____@____:____

____@____:____ ____@____:____ ____@____:____ ____@____:____ ____@____:____ ____@____:____

PERCENT CLOUDY: ____%@____:____ ____%@____:____ ____%@____:____ ____%@____:____ ____%@____:____

DIRECTION FACING & ALTITUDE: ____@____:____ ____@____:____ ____@____:____ ____@____:____

BREAKS: ______________________

COMMENTS: ______________________

NO.	TIME	MAG.	COLOR	TYPE	SPEED	TRAIN	ACCURACY	REMARKS
1								
2								
3								
4								

Figure 1.5. AMS visual meteor observing form. (Used with the kind permission of the American Meteor Society)

Every hour or so, re-check the limiting magnitude and note it on your observing record.

1.3.3 Reducing, analyzing, and submitting your results

Check your observing record for completeness, and then submit a copy of it to your coordinating organization (either AMS or IMO). They will incorporate your observations into their database, and use them for characterization of the meteor shower's activity. Forms should be submitted to the same address as given in Section 1.2.3.

As in the simple meteor count project, if you are mathematically inclined, you may want to use the ZHR equation to estimate the activity profile of the shower you monitored. You may also want to apply the "population index" equation to see where your data falls on that parameter. These calculations are strictly for your interest, education, and amusement. They need not be submitted to the AMS. For the minor showers, where you may gather data on only a few dozen meteors per night, do not be surprised if your estimated ZHR and population index are significantly different from the "generally accepted" values. The reason that your observations alone may not give believable rates is not necessarily a fault in your results, or in your procedures. The issue is small-sample statistics.

The arrival of meteors is a random process. Meteor arrivals, and many other random processes, follow "Poisson statistics". The fundamental feature of this sort of random process is that the probability of seeing an event during time interval dt is $dp = \lambda\,dt$. In this equation, λ is a constant that specifies the average number of events per time period (i.e., for the case of meteor counting the observed number of meteors per hour). Assume that meteor arrivals and detections follow this equation, and that during a one-hour period, you observe N meteors. What does your observation tell you about the underlying "true" rate of meteors (which is a measure of the density of the meteor stream)? Let's do a thought-experiment. Suppose that you could repeat that same hour of measurement, under identical conditions, many times over. Call each repetition of that hour one "realization" of it. Since each "realization" is a sample from a random process, you don't expect to observe exactly N meteors time after time. In some realizations, you would observe a few more than N meteors, and in other realizations you would observe a few less meteors, because of the randomness of meteor arrival and detection.

A common way of describing the variation in such a series of observations is the "standard deviation", σ. The standard deviation is a measurement of the confidence that you can have in your measurement of a random process. You can usually expect that in more than half of your realizations, the observed number of meteors will fall within $\pm 1 - \sigma$ of the mean. In virtually all of your realizations, the observed number of meteors will fall within $\pm 2 - \sigma$ of the mean.

For a Poisson process, it turns out that $\sigma = \sqrt{N}$. So, if you measure 100 meteors per hour (during a major, active shower), you expect that the standard deviation of this measurement is $\sigma = \sqrt{100} = 10$. Using the "two-sigma" criterion, you can be reasonably sure that the "true" underlying average meteor rate is 100 ± 20; that is, the "true" underlying rate is somewhere between 80 and 120 per hour.

Now, consider the case where you measure only 16 meteors in an hour of observing a minor shower. Then, the standard deviation of your measurement is $\sigma = \sqrt{16} = 4$. Using the "two-sigma" criterion, your measurement gives you confidence that the "true" underlying meteor rate is between 16 ± 8; that is, it is somewhere between 8 and 24 meteors per hour—quite a broad range of uncertainty! The general rule is that as the number of events recorded gets smaller, the percentage uncertainty in the estimated rate gets much larger.

This problem of small-number statistics is one reason that it is important to have organizations such as AMS and IMO, who will consolidate the observations from many observers. By combining many observations, they increase the total number of meteors observed, and thereby improve the quality of the statistical data. Of course, this combining of information only works if all observations are made in a consistent way, according to the same observing procedures and all accompanied by relevant descriptions of observing conditions to facilitate merging of data sets from different observers. That is why it is so important to follow standardized observing procedures very carefully. If, for example, one observer neglects to distinguish between "shower member" vs. "sporadic" meteors, then his/her count will not be comparable with those of the diligent observers who did make the distinction. Be a careful, diligent observer!

The population index of a meteor shower

The "population index" of a meteor shower, the parameter "r" in the formula for ZHR, appears in the term that includes your limiting magnitude at the observing site. The population index quantifies how many "faint" versus "bright" meteors a particular stream contains. In order to determine the value of the population index for a given meteor shower, observers record the brightness of each meteor seen. After all observers' data is collated and normalized, the cumulative distribution function of brightness is a plot of the number of meteors brighter than magnitude M on the y-axis, vs. M on the x-axis. When the distribution function is plotted on a log-linear graph, it is a straight line. The slope of this line is the parameter r.

Suppose that a total of 1,000 meteors, down to magnitude 6, are observed in a particular shower. How many "bright" meteors are expected in this sample? The graphs below show theoretical plots for population indices $r = 2$, $r = 2.5$, and $r = 3$. Note that if r is large, then the meteor shower is dominated by faint meteors. Conversely, if r is small, then there is a greater percentage of bright meteors. A low population index is characteristic of an old meteor shower, dominated by large particles.

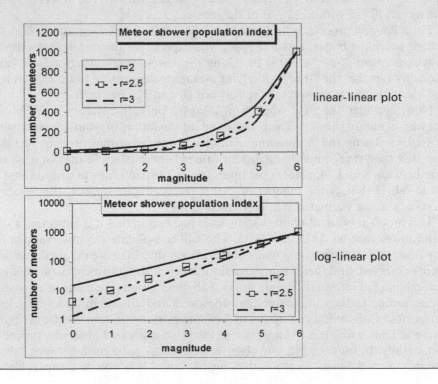

1.4   PROJECT C: CHARACTERIZING SPORADIC METEORS, AND DISCOVERING UNRECOGNIZED SHOWERS

In the previous project, you used a list of known or suspected meteor showers to define the anticipated radiant(s). You used these known radiants as the basis for deciding whether a particular meteor was a shower member or a sporadic.

You might reasonably wonder if there are other, unrecognized, low-rate meteor showers that await discovery. After all, some of the sporadic meteors may very well be the children of long-dead comets, whose debris streams have been dispersed over the eons. They may also represent the outlying fringes of quite-dense meteor streams whose orbits don't quite intersect the Earth's orbit—we just skim through the low-density outer fringe of the meteor stream. The answer is, yes, there probably are low-rate showers hidden in the sporadic meteors [2–4].

These possibilities can be investigated by carefully plotting the paths of the meteors that you observe over several nights. Such a plot, analogous to Figure 1.2, may demonstrate that a sizable number of "sporadic" meteors appear to come from a well-defined radiant: a newly-discovered meteor shower, or confirmation of a suspected shower! If the radiant position on a particular night is identified (from your data), and confirmed on other nights by other observers, then evidence for the hypothesized meteor shower becomes more solid. If the paths of a sufficient number of meteors associated with this new radiant can be plotted, then astronomers can compare the radiant with the orbits of known comets, and (maybe) identify the progenitor comet. Pretty heady territory for a backyard scientist!

The low rate of meteors on "non-shower" nights provides time for the observer to record complete information about each individual meteor. This in turn greatly improves our knowledge about the distribution of particles in near-Earth space. The diligent amateur astronomer can collect information that characterizes infrequently-observed minor showers, as well as the sporadic background. This project has two parts: First, record the complete data set about each meteor (especially its time, and brightness). Second, plot each meteor's path on a map of the stars. With the paths of all the meteors accurately plotted, it should become much clearer which ones (if any) are members of a shower, and their radiant can be estimated.

That is the essence of this project: in addition to noting the time of each meteor (and if possible some of its other characteristics, such as magnitude and duration), you carefully plot its path across the sky onto a star chart. Careful plots will provide several important pieces of information:

- The plotted paths will clearly distinguish "shower" from "sporadic" meteors.
- The combined plots of all "shower" meteors will accurately define the position of the radiant. For most showers, the position of the radiant changes a bit from night to night, reflecting the slightly different relative directions of the Earth's motion and the orbital path of the meteor stream. Refer back to Figure 1.1.
- The statistics of the meteor's brightness (i.e., estimated magnitude) can be used to determine the size distribution of the stream (the population index).

- The data on "sporadics" may show hints of unsuspected radiants whose existence can only be confirmed with a large sample of observations.

This project is best suited to observers who have gained a fair amount of experience doing simple meteor-counts. Even with a modest meteor rate, the rhythm of observing, detecting a meteor, noting its path and brightness, recording the time and magnitude, and plotting its path on a star chart, can get hectic! Your "dead time" will be much greater than in a simple meteor count (probably 15 seconds of down time per meteor, at least, after you've gained some experience at this). The value of the data that you are collecting makes this a very important project.

1.4.1   Equipment needed

Beyond the equipment identified for the previous project, namely:

- Clock or watch, set to an accuracy of at least ±1 minute.
- Clipboard with observation-recording forms, and pencil.
- Dim red-light flashlight.
- Star chart for determining limiting magnitude.

You will also need:

- Several copies of a special star chart used for plotting meteor paths.
- A ruler.
- (Possibly) a length of twine.

The star chart used for plotting meteor paths is a special, somewhat unusual item. The AMS provides plotting charts for its affiliates, but the newly-initiated meteor observer may be reluctant to ask for them. Happily, the use of PC-based planetarium programs has become almost universal among active amateur astronomers, and some of the popular planetarium programs can print out the type of chart that is used for plotting the meteor paths.

The printed charts that you may already use for planning deep-sky observing or teaching your friends the constellations are not quite what you need for meteor-path plots. In much the same way that cartographers have devised many different projections for mapping a spherical Earth onto a flat piece of paper, celestial cartographers use many different ways of putting the "dome of the sky" onto a flat sheet. Most planetarium programs use a "stereographic" projection as their default format. This provides an aesthetically pleasing rendition of the shapes of the constellations, reasonably matching what you see in the sky. Unfortunately, on this type of projection, meteor paths are not necessarily straight lines. The alternate "gnomonic" projection is designed so that any great circle in the sky will appear as a straight line on the star chart. Since the meteor paths are great circle segments, they will all appear as straight lines when plotted on a "gnomonic projection". Hence, meteor plotting always uses gnomonic-projection star charts.

Some popular planetarium programs can create plots in more than one projection. Software Bisque's "TheSky" (version 5 and later), and David Chandler's "DeepSky" both allow the user to select a gnomonic projection for viewing and printing.

1.4.2 Conducting the observations and recording the data

The key challenge in this project is accurately plotting the meteor's path. This requires that you note the starting point of the meteor, relative to the stars, and its burnout point. Doing this will, at first, strain your knowledge of the constellations and your ability to quickly identify points in the sky. Even experienced stargazers are likely to find this to be a tricky task at first! However, like most aspects of amateur astronomy, with a little practice, dedication, and experience, your skill will increase rapidly.

You will want to use either a ruler or a length of twine to assist you in orienting your plot. Immediately after seeing the meteor, hold your ruler or twine up at arm's length, and place it along the path that you saw the meteor follow. This helps you to identify the starting and ending points, and confirm a few milestone stars that it passed by. With this information, you can then accurately plot its path on your star chart.

After plotting the meteor's path, note the time and enter the meteor number and any other information (e.g., magnitude, train, etc.) on your observing log. Be sure to cross-reference the plot to the meteor number in your log! Finally, make an estimate of the amount of time that you spend handling these measurement and recording tasks—that value will be used to calculate your fraction of "dead time", during which you were unavailable to observe other meteors. (Refer back to the equation for ZHR.)

Then return to your gaze to the sky, and await the next meteor.

You can usually use a single chart to plot several meteors. In general, you should begin a new chart when either (1) the chart becomes cluttered, which usually happens after three to five meteor paths have been plotted, or (2) two hours have passed, which will significantly alter the position of the stars relative to the horizon.

This project is not for the faint of heart. It demands a combination of skill and dedication from the observer, and can be either frustrating or boring (or both) while you're learning the skill. However, the great value of the data collected makes it an extremely worthwhile project, and a rewarding way to spend a few nights under the stars. A careful plot of meteor activity is a valuable expenditure of a night's effort, whether or not a meteor shower is predicted.

1.4.3 Reducing, analyzing, and submitting your results

Most nights, the total number of meteors observed and plotted will be modest—a dozen is a very creditable result for a night's effort at this project. Therefore, the problem of "small-quantity statistics" described above is also a significant factor in interpreting the results of your meteor-path plots. Just because the paths of three or four meteors can be extended backward to a near-intersection isn't iron-clad evidence of a meteor shower's radiant. After all, any pair of random meteor paths will have a point of intersection. A little geometric thinking will convince you that it isn't all that rare to have a third meteor path pass within a few degrees of the suspected "radiant". So, you'll want a half-dozen meteors to define a suspected radiant, and you'll want those meteors to betray the other evidence of association with each other (e.g., similar speeds, and path length roughly proportional to distance from the suspected radiant). You'll also want to confirm the suspected radiant on the next night before you get too excited.

Regardless of whether or not you find a new radiant, your data and plots are extremely valuable additions to the statistics of meteor flux. They should be shared with your coordinating organization so that your data can be added to their database, and made available to meteor scientists. Data sheets and copies of your path plots should be sent to the address given above, in Section 1.2.3.

The gnomonic projection for plotting meteor paths

(A) The *stereographic projection* is the "default" projection in most planetarium programs. This projection maintains angular relationships across the field of view. Note that wherever a line of RA intersects a line of Dec, the intervention is a 90-degree angle.

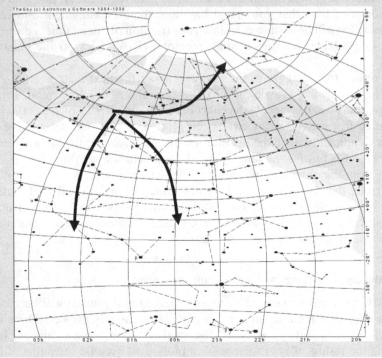

Preserving angular relationships in this way makes the constellations appear most realistic, very similar to what you perceive in the sky. However, meteor paths (great circles) are not necessarily straight lines in this projection. Three meteor paths are shown on this chart. Note that they are strongly curved—hence they would be virtually impossible to accurately plot onto this projection in the field.

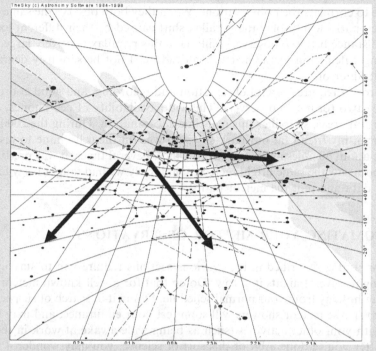

(B) The *gnomonic projection* has the feature that any great circle in the sky is a straight line on the plot. (Note that the lines of RA are all straight.) It is the recommended projection for plotting meteor paths, because meteor paths are great circles on the sky. The same three meteor paths when plotted appear as straight lines, as they would be perceived by an observer.

The gnomonic projection significantly distorts the shapes of constellations near the edges of the plot. The larger the field of view being plotted, the more dramatic this distortion appears.

Screen images from TheSky™ (Software Bisque).

1.5 METEOR PARALLAX AND ALTITUDE DETERMINATION

If you and another observer coordinate beforehand, your plots of meteor paths (or better, your photographs of meteors) offer the possibility of observing parallax to determine the altitude at which the meteor appeared and burned up.

The textbooks will tell you that most meteors flash at an altitude of about 30–60 miles up in our atmosphere. There are good theoretical and observational reasons to accept this estimate, but the actual data is surprisingly sparse. Few good measurements of meteor altitude have been reported in the literature. In principal, a pair of observers, separated by 10–30 miles, both of whom are carefully plotting meteor paths and recording the times and brightness (and any other distinguishing features), will be able to correlate their observations. When they both observe the same meteor, their plotted paths will display the parallax shift caused by their different locations. A sizable distance between observers will make this parallax relatively large, so that the resulting calculation of the meteor's altitude will not be seriously degraded by minor plotting errors.

This project requires careful observation and recording, and good coordination between the two (or more) observers. The wide availability of cell-phones, and cell service in rural areas, simplifies this real-time coordination. During the 2000 Leonid meteor storm, my friend (at our dark-sky site) was on the cell-phone to his wife at their home about 30 miles away, as the crow flies. Several times, it was pretty certain that they both saw the same bright meteor. Oh, if only we had taught his wife how to record the data and plot the meteor paths!

1.6   AUTOMATING YOUR METEOR OBSERVATIONS

The meteor studies described in the previous sections require you to stay up pretty much all night, several nights in a row. Doing that for a well-known meteor shower can be a fun holiday from your normal schedule. Doing it in search of suspected (but not certain) sparse meteor showers will soon test your endurance, and may begin to interfere with your other activities (such as being wide-awake at work in the morning). So, after you've done a bit of this sort of science, you may wonder if you can make an automated meteor-watching system.

It turns out that you can indeed create an obedient robot who will stare at the sky all night and remember what it sees, while you get a good night's sleep. There are several approaches that people have taken, depending on their objectives, technical skill, and budget. In general, they use either a video camera and recorder, or a still camera (film or digital) that can be triggered by an electronic circuit. In Chapter 7 I'll describe some of the things that can be done with such equipment. These are excellent projects for the electronically-oriented amateur astronomer!

1.7   RADIO METEOR MONITORING

Meteors don't cease their activity just because it's cloudy, or the Sun is up in the sky. The background flux of sporadics continues 24 hours per day, and there are meteor streams whose orbital geometry makes them most active during our daytime hours. Obviously, you can't see them visually with the Sun in the sky. It turns out that you can detect them using radio.

Shortwave radio signals travel in straight lines, in the same sense that light waves do: any single photon moves in a straight line unless it is refracted, reflected, or diffracted by intervening material. A radio transmitter is analogous to a light bulb—it sends radio waves in all directions, but each radio photon travels in a straight line. If the radio transmitter is so far away that the curvature of the Earth hides it from you, then you won't receive the signal unless something redirects the rays around the limb of the Earth. That "something" is the ionosphere.

You may know that the Earth is enveloped by discontinuous layers of charged particles, collectively called the "ionosphere". The ionosphere is about 25 to 250 miles above the Earth's surface, and contains a relatively high density of charged particles. A layer of charged particles can act like a radio mirror, refracting radio waves. Some types of long-distance radio communication depend on the existence of the ionosphere [5]. If you place a radio transmitter at one location, and place a receiver several hundred miles away, there is no straight line-of-sight path for the radio waves from the transmitter to the receiver. But if the waves are refracted by the ionosphere, they can still reach the receiver, as illustrated in Figure 1.6.

The ionosphere is not a complete, permanent spherical shell: it is more of a discontinuous patchwork, with patches more-or-less randomly coming and going. It is also a relatively thick shell, with several more-or-less well-defined layers. The density of charged particles in the ionosphere is primarily controlled by the Sun, since it is the Sun's energy that ionizes the atmospheric molecules and creates the ionosphere.

The effect of the ionosphere on radio waves depends on the density of electrons, and the frequency of the radio wave. In general, radio waves of frequency below about 30 MHz are strongly affected. A sizable fraction of their energy is refracted back down toward Earth, where it can be received at great distances from the transmitter. At higher frequencies (above 30 MHz), the radio waves are weakly refracted, and essentially all of the energy is transmitted into space, with almost

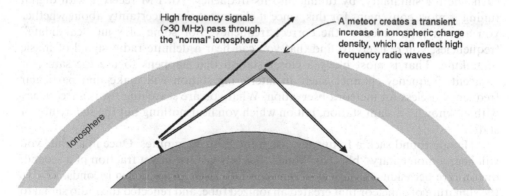

High frequency signals
(>30 MHz) pass through
the "normal" ionosphere

A meteor creates a transient
increase in ionospheric charge
density, which can reflect high
frequency radio waves

Ionosphere

Figure 1.6. The ionosphere reflects/refracts low-frequency short-wave signals, but (mostly) passes high-frequency signals. A meteor creates a transient increase in ion density that is a good reflector of high-frequency radio signals.

none being refracted back downward toward the Earth. Almost none, that is, unless something happens to dramatically increase the electron density in the ionosphere. That "something" can be a meteor.

Earlier, I mentioned that when we see a meteor, what we're really seeing is the glowing tube of ionized gas that was created by the meteor's fiery demise as it was vaporized in the upper atmosphere. That tube of ionized gas is a tube of very high charge density, and it will quite effectively reflect high-frequency radio waves. So, in order to detect meteors by radio, you need a high-frequency transmitter, and a receiver located far enough away that it can't get the transmitter's signal by direct (line-of-sight) propagation. And you need a way to detect, recognize, and record the momentary signals that will be received when a meteor creates an ionized region that reflects the signal into the receiver.

In general, to eliminate direct (line-of-sight) propagation, the transmitter and receiver should be more than 300 miles apart. Distances of up to about 600 miles will be satisfactory. If you are relatively close to the transmitter, even if you are shadowed from it by an obstruction such as a mountain, energy that is diffracted around the obstruction still reaches you with sufficient intensity for your receiver to detect it. After all, you can hear nearby radio stations that aren't perfectly visible to your line-of-sight. That is why for radio meteor monitoring, you want the transmitter to be a few hundred miles from your location—you avoid the confusing effect of diffraction from a local transmitter.

There is a convenient source of high-frequency transmitters available for free in most parts of the world: commercial FM radio stations. These stations broadcast on frequencies between 88 MHz to 108 MHz, which generally are not refracted by the "normal" ionosphere.

You can locate suitable transmitters by drawing two circles on a map, with radius of 300 miles and 600 miles. Select a few large cities that lie between the two circles, and find out the call letters and frequencies of the FM radio stations that broadcast in those cities. Those stations/frequencies are your candidate transmitters. Then, check each one for suitability, by tuning into its frequency. An FM receiver with digital tuning is most convenient for this, since it eliminates the uncertainty about whether you are, in fact, tuned to the correct frequency. For some of your "candidate" frequencies, you're likely to find that you can hear a definite radio signal of music or talking. That is most likely a closer station that happens to use the same, or adjacent, frequency channel; such an interfering station will make that particular frequency useless for meteor observation. What you are searching for is a frequency with a "known" distant station, but on which you hear nothing but the steady hiss of static.

Having found such a frequency, listen for 15 to 30 minutes. Once in a while you will hear a momentary "blip" or "tone" typically lasting only a fraction of a second, or (on rare occasions) a snippet of music or talk lasting one or two seconds. That is the signature of a meteor that created an ionized tube, and reflected the radio signal to you during its brief existence.

If you have some knowledge of electronics, you can probably imagine how to connect a recording device (computer or strip-chart recorder) to your receiver, and trigger it by the momentary signal, so that you can record the onset time, intensity, and duration of the signal. Such a record provides information about the rate of meteors (analogous to meteor-counting at night), and the distribution of signal durations (analogous to magnitude estimates).

Unfortunately, evaluation and analysis of meteor radio observations is still in the "experimental" stage. There are no well-defined observing procedures. Therefore, there is no commonly-accepted method to combine observing results from different methods and equipment. There is active work going on to put radio-meteor studies on a firmer theoretical and procedural basis, so if you are a radio-savvy amateur astronomer, you may want to experiment with this technique. The most active radio-meteor group operates under the auspices of the International Meteor Organization (*http://www.imo.net*).

A meteor shower without a radiant: the "Cyrillids" [3, 4]

As shown in Figure 1.2, meteor showers have a "radiant" because the meteor stream is composed of particles whose orbital paths are (nearly) identical, and the cross-section of the meteor stream is substantially wider than the diameter of the Earth. No matter where you are on Earth, when we're passing through such a stream the meteors' paths appear to radiate from a single point in the sky.

There is one recorded instance of a meteor shower without a radiant. On February 9th, 1913, an unusual display of meteors was observed in Toronto. The meteors all tended to travel toward the southeast. They all appeared to be in level flight, following horizontal paths. They were astonishingly slow: some reports claimed that individual meteors could be followed for as long as a minute before they burned out. One observer described the event as a "procession" of meteors, rather than a "shower". The meteors seemed to be associated with each other, but there was no definable radiant.

Similar reports were later found from several locations. The curious thing was that these meteors were only seen from locations that were located (roughly) along a great circle that ran from Toronto to Bermuda. Locations far from this great circle path didn't report seeing anything unusual.

What was going on?

One plausible reconstruction of the event is that a fairly large object, or a group of objects, was somehow captured by Earth's gravity, and became a short-lived satellite of our planet. The demise of this object or group, as it entered our atmosphere and was burned up on entry, created the unusual meteor display.

This event is referred to as the "Cyrillid" meteor shower. It was apparently a unique event; at least it was a unique observation.

Who knows? Considering the relatively few people who are diligent about observing meteors, adding your efforts to theirs may be just what's needed to discover another, similar event.

1.8 COORDINATING ORGANIZATIONS: AMS AND IMO

An excellent reference for meteor observers is Neil Bone's book *Meteors* (Sky Publishing, 1994).

The two most active meteor coordinating organizations are the American Meteor Society (AMS), and the International Meteor Organization (IMO). The AMS, as its name indicates, primarily maintains a North-American focus. The IMO maintains a primarily European focus. Every active meteor observer should be a member of one or both of these fine organizations. Their websites are wonderful sources of meteor-observing methods, advice, forms, and reports from meteor-observing campaigns. You can reach the AMS main page at *http://www.amsmeteors.org* The IMO main page is located at *http://www.imo.net*

Meteor observers also correspond with each other via the Yahoo group called "imo-news" at *http://groups.yahoo.com/group/imo-news/* Most active meteor observers will want to participate in this on-line forum.

1.9 REFERENCES

[1] Chesser, H., Gandhi, A., and Hiemstra, D., *Space Engineering Materials*, lecture charts for ENG 3330, York University.
[2] Poole, L.M.G. and Kaiser, T.R., "The detection of shower structure in the sporadic meteor background", *Monthly Notices of the Royal Astronomical Society*, vol. 156, p. 283 (1972).
[3] O'Keefe, John A.: "Tektites and the Cyrillid Shower", *Sky & Telescope*, vol. 21, no. 1, p. 4 (January 1961).
[4] O'Keefe, J.A.: "The Cyrillid Shower: Remnant of a Circumterrestrial Ring?" Lunar and Planetary Science Conference XXII.
[5] Lusis, D.J., "HF Propagation: The Basics", *QST* (journal of the American Radio Relay League), December, 1983, p. 11.

2

Occultations

Astronomers attempt to make precise measurements of the size and shape of astronomical objects for a variety of reasons. Let's suppose that we want to know the size and shape of a particular object. One way would be to try to carefully photograph the target at high magnification. We'd quickly discover that this technique doesn't work for many types of objects. In a well-done astronomical image, all of the stars will have blur circles that are essentially the same size. This size is set by optical diffraction or (more commonly) atmospheric "seeing". Typically, "seeing" restricts you to no better than 1–2 arc-sec resolution. The image of a star does not tell us anything about the star's actual size.

Stars aren't the only things that are too small to measure directly. Even in this age of space telescopes and ground-based adaptive optics, almost all asteroids appear as star-like points of light. Yet it's important to be able to determine their sizes and shapes so that we can learn more about their composition. Closer to home, the topography of the Moon is surprisingly poorly known, despite a variety of lunar-orbiting satellites and the manned missions of the 1970s.

Happily, under the right circumstances we can use the motion of solar-system objects as a way of making very precise measurements of the position and size of our target. The "right circumstances" occur when one object passes directly in front of another. Such an event is called an "occultation". Two projects where amateur astronomers can make important observations are "lunar occultations" (when the Moon passes in front of a star), and "asteroid occultations" (when an asteroid passes directly in front of a star).

2.1 PROJECT D: LUNAR OCCULTATION TIMING

Have you ever been looking at the Moon through your telescope, and noticed a star hanging next to the lunar limb? The first time I saw such an occurrence, I was using a

6-inch F/5 Newtonian at about 25×. The star was Regulus, and I had just watched it pop out from behind the bright limb of the crescent Moon, set against the twilight sky. I remember being struck by two things. The first was how bright the star was, compared with the lunar surface. We are used to the Moon's glare making it by far the brightest object in the evening sky. That common impression usually outweighs our scientific knowledge that most of the lunar rocks and regolith are not much brighter than asphalt. That evening, the twilight sky veiled the Moon's glare, and the intensity of the star—a brilliant spark set against the deep-blue sky—made the sunlit lunar surface appear as a faded gray by comparison. It was a beautiful, magical scene.

The second thing I noticed was the speed with which the Moon slid away from the star. You know that the Moon moves eastward relative to the "fixed" stars, and some quick math shows that it moves through a distance equal to its diameter in about an hour:

$$\text{Moon's orbital rate} = (360 \text{ degrees}/27.3 \text{ days}) \times (1 \text{ day}/24 \text{ hours})$$

$$\approx 0.56 \text{ degrees/hour}$$

This is just another dry fact until you've actually watched it happen. The presence of a visible star adjacent to the Moon provides the benchmark you need to observe the motion of the Moon, and it is a sublime way to spend a half-hour.

I had selected that particular night to look at the Moon because I knew that the reappearance-phase of an occultation was predicted. (I'll tell you how to get such predictions later.) Careful observation and timing of lunar occultations provides information that is of great value in several fields. Even if you decide that this area of research isn't your cup of tea, I strongly encourage you to take the trouble of observing at least one lunar occultation of a bright star just for the experience of witnessing it. The star's gradual approach to the lunar limb, the way it seems to hang on the edge of the Moon for a minute, and its instantaneous disappearance are visual treats that will live in your memory for a long time.

The scientific value of lunar occultations is surprisingly wide-ranging [1, 2], and provides a case-in-point of how scientists sometimes succeed in pulling themselves up by their own bootstraps. Start with a single careful timing of a lunar occultation, such as the one I've described. That gives you precise knowledge of the Moon's location at a single point in time (assuming that you know the precise coordinates of the star!). If the real world were as simple as the world of your freshman physics class, that would be all the information you'd need to predict the Moon's position for any time in the future. Alas, the real world isn't that simple.

The theory of the Moon's motion is a rock on which many careful observers and theoreticians have broken their picks. The freshman-physics of a small satellite orbiting its spherically-symmetric planet in a Newtonian gravitational field barely scratches the surface of the problem. In real life, the gravitational fields of Earth and Moon aren't quite perfectly spherically symmetric, and the effect of the Sun's gravity cannot be ignored (i.e., it's really a three-body problem). Both the Moon and the Earth are compliant bodies, and the Earth has oceans, so that the Moon's gravitational force raises tides, that in turn create a sort of "drag" that tries to slow the

Moon's velocity in its orbit. The exchange of angular momentum between the Moon's orbital motion and the Earth's rotation has a deceleration term that is predicted to amount to 23 arc-sec per century per century—assuming that the gravitational constant is truly constant. If, as some people occasionally speculate, the gravitational "constant" actually changes with time, then such a change could be observed as an additional deceleration in the Moon's motion, amounting to a few arc-seconds per century per century [3]. That's a small number, but should be detectable in a long record of lunar position measurements derived from occultation timing.

In order to compare their theories with the "real world", astronomers need a continuous stream of data points that anchor the location of the Moon. That stream of data includes precise occultation timings. The question of "where is the Moon" is answered with reference to the stellar reference frame, which is itself a somewhat idealized concept (see Appendix B). Real stars do, after all, move around (due to proper motion, parallax, and the aberration of light), and one way to cross-check their positions is to refer back to lunar occultation timings.

The precise moment of a star's disappearance or reappearance in a lunar occultation is affected by the topography of the lunar profile. Consider the two occultation disappearances illustrated in Figure 2.1. If a star's path toward the lunar limb happens to carry it toward a lunar valley, then its disappearance will be delayed,

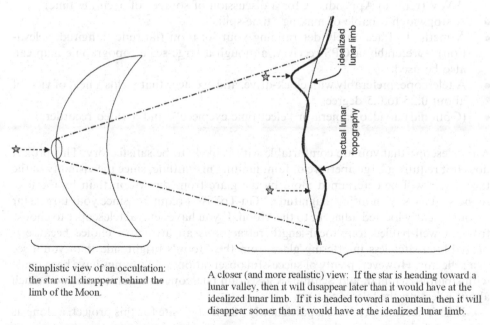

Simplistic view of an occultation: the star will disappear behind the limb of the Moon.

A closer (and more realistic) view: If the star is heading toward a lunar valley, then it will disappear later than it would have at the idealized lunar limb. If it is headed toward a mountain, then it will disappear sooner than it would have at the idealized lunar limb.

Figure 2.1. The time of stellar disappearance and reappearance in a lunar occultation is sensitive to details of lunar-limb topography.

compared with what you would have expected if the Moon were a perfect sphere. Conversely, if the star happens to be headed toward the peak of a lunar mountain, then its disappearance will happen sooner. Therefore, accurate interpretation of occultation timings requires that we have a good knowledge of the lunar-limb topography. Surprisingly, despite all of the spacecraft that have visited and orbited the Moon, there are significant uncertainties in the limb's topography ("significant" in terms of the accuracy requirements for lunar occultation interpretation, where a few hundred feet of uncertainty in the limb position is noticeable). The details of the limb's topography can be very accurately measured by teams of observers monitoring "grazing occultations", which will be described in the next section.

Another research area that benefits from accurate records of lunar occultations is that of solar studies. There are occasional, tantalizing hints (primarily from observations of solar eclipses) that the Sun's diameter may not be constant. Precise knowledge of the size, position, and topography of the Moon is required in order to interpret measurements of the duration of solar eclipses in terms of the Sun's diameter, and these are best determined from lunar occultations.

2.1.1 Equipment needed

The equipment required for lunar occultation timing is:

- A portable shortwave receiver tuned to one of the standard time services such as WWV (refer to Appendix A for a discussion of sources of accurate time).
- A stopwatch capable of making "time-splits".
- A method of accurately determining your location (latitude, longitude, elevation)—preferably a GPS receiver, although a large-scale topographic map can also be used.
- A telescope, preferably with clock-drive, and eyepiece that yields a field of view of about 0.25 to 0.5 degree.
- (Optional) a video camera or "electronic eyepiece", and a video recorder.

Any telescope that you are comfortable with is likely to be satisfactory. This project does not require a large aperture or faint limiting magnitude, since the visibility of the target star will be determined more by the glare from the Moon than by the telescope's "dark sky" limiting magnitude. "Go-To" isn't required, since your target star is conveniently located adjacent to the Moon. If you have several telescopes to choose from, a well-baffled long-focal-length refractor is an attractive choice because it probably creates less interfering glare from the Moon's bright side than you'll see in a reflector. However, plenty of successful observations and timings have been made with small Newtonian and Schmidt–Cassegrain telescopes, so don't worry too much about selecting the telescope type.

It isn't necessary to have a particularly "dark-sky" site for this project, as long as you can reliably monitor the target star. However, sky clarity and transparency are advantageous. The slightest haze or cirrus will dramatically increase the amount of atmospheric scattered light from the brightly-lit portion of the Moon, and this veiling glare may make it difficult to see your target star.

2.1.2 Observing lunar occultations and recording the data

The geometry and key terms involved in a lunar occultation are illustrated in Figure 2.2. The broad-brush concept of lunar occultation timing is quite simple. Ten or fifteen minutes before the predicted occultation, you locate the target star. That's not too hard, considering that it's right next to the Moon. You record the precise time of the star's disappearance, wait until the star is about to pop out the other side of the Moon, and record the precise time of its reappearance. If you can't reliably time the event that occurs on the bright limb, that's OK—report only the time of the "dark limb" event. Either before or after the occultation, you also record your precise location (latitude, longitude, and elevation). Those data points—one or two event times, and one location—comprise the raw data of your observation.

There are, of course, a few details that require meticulous attention if your observation is to be of scientific value. These relate to the accuracy and precision of your measurements, and are to some degree affected by the method you use to observe the occultation and record your observations.

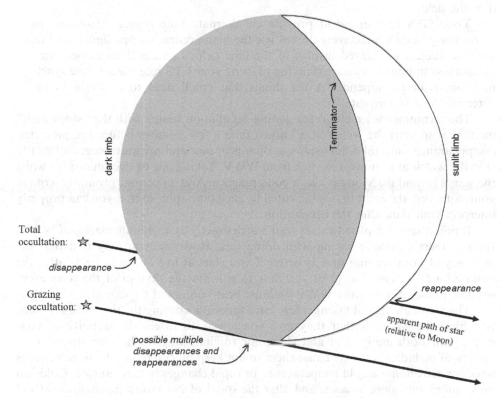

Figure 2.2. Geometry of a lunar "total occultation" and a "grazing occultation".

2.1.2.1 Visual/stopwatch timing

The required accuracy of your timings is at least ± 0.5 sec, and preferably ± 0.2 sec. With practice, you can achieve this accuracy by visually monitoring the scene in your eyepiece, and clicking a stopwatch at the disappearance/reappearance of the star. Timing the disappearance is most reliable (especially if it occurs on the dark limb), because you can watch the star approaching the Moon's limb. The reappearance will come as a complete surprise, and so you need to do everything you can to minimize the inevitable delay in your response to this event. As a minimum, you should have carefully identified the location on the Moon's limb where the star will appear. An eyepiece reticle can be helpful in this regard.

Your stopwatch will only tell you the interval between time-splits. For occulta-tion timing, you'll also need a precise time reference. The best time references are the National Institute of Standards and Technology (NIST) radio stations WWV, or (in the Pacific) WWVH. Most portable shortwave receivers can tune in WWV (standard time broadcasts are on 5, 10 and 15 MHz). You'll recognize the station by its distinctive "tick" sounds at exactly 1-second intervals and the hour/minute identifier at the beginning of each minute. Spend some time at home with your receiver to find and listen in on WWV. Once you've heard it, you'll easily recognize it in the field.

Your GPS receiver *might* provide an alternate time source. However, most consumer-grade GPS receivers do not use the high-accuracy 1-pps signal, and most are also subject to delayed display of the time (which makes them inadequate for occultation timing). If you are thinking of using your GPS receiver for time synchro-nization, refer to Appendix A for things that you'll need to check in order to determine if it is adequate.

The recommended method for getting occultation timing with the "stopwatch" method is to start the watch at a known time a few minutes before the predicted disappearance time, take time-splits at disappearance and reappearance, and finally stop the watch at a known time tick from WWV. Taking one or two time-splits while the star is behind the Moon is also a good practice. And, of course, promptly write in your notebook the event that is indicated by each time-split, so that you can properly interpret your data after the occultation.

It might seem simplest to start your watch exactly at a "minute mark" of WWV. Indeed, there's nothing wrong with doing this. However, you may find that the accuracy of your starting time is better if you start at five or ten seconds after the minute beat, because that gives you time to get into the rhythm of the ticks every second. Remember that the WWV "minute beat" occurs at 00 seconds.

The rationale behind taking a few extra time-splits during the time when the star is behind the Moon is that they give you a way to assess the stability of your stopwatch. Both mechanical and electronic (digital) stopwatches are subject to a variety of maladies that may cause them to not keep precise time. The most obvious issue for astronomy is cold temperatures, or rapid changes in temperature. Cold can make lubricants more viscous, and alter the speed of clockwork mechanisms. Cold can also reduce the voltage of the batteries, and alter the characteristics of some electronic components. So, you'll want a way to check the stability of your stopwatch under realistic field conditions. I know people who recommend doing special experiments (e.g., leaving the watch in the refrigerator for a while), to qualify the watch for occultation observations. Even with a "qualified" watch, the belt-and-suspenders aspect of having a few extra time-splits is attractive as a way of confirming that the watch is operating properly in the field.

During the "idle time" while the star is behind the Moon, you can also determine if there is a fraction-of-a-second correction needed to your stopwatch readings. The concept goes as follows. You tried to start the watch exactly at a "second" beep-tone of WWV. If you were perfectly, precisely successful, then you will see your watch's display increment by 1 second with each subsequent "second-beep" of WWV. Odds are, however, that what you'll see is that the watch leads or lags the WWV beep-tone by a fraction of a second. That lead or lag represents your error in starting the watch. By monitoring this for a minute or so, you will be able to estimate the amount of lead/lag quite accurately—certainly to within 0.2 seconds. Write this value in your notebook, and note whether your watch "leads" WWV (i.e., watch is a fraction of a second "ahead" of WWV) or "lags". This correction increment will be used during your data reduction.

There is clearly a difference between your ability to make a time-split when you know the event is going to happen and you're in sync with the rhythm (e.g., the "second" beep of WWV), vs. your ability to react when a surprise event (such as the disappearance of the star) occurs. See Section 2.3.3.1 for a discussion of reaction time, and how to determine yours.

2.1.2.2   Video recording

There are, of course, a myriad of things that can corrupt your visual timings (see the next project—Asteroid occultations—for some real-life examples). Therefore, your results will be more secure if you can make a permanent record of the occultation. By far the most common method is video recording. The easy availability of low-cost video cameras, camcorders, and "electronic eyepieces" puts video well within the capability of many amateur astronomers. A video record offers two wonderful features. First, you can re-play the event over and over, so that your timing is more accurate and free of corrupting surprises (especially at the challenging "bright limb" and "reappearance" timings). Second, by recording the WWV time ticks directly onto the audio channel of the video, you have a positive record of the absolute time of each event. Your data reduction may still rely on observation of the video and stopwatch time-splits, but with the audio record you can easily correlate your stopwatch to precise time.

If you are willing to buy one more piece of equipment, you can get a time-insertion box that writes accurate time as text onto the video itself, so that each frame contains a visual time-record. The one I'm aware of is the Kiwi Time Inserter, available from PFD Systems. With this device, it is possible to replay and analyze the occultation frame by frame. This presents several opportunities. Obviously, the occultation timing can now be accurate to ±0.03 second (assuming standard 30 frames/sec video). Of additional interest is the possibility of measuring the light curve of the occultation. If the target star is a close double, the "disappearance" may actually be a two-step affair. Accurate analysis can determine the separation of the components, to an accuracy that is competitive with speckle interferometry (about 0.01 arc-sec). For particularly large stars, the disappearance may be a gradual fade, rather than a sudden "blinking out". Analysis of the duration of the fade is one of the few ways of directly measuring the diameter of a star.

2.1.2.3   *Determining your location*

The value of your occultation timings is critically dependent on the accuracy with which you determine your location. An error of 0.3 mile in establishing your location causes the same-sized error as a 1-second timing error. Either of these is a totally unacceptable level of error for useful lunar occultation timing. Nearly an order of magnitude better is needed: that is, timing to an accuracy of ±0.2 sec, and position accuracy to 0.05 mile or better (i.e., about 250 feet) in all three dimensions (latitude, longitude, elevation).

This degree of positional accuracy is difficult to achieve using a topographic map. Although doing so was once part of the occultation observer's routine, it is not a simple task for an inexperienced navigator. Therefore, most modern occultation observers use a GPS receiver to determine their location. Thankfully, this very handy tidbit of technology is relatively inexpensive, and many people (astronomers, boaters, campers, hunters) already have one.

If you have some experience with GPS receivers, the following notes will be familiar to you. First, be aware that if a GPS receiver has been moved more than a couple of hundred miles (e.g., on an expedition to an occultation occurring at a remote site), you may need to "re-initialize" it in order to get an accurate lock on the satellite signals. Second, at your observing site, you should allow the receiver to conduct a 15-minute "average" of its position estimate before accepting the results. Especially during the first minute of this "averaging", you are likely to be able to watch the least-significant digits of the position display change as the receiver updates its calculation of your position. This averaging is very important in order to get a good elevation reading, because GPS tends to have a tougher time settling in on its elevation estimate than it does in latitude and longitude.* Third, be sure that you have spare batteries if you are using a hand-held GPS receiver: they are real power-hogs!

You should also confirm the settings of your receiver, particularly the "datum" that it is using. The normal setting is "WGS84" (World Geodetic System 1984), which is the preferred datum reference. However, most commercial receivers can use several other datum references (e.g., my old Magellan contains about 75 optional

---

* You may find that a topographic map, or comparable data from the website *topozone.com*, is a useful source of elevation data for initializing your GPS receiver. Then, a 5-minute "averaging" with your GPS receiver is likely to be sufficient.

datums). So, do check your settings, and either use WGS84, or be able to report the datum that your receiver uses.

2.1.3   Reducing and analyzing your results

Whether you used the "visual/stopwatch" or "video" method of gathering data, your data reduction process has two essential steps: determining the timings themselves, and documenting the details of the observing conditions.

2.1.3.1   Timing analysis

The method of translating your stopwatch readings into timings is easiest to explain by example. In this example, "true" UTC times are indicated by capital "T", and "stopwatch" time-split readings are indicated by lower-case "t".

Assume that you started your stopwatch at T_{start}, a few minutes before the predicted disappearance, and precisely on the minute mark of WWV. For example, assume that $T_{start} = 08:12:00$ UT. Your stopwatch reading was $t_0 = 00:00:00.00$ (HH:MM:SS.ss) at T_{start}. You took your first time-split when the star disappeared; call that time-split reading t_1. For our example, assume that $t_1 = 00:02:37.45$.

During the time that the star was behind the Moon, you did three things. First, you took a stopwatch time-split at exactly a WWV minute-mark. For our example, suppose that this time mark was $T_{check} = 08:25:00$ UT (by WWV). The stopwatch time-split at this mark is t_2. For our example, $t_2 = 00:13:00.15$. Second, you spent a few minutes monitoring the watch display while listening to WWV, and you noted that the stopwatch turned from one second to the next about 0.1 second *after* the WWV second-tick. Third, you adjusted your telescope pointing to aim it at the position on the lunar limb where reappearance was predicted to occur.

When the star reappeared, you took a time-split t_3. For our example, assume that your stopwatch reading at this time split was $t_3 = 00:19:56.35$. Lastly, at the WWV minute-tone after reappearance, you stopped the watch. Assume that this occurred at $T_{stop} = 08:33:00$, and that your stopwatch reading was $t_4 = 00:20:59.92$.

All of this data can be entered into a table, such as that shown in Figure 2.3. The purpose of the calculations is to determine the unknown times of disappearance,

$$T_{disappear} = T_{start} + t_1$$

and of reappearance

$$T_{reappear} = T_{start} + t_3$$

The worked-out example in Figure 2.3 also shows how to use the "check" and "stop" times to confirm the stability of the stopwatch and the observer. In the example given, everything looks good: the calculated times of T_{check} and T_{stop} match the actual times, to within a small fraction of a second.

In addition to recording the time-splits, our example assumes that you also noticed that the stopwatch ":00" second turnovers happened about 0.1 second *after*

Event	T	WWV UTC		t	stopwatch reading
start	T_{start}	08:12:00		t_0	00:00:00.00
disappearance				t_1	00:02:37.45
time check	T_{check}	08:25:00		t_2	00:13.00.15
reappearance				t_3	00:19:56.35
stop	T_{stop}	08:33:00		t_4	00:20:59.92

•Raw calculations for time of disappearance and reappearance:

$T_{disappear} = T_{start} + t_1 =$
  08:12:00
  + 00:02:37.45
  = 08:14:37.45 (uncorrected)

$T_{reappear} = T_{start} + t_3 =$
  08:12:00
  + 00:19:56.35
  08:31.56.35 (uncorrected)

•Check consistency & stability of stopwatch using the "time check" and "stop" splits:

$T_{check} = T_{start} + t_2 =$
  08:12:00
  + 00:13:00.15
  08:25:00.15 → observed – calculated = 0.15 sec

$T_{stop} = T_{start} + t_4 =$
  08:12:00
  + 00:20:59.92
  08:32:59.92 → observed – calculated = -0.08 second

Figure 2.3. Calculation of event times from stopwatch time-splits.

the WWV second-beep (or, equivalently, when WWV beeped, the watch read :59.9 seconds). This means that there was a slight error in synchronizing the watch to WWV when you started the watch. The effect is illustrated in Figure 2.4. The synchronization error Δt is positive if the watch lags WWV (as in our example), and negative if the watch leads WWV. This systematic error can be compensated in the disappearance and reappearance times by using:

$$T_{dis, corrected} = T_{start} + t_1 \pm \Delta t$$

and

$$T_{re, corrected} = T_{start} + t_3 \pm \Delta t$$

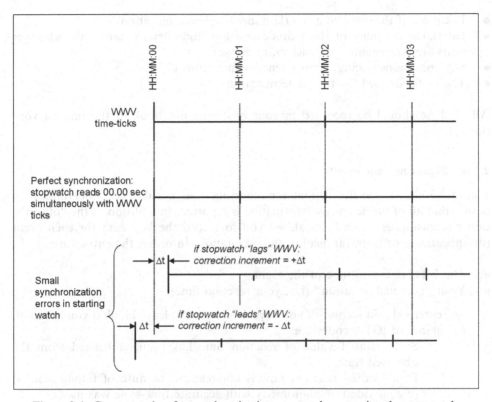

Figure 2.4. Compensation for synchronization errors when starting the stopwatch.

In our example, since Δt is in the positive sense, the corrected values are:

$$T_{\text{dis, corrected}} = 08:14:37.45 + 0.1 = 08:14:37.55 \text{ UTC}$$

and

$$T_{\text{re, corrected}} = 08:31:56.35 + 0.1 = 08:31:56.45 \text{ UTC}$$

Both the "uncorrected" and "corrected" times should be entered into your occulta-tion report. This provides a basis for consistency checking between observers, and a way to detect simple mistakes (such as incorrect sign in the correction term).

2.1.3.2 Observation circumstances

As with any visual observation, a variety of circumstances can affect the reliability and accuracy of the result. Therefore, any lunar occultation timing must be accom-panied by a complete description of the circumstances of the observation, including:

- The characteristics of the telescope (type, aperture, focal length, magnification used).
- The characteristics of the mount (driven or manually guided).

- Location of the observing site (latitude, longitude, elevation).
- Estimated accuracy of the latitude and longitude determination (in whatever units are convenient: feet, meters, or arc-sec).
- Sky conditions (seeing, transparency, temperature).
- The methods used for time determination

All of these should be recorded in your observing notebook at the time of your observations.

2.1.4   Reporting your results

Figure 2.5 shows a useful format for reporting your occultation observations, to ensure that all of the necessary information is reported. In addition to the "observation circumstances", this form allows you to record the key data for each event (disappearance or reappearance) of the occultation. In order, the entries are:

- The UTC date and time of the event.
- Your "personal equation" (i.e., your reaction time):

  - enter value in seconds if you have determined it; leave blank if you have not;
  - standard IOTA codes are:
    - S: the stated value of reaction time has been subtracted from the observed time;
    - E: individual reaction time is not relevant because of timing method (e.g., if video or photometry with accurate time-sync was used);
    - N: reaction time is not known;
    - U: the stated value of the reaction time is known, but has not yet been subtracted in the timings.

- Estimated accuracy of timing (the consistency check made using T_{check} and T_{stop} shown in Figure 2.4 give you a basis for making this estimate; although your uncorrected reaction time is likely to be the largest error source).
- Identity of the star (e.g., HD number or DM number).
- Phenomenon observed (e.g., D = disappearance or R = reappearance).
- Limb location (e.g., B = bright limb, D = dark limb).
- Method used to measure event times (e.g., R = WWV radio signal; C = clock adjusted by standard time signal; O = other method described in "remarks").
- Identify which component the event relates to, in the case of a known or suspected double/binary star.
- Sky conditions:

  - seeing (1 = good, 2 = fair, 3 = poor);
  - transparency (1 = good, 2 = fair, 3 = poor).

- Ambient temperature.

This form is available at the IOTA website.

LUNAR OCCULTATION OBSERVATION REPORT
PLACE NAME: _____ *(name of closest city)*
ADDRESS: _____ *(Mailing address for observer or team leader)*
E-MAIL ADDRESS: _____ *(e-mail address for observer or team leader)*
REPRESENTATIVE: _____ *(name of observer or team leader)*
REPORTED TO: _____ *(name, address of org or person receiving the report, e.g. IOTA)*

Telescope information:
    Type= ___ *(R, refractor; N, Newtonian; C, catadioptric or Cassegrain; O, other)*
    Mount= ____ *(E, equatorial; A, alt-az)*
    Drive= _____ *(D, motor-driven; M, manual)*
    Aperture (cm)= _____
    Focal Length (cm)= _____

Observing site location:
    Longitude (d-m-s)= _____ *(specify E or W)*
    Latitude (d-m-s)= _____ *(specify N or S)*
    Accuracy (meters)= ____
    Elevation (ASL, meters)= _____
    Geodetic datum= ____
Year: _____

Timings:

	event #:	1	2	3	4
date (UTC)	Yr				
	Mo				
	Day				
Time UTC	hr				
	min				
	sec				
PE	sec				
	code				
	Accuracy (sec)				
	Star ID				
	Phenomenon				
	Limb				
	Time Method				
	Time comp				
	seeing				
	transparency				
	temp(°C)				
	Remarks				

For a total lunar occultation, event #1 will be D ("disappearance"), and event #2 will be R ("reappearance").
    For a graze occultation, you may see a series of events D-R-D- ... as the star passes behind lunar hills and valleys.

Figure 2.5. Data and report format for lunar occultation timing. (Used with the kind permission of the International Occultation Timing Association)

The main collecting, analysis, and disseminating body for total lunar occultations is the International Lunar Occultation Center (ILOC). Total lunar occultation reports should be sent to:

International Lunar Occultation Center (ILOC)
Geodesy and Geophysics Division
Hydrographic Department
Tsukiji-5, Chuo-ku
Tokyo, 104-0045 Japan
E-mail: *iloc@jodc.go.jp*

This address was correct as of the date of publication of this book. I recommend that you check the International Occultation Timing Association (IOTA) website at *http://www.lunar-occultations.com/iota* for any updates to the mail and e-mail address.

2.2   PROJECT E: GRAZING LUNAR OCCULTATIONS

A very special circumstance occurs when you place yourself at a position where the star skims across the limb of the Moon. If you are in just the right place, you will see the star blink out and reappear several times as it passes behind mountains silhouetted on the lunar limb. This is called a "grazing occultation", and the situation is summarized in Figure 2.6. As you can imagine, this is a fascinating thing to watch: a star hangs right on the lunar limb, it blinks out, then it flashes back into view, and blinks out again, reappears ...

The grazing occultation presents an opportunity to gather very precise information about lunar topography in the region being "grazed" by the star. This is a group activity, and can be a very interesting, entertaining, and educational project for the members of your local astronomy club. You find the predicted path of the graze (i.e., the very edge of the occultation zone), and position your observers along a line that is perpendicular to the graze path. You'll want to space the observers so that the entire graze zone is spanned—that is, the observer on the inner edge is predicted to see a total occultation, and the observer at the outer edge is predicted to see no occultation. The more observers, the better, since the spacing of the observers sets the resolution of your determination of lunar topography.

2.2.1   Equipment needed

For this project, you will need a group of observers, and each observer will need everything that would be used for observing a total lunar occultation:

● A stopwatch capable of making "time-splits".
● A means of starting and stopping the watch at a precisely known time (this can be a shared WWV receiver).

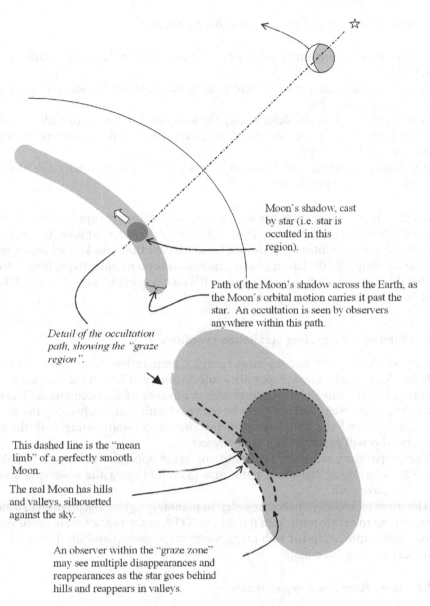

Moon's shadow, cast
by star (i.e. star is
occulted in this
region).

Path of the Moon's shadow across the Earth, as
the Moon's orbital motion carries it past the
star. An occultation is seen by observers
anywhere within this path.

*Detail of the occultation
path, showing the "graze
region".*

This dashed line is the "mean
limb" of a perfectly smooth
Moon.

The real Moon has hills
and valleys, silhouetted
against the sky.

An observer within the "graze zone"
may see multiple disappearances and
reappearances as the star goes behind
hills and reappears in valleys.

Figure 2.6. Geometry of a lunar "grazing occultation".

- A telescope, preferably with clock-drive, and eyepiece that yields a field of view of about 0.25 to 0.5 degree.
- A digital tape recorder, or (preferably) a video camera or "electronic eyepiece", and a video recorder. Video recording is preferred, because it's easy to get confused with rapid-fire events when using visual monitoring and audio recording.

As the coordinator of the graze expedition, you'll also need:

- An accurate prediction of graze opportunities and graze paths (available from IOTA).
- A topographic map (or equivalent web-based map) of the location of the graze that you will observe.
- A method of accurately determining the location (latitude, longitude, elevation) of each observer—preferably a GPS receiver, although the topographic map can also be used.
- (Optional) a planetarium program to assist you in determining the observing circumstances of the graze.

In the USA, topographic maps are available at most camping supply stores, or direct from the US Geological Survey. The "7.5 minute" series is appropriate for planning your graze occultation expedition. I will assume that you know how to read a topographic map. If you haven't been exposed to topographic maps, then a trip to your local library or book store is in order. It's an easy skill to gain, and you'll find it useful for a variety of activities.

2.2.2   Planning for a grazing occultation expedition

Grazing occultations require the most complex preparation of all of the projects in this book. Most likely, these preparatory calculations will fall on the shoulders of the expedition leader, who should do them well in advance of the occultation. The week prior to the graze event will probably be occupied with double-checking the calculations, selecting and checking the observing sites, and coordinating with the other observers who will participate in your project.

The preparatory steps are: (1) identifying graze opportunities in your neighborhood, (2) determining the graze path, and (3) establishing the observing stations across the graze path.

The most technically challenging step in planning a graze occultation expedition is determining the graze path. You'll need the IOTA graze-prediction for your region, a good topographic map (or web map), some map-reading and drafting skills, and willingness to do a little math.

2.2.2.1   *Identifying graze opportunities*

If you're not already a member of the IOTA (International Occultation Timing Association), your starting point will be the January issue of *Sky & Telescope* magazine. There is usually a summary article discussing the lunar occultation opportunities for the coming year, including a map that shows the best and brightest graze occultation paths. This summary map of graze occultations is also usually available on the magazine's website *http://skytonight.com* An alternative starting point is the annual *Handbook of the Royal Astronomical Society of Canada*, which includes details of many more lunar occultations that are visible from within Canada, USA, and

```
2006: BUCHHEIM~ROBERT          , COTO DE CAZA, CA        STATION: LAT.  33.5918N
                                 TRAVEL RADIUS  100 MI.            LONG.-117.5822E

OVERVIEW OF GRAZING OCCULTATIONS WITHIN TRAVEL RADIUS:
-----------------------------------------------------

DATE     USNO   H/P/S#   UT H M S   MAG %SNL D(MI)  ALT    AZ     SUN    PA   CUSP

JAN  3  X 30038 H107454   2  8 16   8.5  12+   52   19.7  230.4 -14.6 145.6 19.8D
JAN 24  X 21509 P264468  14 33  5   9.0  31-   31   31.1  168.6  -4.9 208.4 17.1D


JUL 19  ZC  435 H 13834  10 32  5   5.8  33-   98   30.7   83.7 -24.4 329.7 15.1D
JUL 22  X  7788 P 95030  11 30 48   8.2   8-   13   12.6   64.2 -16.0 349.5 17.5D
JUL 29  X 17073 H 55467   4 11 36   8.1  14+   30    9.5  268.6 -14.7  31.4  8.1D
```

A potentially-attractive graze occultation opportunity:
within my travel radius, a "bright-enough" star for my
portable 6-inch Newtonian telescope, and the path is only 13
miles from my home!

Figure 2.7. Example IOTA "Overview of Grazing Occultations" for an observer's location. (Used with the kind permission of the International Occultation Timing Association)

Mexico. From this information you can identify graze paths that are within your traveling radius.

After selecting one or two candidate occultations whose graze paths pass near your location, you will need to get accurate, detailed path predictions in order to plan your expedition. These are available from the IOTA.

Each IOTA member receives an annual e-mail that provides an "Overview of Grazing Occultations" that are observable from his/her location during the year. An example of this report is shown in Figure 2.7. This e-mail identifies graze opportunities that are nearby to your location. The default definition of "nearby" is a 100-mile radius, but you can adjust that on your IOTA membership information.

The next level of detail, also provided to IOTA members in their annual occultation predictions package, is given in the "Occultation Predictions for ... Observer Location" report. This report provides a variety of information about each occultation that is observable from your location. An example is shown in Figure 2.8, annotated with the definitions of the data columns. The most important planning information is:

- The date and time of the graze.
- The Moon's position, in altitude and azimuth.
- The star's magnitude.

The Moon's position is likely to influence your choice of observing site. For example, if the graze occultation happens when the Moon is at a low elevation angle, you'll need to select a site that has an unobstructed horizon in the direction of the Moon. Although this information is listed in the "Occultation Predictions" report, you will probably want to use your planetarium program to study the situation in the sky for

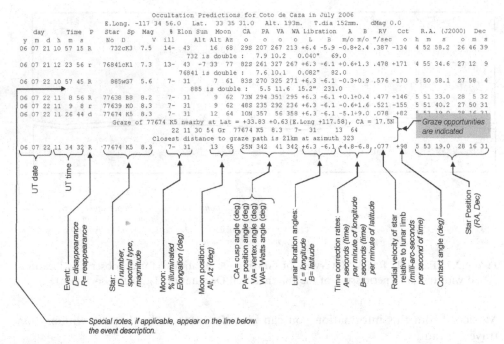

Figure 2.8. Explanation of the IOTA "Occultation Predictions" report. (Used with the kind permission of the International Occultation Timing Association)

the time of the occultation, confirming the elevation and direction of the Moon, and following the occulted star to see where on the lunar limb the graze will occur. The star's magnitude and the phase of the Moon will establish the minimum telescope aperture to use. Faint stars next to the lunar limb are quite a bit harder to see—and hence require larger aperture—than they would under a dark sky.

Figure 2.8 was prepared for my home location, and it confirms that the July 22 (UT) graze is a good candidate for an expedition. The target star will disappear at 11:26:44 UT, and will reappear at 11:34:32 UT. The period of invisibility is short (only $7\frac{1}{2}$ minutes) because the star is just skimming behind the lunar limb. As indicated in the "Occultation Predictions" listing, the graze path passes only 13 miles from my backyard. At my home's longitude ($-117°34'55.8'' = 117.58$ deg) the graze path is at latitude $= +33.83$ degrees. The number "+0.63" in this record is used to indicate the orientation of the graze path. For every 1-minute change in longitude, the graze path moves +0.63 minutes of latitude.

2.2.2.2 Determining the graze path

The information in Figures 2.7 and 2.8 identify attractive graze opportunities, but in order to select a good observing site, you need to accurately determine the graze path—the line on the Earth where a graze will occur—and plot it on a good-quality map. A graze occultation is observable only within about ±1 mile of this "graze line".

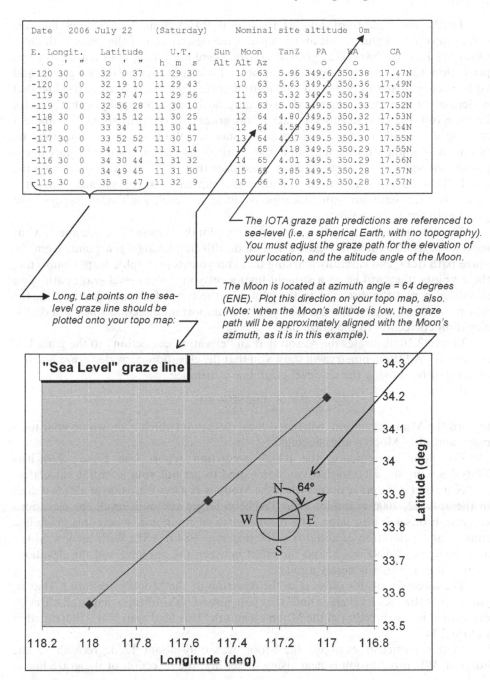

```
Date    2006 July 22    (Saturday)     Nominal site altitude  0m

E. Longit.     Latitude       U.T.    Sun  Moon   TanZ   PA     WA      CA
   o   '   "    o    '   "    h  m  s  Alt  Alt Az   o      o      o       o
-120 30   0   32  0  37   11 29 30     10   63   5.96 349.6 350.38   17.47N
-120  0   0   32 19 10   11 29 43     10   63   5.63 349.5 350.36   17.49N
-119 30   0   32 37 47   11 29 56     11   63   5.32 349.5 350.34   17.50N
-119  0   0   32 56 28   11 30 10     11   63   5.05 349.5 350.33   17.52N
-118 30   0   33 15 12   11 30 25     12   64   4.80 349.5 350.32   17.53N
-118  0   0   33 34  1   11 30 41     12   64   4.58 349.5 350.31   17.54N
-117 30   0   33 52 52   11 30 57     13   64   4.37 349.5 350.30   17.55N
-117  0   0   34 11 47   11 31 14     13   65   4.18 349.5 350.29   17.55N
-116 30   0   34 30 44   11 31 32     14   65   4.01 349.5 350.29   17.56N
-116  0   0   34 49 45   11 31 50     15   65   3.85 349.5 350.28   17.57N
-115 30   0   35  8 47   11 32  9     15   66   3.70 349.5 350.28   17.57N
```

The IOTA graze path predictions are referenced to sea-level (i.e. a spherical Earth, with no topography). You must adjust the graze path for the elevation of your location, and the altitude angle of the Moon.

Long, Lat points on the sea-level graze line should be plotted onto your topo map:

The Moon is located at azimuth angle = 64 degrees (ENE). Plot this direction on your topo map, also. (Note: when the Moon's altitude is low, the graze path will be approximately aligned with the Moon's azimuth, as it is in this example).

Figure 2.9. IOTA report of graze path (top), and example of how to plot it onto your topographic map (bottom). (Graze path report is used with the kind permission of the International Occultation Timing Association)

That's where the third data set from your IOTA annual predictions message comes into play. Figure 2.9 is an example of the description of the graze path data, using the July 22, 2006 (UT) graze example. There are a couple of features of this prediction that are very important to note. First, it provides the longitude and latitude coordinates of points on the "central graze line". That is, the graze line that is defined by the mean lunar limb, assuming no lunar topography. (That's useful because you'll place your observers across the graze line, typically spread out about a mile on either side of the line, in order to measure the true lunar topography by timing the occultation.) Second—and more critical—is that this predicted graze line is plotted for sea level on Earth. That is, it assumes that there is no topography on the Earth. This assumption is strictly a matter of computational convenience. Since it is almost certain that your actual location is not at sea level, you'll need to adjust the graze path for your elevation.

The first step in locating the graze path is to plot the "sea level" graze line on your topo map, as illustrated in Figure 2.9. You do this by picking a few points from the graze path data, and carefully marking them on your topographic map. Connecting these points (they will lie on a straight line) gives you the sea level graze path. You should also plot the direction to the Moon at the time of the occultation (i.e., the moon's azimuth angle), because that will be an important direction when you do the elevation correction to the graze line.

Figure 2.10 illustrates the reason that an "elevation correction" to the graze line is required. A little geometry will show you that the elevation-corrected graze line will be created by moving the sea level graze line a distance

$$d = h \cdot \tan(90 - \text{alt})$$

toward the Moon's azimuth, where $h = $ local elevation (which is shown on your topo map) and alt $=$ Moon's altitude angle.

For your convenience, the IOTA prediction report (see Figure 2.9) lists TANZ $= \tan(90 - \text{alt})$, so that you don't need to get out your scientific calculator.

Note that if the graze occurs when the Moon is at a low elevation angle (as it does in the July 22, 2006 example), then TANZ is large, and as a result the elevation-correction offset is also large. The region north of my home, where this graze line runs, is at an elevation of about $h = 300$ meters $= 984$ ft ASL. With the low Moon altitude angle, TANZ ≈ 4.5, and the offset between the sea level and the elevation-corrected graze lines is nearly a mile.

The direction of this offset is in the direction of the Moon's azimuth.* That is, you start at the "sea level graze line" that you plotted on your topo map, and offset by distance d, in the direction of the Moon's azimuth. This final map plot is illustrated in Figure 2.11.

In this particular example, the Moon was only about 12 degrees above the horizon. When the Moon is near rising or setting, the direction of the graze line is

---

* Since most land areas of Earth are above sea level, the offset will be *toward* the Moon. If you are observing from a site that is below sea level (e.g., California's Death Valley, or Israel's Dead Sea), then the offset will be *away* from the Moon.

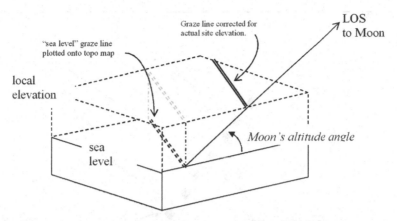

Figure 2.10. The "sea level" graze line must be corrected for the Moon's altitude angle and the local elevation, in order to find the elevation-corrected graze line.

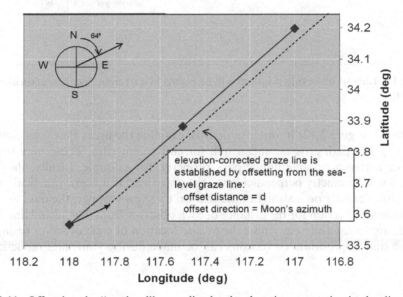

Figure 2.11. Offsetting the "sea level" graze line by the elevation correction in the direction of the Moon's azimuth defines the elevation-corrected graze path.

nearly parallel to the Moon's azimuth. When the Moon is near its culmination, the graze line will run nearly perpendicular to the direction of the Moon's azimuth.

2.2.2.3 Establishing the observing stations

Now that you know the elevation-corrected graze line, you need to carefully examine the path in order to find convenient locations for setting up your observing stations. Ideally, the stations will be established along a line perpendicular to the graze line, as

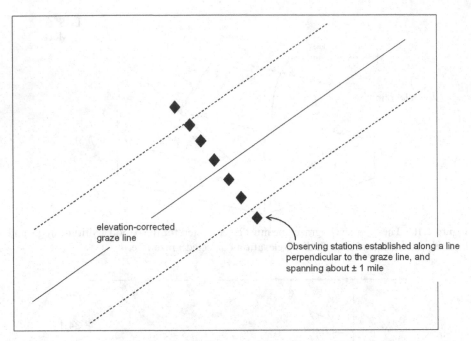

Figure 2.12. Ideally, observing positions will be placed in a straight line, perpendicular to the graze path.

illustrated in Figure 2.12. In order to do this, examine the areas that your elevation-corrected graze path crosses, and search for convenient sites. These may be along a road or across accessible public park property, for example. Ideally, the line of observers will be exactly perpendicular to the graze line. If it is not practical to place the observers exactly on a straight line, that isn't a show-stopper: the data reduction will include a correction for the distance of each observer from the "ideal" line. In any case, be sure to carefully determine the precise location of each observer (using your GPS), so that appropriate corrections can be made during your data reduction.

2.2.2.4   WinOccult3 software

A special piece of software called WinOccult3 (written by David Herald) can be downloaded from the IOTA website. This software package enables you to calculate specific information that will be useful in planning graze expeditions. WinOccult3 enables you to calculate your own graze profiles (replicating the reports illustrated in Figures 2.7, 2.8, and 2.9). In addition, it gives you a large-scale plot of the graze path, and displays the results of previous measurements/calculations of the actual lunar limb in the region of your graze. An example of the large-scale plot of the (sea level) graze line is shown in Figure 2.13. The primary value of having WinOccult make such a plot is that it offers a simple check that you didn't make an arithmetic error in

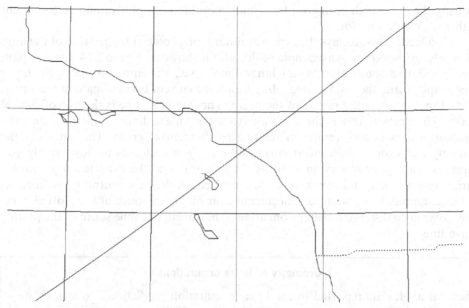

Figure 2.13. Example of WinOccult's summary graze path plot. (Used with the kind permission of David Herald and the IOTA)

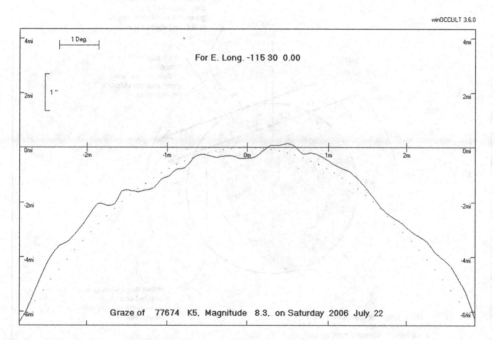

Figure 2.14. Example of WinOccult's plot of estimated lunar limb profile. (Used with the kind permission of David Herald and the IOTA)

plotting the graze path onto your topo map. Your detailed plot should be consistent with the WinOccult plot.

WinOccult also displays the expected lunar limb profile at the portion of the limb where the graze occurs. An example of this plot is shown in Figure 2.14. On this plot, the dotted line shows the "mean lunar limb" (i.e., the limb assuming no lunar topography), and the solid squiggly line shows the current best estimate of the actual lunar topography in the region of the graze. The "0 miles" line is the line of central graze. The vertical axis is the distance from the "mean limb" limit line, and the horizontal axis is time (minutes) relative to the central graze. The range of the squiggly line around the central graze line gives you a clue as to how widely you want to space your observing stations. In this example, the expected topographic variation is modest, and seems to be concentrated south of the central graze line. So, in this example, you'd want to concentrate your observers south of the plotted graze line, covering a distance ranging from about $\frac{1}{2}$ mile north to 1 mile south of the plotted graze line.

Geometry of lunar occultations

Several angles are reported in the lunar occultation predictions, to specify positions on the lunar limb. You may see them on IOTA occultation reports, and wonder, "What do these mean?" They are defined in this sketch.

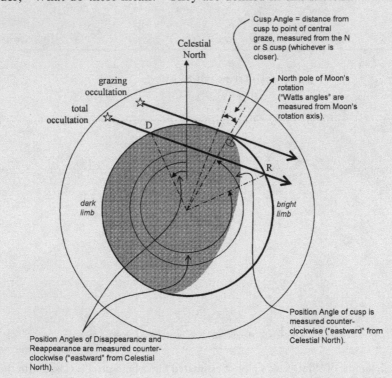

2.2.3   Conducting the observations

This is one project where the leader's organizing and planning skills will be critical to success! Observers must be recruited, the purpose and strategy explained, and detailed plans prepared. The leader will need to determine the availability of equipment for everyone, and perhaps arrange for necessary borrowing and lending (e.g., not everyone has a stopwatch, some people may need to share transportation, etc). The leader will also want to examine the planned observing location at least a few days in advance of the event (especially if it is in an unfamiliar area) to select observing sites, establish the inner and outer boundaries of the predicted observation path, discuss access permission with private property owners, evaluate the safety and security of roadside or wilderness territory, etc. These are all of the things you'd do for a total occultation or asteroid occultation expedition, multiplied by the fact that you're planning for multiple observing sites that will be scattered over a line that's a mile long.

At the observing area, each observer should be equipped with a telescope, an accurate time reference, an audio or video recorder, and a notepad. The time reference can be achieved as simply as starting each stopwatch at a known time (e.g., a WWV time tick) before the occultation. Then each observer proceeds as he/she would for a total occultation, with the added excitement that multiple disappearances and reappearances will be expected. A few minutes before the predicted start of the action, make an accurate audio time mark on the recording. When events begin, the audio recording becomes your time record. Simultaneously with each event, the observer says "on" or "off" into the audio recorder ("off" meaning that the star has disappeared, "on" meaning that it has reappeared). At instantaneous events, the observer can say "blink" (for a momentary disappearance) or "flash" (for a momentary appearance). Continue this process of watching and announcing events into the audio recorder until the series is completed. Then, make a final accurate time announcement based on either WWV or a synchronized stopwatch.

As many of the observers as possible should attempt to make video recordings of the events. The video records will be invaluable in sorting out problems or confusion during data reduction. For example, if the star blinked on and off rapidly, a visual observer might be late in announcing the "flash", but the accurate video record from a different observer might make it possible to reconstruct the event timing. They may also make for a fun focus of discussion at the next meeting of the group!

Each observer should transcribe his or her notes promptly after the event. These notes should record the WWV-time of the accurate time announcements made on the audio record, and should also explain any mistakes, ambiguities, or unusual events that occurred during monitoring of the star and lunar limb (e.g., "I think I failed to record one blink ... too slow in reacting").

Either before or after the graze, the precise location of each observer must be measured. Your GPS receiver will be a great help with this. Be sure to remain a sufficient length of time at each observer's site to get a good "average" fix on the location before moving on to the next observer's location.

Each observer then prepares an individual report, following the IOTA format (refer back to Figure 2.5). All of the details and observers' comments should be included, since they may help in weighing discordant observations. The leader (or designated data-reduction analyst) collects all observer reports, and all observing records (timing results, audio and video recordings). With that data set, he or she can prepare a consolidated graze report, and can also conduct an initial data analysis.

2.2.4 Reducing, analyzing, and reporting the results

You do not need to do any specific data reduction or analysis on your graze occultation data. The actual data sheets for each observer location are what you will submit to IOTA.

However, you will probably want to do a simple assessment of your results, both to see what you found, and as a way of confirming the consistency of your data. This assessment consists of three steps: plotting each observer's location, calculating time-offsets for each observer, and combining the data from all observers.

Using your measured coordinates, make a map-like plot that shows each observer's location on a latitude/longitude grid. Plot the direction of the graze line on this grid, also. An example of this sort of plot is shown in Figure 2.15. Select one observer's location to be the "zero-point" of coordinates for your timing analysis. If the observer locations are set perfectly on a line perpendicular to the graze path, you don't need to worry about time-offsets. However, in the usual case where the

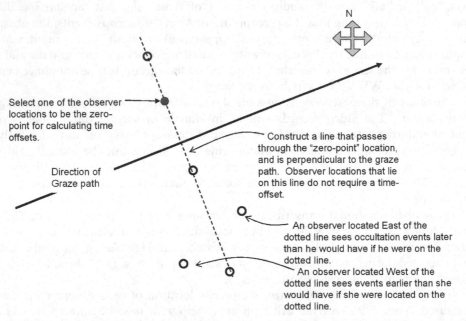

Figure 2.15. Assessing the need for time-offsets when combining the graze results from your observer stations.

observers aren't located exactly on a line perpendicular to the graze path, you need to compensate for the fact that observers that are a bit eastward of the "perfect" line will see events a little later than they would have on the "perfect" line.

The IOTA "Graze Path" report (Figure 2.9) contains the information that you need to calculate the time-offsets for observers that were east or west of the "perfect" line passing through your "zero point" observer. Take any two points on the table of the graze path prediction, and calculate the speed of the Moon's motion along the graze path:

$$V = D/\Delta t$$

where D is the distance between two points on the predicted graze path and Δt is the time interval of predicted central graze for these two points.

For each observer location that falls off the "perfect" line, measure this observer's distance from the "perfect" perpendicular line. Call this distance "d". The time adjustment for this observer is $\delta t = d/V$. Be sure that you use a consistent set of units, and that you apply the time correction in the proper sense (i.e., earlier or later than "zero-point", depending on whether the observer is east or west of your "zero point" observer's location).

With time-offsets determined and applied to observer locations that are east or west of the "perfect" line, you can make a preliminary plot of your team's results. Plot each observer's location along the "perpendicular line", vs. the timing of occultation events. Make the line thick when the star was visible, and thin/dotted for times when the star had disappeared. A somewhat idealized example of this plot is shown in Figure 2.16. With this plot, it isn't too hard to envision what you've learned

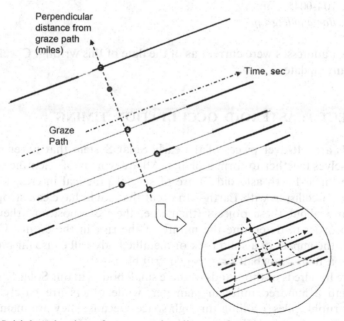

Figure 2.16. Initial evaluation of graze results: distance from graze path vs. time of events.

about the lunar topography in this region. By "connecting the dots" of disappearance and reappearance, as shown in the inset, you get a nice picture of the hills and valleys on the Moon's limb.

This effort of plotting your results is a useful thing to do, since it may highlight minor troubles with your observers' reports (such as having missed an event). You can then re-examine the raw data (tapes and timings) to see if you can sort out and correct the mistake.

Your graze report to the IOTA should contain all observers' data sheets, along with a description of any special features of the plan or the instrumentation. Graze reports should be sent to the IOTA's Coordinator for Grazing Occultations:

> Dr. Mitsuru Soma
> V.P. for Grazing Occultation Services
> National Astronomical Observatory
> Osawa-2, Mitaka-shi
> Tokyo 181 -8588, Japan
> E-mail: *somaMT@cc.nao.ac.jp*

Also send a copy of your graze report to:

> International Lunar Occultation Center (ILOC)
> Geodesy and Geophysics Division
> Hydrographic Department
> Tsukiji-5, Chuo-du
> Tokyo, 104-0045, Japan
> E-mail: *iloc@jodc.go.jp*

(These contact addresses were current as of the date of this writing. Check the IOTA website for any updates.)

2.3   PROJECT F: ASTEROID OCCULTATION TIMING

The asteroids are left-over pieces of the early Solar System that never managed to gather themselves together to form a planet. All are worthy of scientific study. Some of them (the "near-Earth asteroids") are of special practical interest because they present a risk of collision with Earth—an event that could have catastrophic effects. Learning more about these objects (their size, their composition, their orbits) is important in order to determine the nature of the risk in the event of a collision. For example, the impact of a solid rock or metallic body will cause far more damage than a hit by a fluffy snowball or a fragile "rubble pile".

There are hundreds of thousands of these small bodies in our Solar System. A few are more than a hundred miles in diameter, while others are barely more than boulders (or rubble piles?) rolling through space. Because they are members of our Solar System, they are much closer to us than the stars are. Given the huge number of

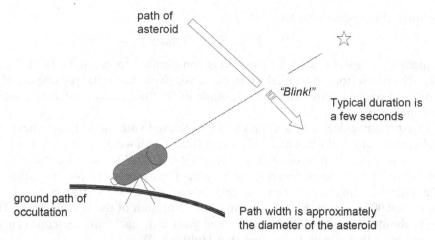

Figure 2.17. Geometry of an asteroid occultation.

asteroids, and the even more numerous stars, it is reasonable to guess that occasionally an asteroid will pass directly between us and some star. The geometry of such an event is illustrated in Figure 2.17. Let's imagine that you were staring into the eyepiece of your telescope, looking at that particular star, when an asteroid passed directly between you and the star. What would happen?

You may have been able to detect the asteroid—as a star-like point of light— gradually approaching the target star. The odds are that the asteroid will be very much fainter than the star. Whether you saw it coming or not, if the asteroid passed directly between you and the star, the star would have blinked out for a few seconds when the asteroid blocked its light.

An asteroid occultation can be seen only along a narrow strip of land, analogous to the narrow "path of totality" of a solar eclipse. The width of the occultation path is somewhat larger than the diameter of the asteroid (due to the projection of the asteroid's diameter onto the surface of the Earth). If you position your telescope in that narrow path, and observe the target star, you'll see the star "blink", disappearing for a few seconds when the asteroid blocks the starlight.

If you carefully measure the duration of the "blink", a surprisingly simple calculation gives you the size of the asteroid. It works like this: The asteroid has a physical diameter D (in miles), and it is at distance R (in millions of miles, Mmi) from the Earth. The asteroid's angular diameter is then

$$\theta = 0.21 \cdot (D_{mi}/R_{Mmi}) \text{ arc-sec}$$

(the factor 0.21 converts the angular size from micro-radians to arc-seconds).

The star is, for practical purposes, a point-source of light. The asteroid moves past this point-source at an angular velocity ω (arc-sec per second), so the duration of the blink is

$$\Delta t = \theta/\omega$$

Re-arrange these equations, and you get:

$$D = R \cdot \omega \cdot \Delta t / 0.21 \text{ miles}$$

The apparent angular rate of the asteroid is complicated to calculate, but it is very accurately known from the orbital parameters. So, given the orbital parameters of the asteroid and your measurement of the duration of the "blink", you can determine the size of the asteroid.

An astounding degree of accuracy can be achieved with modest equipment. The rate of motion of a main-belt asteroid as seen from Earth will be about 30 arc-sec per hour (give or take a factor of two), and it is not too hard to measure the duration of the "blink" (Δt) to an accuracy of 0.3 second. The accuracy of your measurement of the asteroid's angular diameter is thus about $\varepsilon = 0.3\,\text{sec} \times 30\,\text{arc-sec/hr} \times 1\,\text{hr}/ 3{,}600\,\text{sec} = 0.0025$ arc-sec. For comparison, the resolution of the Hubble Space Telescope is about 0.1 arc-sec. In this application, your little backyard telescope can give you resolution that's forty times finer that Hubble's. Wow!

Amateur astronomers have always been at the forefront of asteroid occultation studies. The very first successful observation and timing of an occultation [4] occurred in 1958, when Bjorklund and Muller watched an 8th magnitude star momentarily disappear, occulted by 3 Juno. This was a visual observation—eyeballs peering into an eyepiece.

An amateur astronomer also made the first photographic record of an occultation, when SAO 80950 was occulted by asteroid (9) Metis, in 1979 [5]. Despite being taken twenty-one years after the first-ever observation of an asteroid occultation, this historic photograph represented only the 14th successfully measured asteroid occultation.

Very few occultations were observed during that 21-year interval because of two factors. One was that very few people made the attempt. The second was that it was very difficult to accurately predict the occultation's ground track: the uncertainty in the ground track's location could be several times larger than the width of the path. Hence, even if an observer was positioned at the centerline of the predicted ground track, he only had a small chance of seeing an occultation (because the *predicted* ground track could be quite different from the *actual* ground track!) Plus, using topographic maps and an odometer, it was tricky for the observer to get an accurate reading of his actual location.

Today, modern star charts based on catalogs derived from space-based astrometry make the predictions more accurate, and GPS receivers make it a lot easier to determine your location and get into the occultation path. Hence, it is no longer a "long shot" for an amateur astronomer to be able to plan for, and successfully observe an asteroid occultation. A few minutes before the predicted occultation time, you begin staring into your eyepiece, monitoring the target star, when ... "blink" ... the star disappears, and after a few seconds reappears. That's it!

You did have the presence of mind to click your stopwatch when the star disappeared, and again when it reappeared, didn't you? Because that's where the science data is hidden. For the scientist, your observation of the occultation has several valuable features:

- The mere fact that you saw the occultation from a particular point on Earth enables astronomers to confirm the orbit of the asteroid and the position of the star.
- The duration of the "blink", times the known velocity of the asteroid, is a very precise measurement of the size of the asteroid—about the only way to directly measure an asteroid's size, without borrowing a spaceship from NASA, or being offered a night's use of the Arecibo radar telescope.
- If several observers at different locations see the occultation, their observations can be combined to get a reasonably accurate model of the shape of the asteroid.

And, if you're really lucky, you might discover that the target star is a very close double-star, or find evidence that the asteroid may have a satellite of its own.

From a purely personal and aesthetic standpoint, if you observe an asteroid occultation, you will have seen something that very few people have witnessed. As of November, 2005, there had been just over 700 successful observations of asteroid occultations. Chasing asteroid occultations is also a grand excuse to take your telescope and your spouse for a short holiday to an exotic location. For example, when I told my wife that I was leaving for the weekend to go see Ursula, she insisted on being our chaperone. That turned out be a fine weekend. I saw a 10th magnitude star disappear when it was occulted by asteroid (375) Ursula—my very first success!

Asteroid occultations are most valuable if they're observed and timed by more than one astronomer. Imagine if you could position a group of observers across the predicted occultation path. Obviously, that increases the odds of at least one of the observers seeing something. More importantly, if each observer sees the "blink" and accurately measures its time and duration, the result can be plotted as shown in Figure 2.18. There is one line per observer location, spaced relative to their offset from the predicted occultation path centerline. Time increases to the left. The light line indicates the star was visible, and the absence of a line shows the time period of the "blink" recorded by each observer. This plot gives fine resolution of the asteroid's size and shape. In the example of Figure 2.18, it isn't too hard to visualize how this result shows that the asteroid is elliptical in shape.

When you are monitoring a star in anticipation of an occultation, you expect a single, clean blink, making it easy to accurately time the beginning and end of the occultation. However, it's not always so simple. Consider the "lightcurves" illustrated in Figure 2.19. If the target star is a close double, the occultation presents a confusing combination of brightness changes, as the target star is occulted, then the companion, then the target star emerges from occultation, and finally the companion star emerges. This sort of event is one reason that an electronic eyepiece and a video recorder are preferred for occultation observations. There's a permanent record, and you can play it over and over while you unravel what was happening. It's also a good idea to watch the target star for a few minutes before and after the predicted occultation event. There are few cases of suspected asteroid satellites, indicated by secondary "blinks" of the target star.

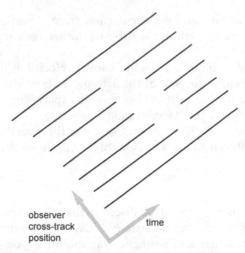

Figure 2.18. If multiple observers measure an asteroid occultation, the size and shape of the asteroid can be determined.

2.3.1   Equipment needed

The equipment required for observing and timing asteroid occultations is essentially the same as that required for lunar occultations:

- A portable shortwave receiver tuned to one of the standard time services such as WWV (refer to Appendix A for a discussion of sources of accurate time).
- A stopwatch capable of making "time-splits" (although one of the "optional" methods of recording the results is preferred).
- A method of accurately determining your location (latitude, longitude, elevation) to within about 100 meters—preferably a GPS receiver, although a topographic map or its web-based equivalent can be used.
- A telescope, preferably with clock-drive, and eyepiece that yields a field of view of about 0.25 to 0.5 degree.
- A good PC-based planetarium program (e.g., TheSky, SkyMap Pro, DeepSky, or one of the many other programs that contains the Guide Star Catalog).
- An accurate finder chart for the target star. If you are star-hopping, make a finder chart for your finder-scope, for your low-power eyepiece, and for your high-power eyepiece. If you are using a "Go-To" mount, make the finder charts anyway: they'll be invaluable in identifying the target star.
- A good map (if you are navigating to an observing site within the occultation path, rather than observing from your "home" site).
- (Optional) an audio recorder (for recording your observation notes while keeping your eye to the eyepiece).
- (Optional) a video camera or "electronic eyepiece", and a video recorder.
- (Optional) a CCD imager and associated PC and software.

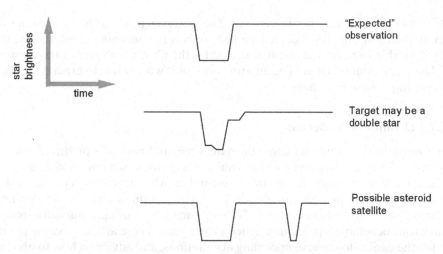

Figure 2.19. Possible asteroid occultation "lightcurve signatures".

The choice of telescope is driven primarily by the brightness of the target star—you need to be able to confidently and reliably watch the star for several minutes. The fainter the target, the larger the instrument you need. The FOV should be large enough that you can reliably match your eyepiece view to the star map that identifies the target star, and be sure that you're monitoring the correct star. A "Go-To" telescope is a definite plus, because it simplifies the process of finding the target star. It is most emphatically not a requirement, however. I suspect that a majority of successful asteroid occultation observations were made the old-fashioned way—by star-hopping.

Effective monitoring of asteroid occultations does not require a "dark sky" site or perfect skies. As long as conditions permit you to reliably monitor the target star, a bit of light pollution or thin haze is not an impediment. However, scattered clouds or unstable atmospheric conditions are problematic. Theoretically, you can monitor the target star through a "hole" in the clouds, but if the star does disappear, you may be left with a nagging uncertainty about the cause: Was it an asteroid, or a wisp of cloud that caused the star to blink out? If your target star disappeared while the other stars in the field continued to shine, then odds are that you observed an occultation event; but you should record the unstable sky conditions in your notebook, in case your observation is discordant with other observers'.

2.3.2 Preparing for the observation

Observing asteroid occultations is a tricky business because of the multiple constraints. The observer has to position himself within an occultation path that is only about as wide as the asteroid, typically less than 50 miles. The observer has to be lucky enough that his chosen location turns out to be within the *actual* occultation path (which unfortunately may or may not match the *predicted* path). The observer needs to find and identify the target star. The observer needs to be monitoring the target at the right time (the occultation lasts only a few seconds). And, finally, it is highly desirable to have equipment that enables the observer to record the observation. The great value of asteroid occultations makes it worthwhile to expend the effort in attempting to observe them!

2.3.2.1   Occultation predictions

Of first importance is that you know the date, time, and path of a predicted asteroid occultation. They are rare enough that you're not going to see one by accident! For the casual observer—and the amateur researcher who wants to try one or two occultation projects in order to decide if this is a subject that he or she finds interesting and entertaining—*Sky & Telescope* magazine usually publishes one or two asteroid occultation-prediction articles each year. These articles include predictions for the easiest-to-observe upcoming occultations, and advice on how to observe, record, and report your observations.

The best sources of detailed occultation predictions are Steve Preston's website (*www.asteroidoccultations.com*) and the IOTA website (*www.lunar-occultations.com/ iota*). At those websites you will find a listing of all the predicted events for the next few months (usually updated quarterly), along with hyperlinks to maps showing the predicted occultation paths. It is my habit to plan a month in advance, and note on my calendar the dates of those occultations whose paths are near my home.

Figure 2.20 is an example of the "star's-eye" view of the occultation. If the imaginary inhabitants of a hypothetical planet orbiting the target star were to care-

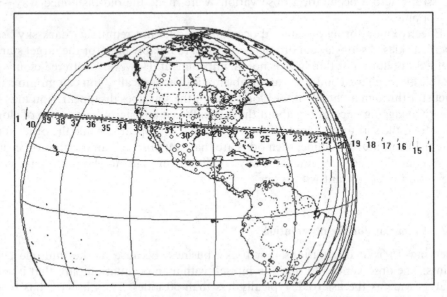

Figure 2.20. "Star's-eye" view of an asteroid occultation prediction. (Used with the kind permission of the International Occultation Timing Association)

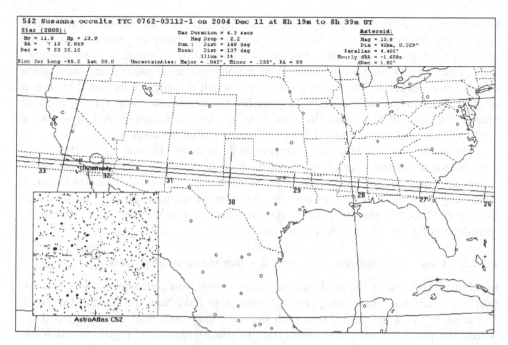

Figure 2.21. Example of a detailed prediction of an occultation path. This scale of prediction is useful for identifying occultations within your travel radius. (Used with the kind permission of the International Occultation Timing Association)

fully watch the Earth (with their impossibly-magnifying telescopes), they would see the asteroid traveling along the indicated path, as it passed between the star and the Earth. Any point within the occultation path will see the target star "blink" as the asteroid blocks its light. The time ticks on the plot show the position of the asteroid—and the predicted time of the occultation—at 1-second intervals. This scale of chart is useful for deciding if the occultation is within your range of travel.

Figure 2.21 is a closer view of the predicted occultation path. This scale is useful for selecting candidate observing locations. The IOTA website will also contain links to "detailed circumstances" of the occultation, including tabulation of the predicted centerline coordinates (latitude and longitude) at intervals of 1 degree in longitude. This tabulation, along with a good topographic map, will guide you in selecting your precise observing location. Remember that the occultation path width is comparable with the diameter of the asteroid, so most paths are less than 50 miles wide; you'll want to take some care to position yourself well within the predicted path, if you are traveling to observe the occultation. On the other hand, if the predicted path passes near your home it is worthwhile to make the observation even if you're 50 miles or so outside of the path—many path predictions are uncertain by that amount.

The odds of having an occultation path go directly over your backyard or observatory are pretty small. Hence, a necessary step in planning is to determine which occultation paths are close enough that you can travel to a point within them.

How long a journey you're willing to make depends on many things: your travel budget, the flexibility of your work schedule, the portability of your telescope, the sufferance of your spouse, etc. For most amateur astronomers, "close enough" usually means "within a 4-hour drive from home" (perhaps longer if the occultation occurs on a weekend). This is, of course, very much a matter of personal taste and interest. I know people who have a portable telescope and video set-up, and are prepared to fly halfway across the country to monitor an occultation. I myself have made a 3-day journey to central Baja California for an occultation.

The accuracy of predictions has improved dramatically in recent years, thanks to the space-based astrometry of target stars. As a result, more and more targets are being identified, and the path predictions are generally quite good. This has led to an increasing number of opportunities for observers to monitor asteroid occultations from their "home" observing sites, and to travel to remote sites with reasonable confidence of being in the occultation path.

2.3.2.2   Confirming that you have arrived within the occultation path

In addition to knowing the predicted occultation path, you'll also need a way to know that you've arrived. Historically, topographic maps have been the standard method: you carefully plot the occultation path onto the map, and find a convenient point within the predicted path that meets the obvious astronomical requirements (clear line-of-sight, minimal light pollution) and prudent safety considerations (easy to get to, comfortable location for you and your instruments). In the southwestern USA, it is often possible to find locations on public land (national parks, etc). If your selected location is on private property, be sure to get the owner's permission! If you're using the "map" method of putting yourself into the occultation path, you'll want to be very careful to identify landmarks (such as road or highway intersections) that you can use to confirm that you are at your target location. If you aim for the centerline of the occultation path, then an error of a mile or so is usually of no consequence, but an error of 10 or 20 miles cross-track might put you completely out of the predicted occultation path. Plus, wherever you do set up, you'll need to be able to accurately locate your observing position on the map, so that your report can be combined with other observers at other locations (the IOTA goal will be to prepare a plot similar to that shown in Figure 2.18).

Today the easy availability of portable GPS receivers simplifies the determination of your location, but doesn't really make the map-reading method obsolescent. You still have to figure out how to get from your home location to the occultation path. Once you've set up at the observing site, your GPS will tell you precisely where you are.

2.3.2.3   Accurate time

In addition to the duration of the "blink", it is very important to measure the actual times of disappearance and reappearance. These data points facilitate combining your results with those of other observers, at other positions along the occultation path. Therefore, you must have a way of knowing the precise time (preferably to within ± 0.2 sec.) The best way to do this is with a portable shortwave receiver that can tune into a standard time broadcast (WWV). Refer to Appendix A for a discussion of methods and challenges of determining the correct time.

2.3.2.4   Finding the target star

When you are planning your trip to the occultation path, be sure to factor into your plan a sufficient amount of time to locate the target field of view, and identify the target star.

For your first few occultations, I recommend that you plan this step very carefully, even to the extreme of over-planning. If you're going to be star-hopping, prepare a "finder chart" showing the field at the scale of your finder scope, plus a detailed chart at the scale of your main-scope eyepiece view. Observe the target field the night before the occultation, to be sure that you can find the target star, and that it is bright enough for you to monitor it with high confidence. If you have digital setting circles or a "Go-To" telescope, do all of that anyway, just in case.

After doing all of that preparation, I still budget a minimum of a half-hour for finding and centering the target star when I get to the observing site—longer if I had any trouble finding it during my "dry runs". It can be a real test of your star-hopping skill and your ability to remember star-patterns as you glance from eyepiece to star-map and back! You'll be amazed at the things that can go wrong when time is of the essence.

2.3.3   Conducting the observation

Now that you have a method for accurately determining your location (map or GPS receiver), a method for accurately noting the time (WWV receiver), and confidence in your ability to find and monitor the target star, the other equipment you'll use and the procedure you'll follow depend on your approach to monitoring and timing the occultation. There are three ways to determine the time and duration of an asteroid occultation.

First is the "eyeball and stopwatch" method. You visually monitor the target star in your eyepiece, keeping a trembling finger on the stopwatch, and take time-splits at the disappearance and reappearance of the target star. Time-splits are also taken at precisely known times (e.g., from a WWV receiver). From these you can determine the start time and duration of the occultation. The danger with this method lies in the variety of potential misfortunes: you may blink at the wrong time, or your finger may slip, or your reaction time may be slow due to cold or discomfort, or the occultation may have unusual features.

Second is the "video camera" method. Small, high-sensitivity video cameras are surprisingly inexpensive these days (SuperCircuits sells several models). You slide one of these into your telescope, and record the video of the target star, while also recording WWV on the audio track. This provides a permanent record, so that you can carefully analyze the timing by re-playing the tape. The drawbacks that I've seen in this method are that a too-small FOV of the video camera may make it tough to confirm that you're actually watching the correct target star, and that video sensitivity may not be sufficient for faint target stars.

Third is CCD drift-scan imaging. In this method, you use your CCD imager, with its excellent sensitivity, to record the occultation. The target star is set at the eastern edge of the FOV, and allowed to drift across the imager chip (with your clock-drive turned off). An occultation will cause a "break" in the drift-image of the target star. This method is often capable of detecting occultations of fainter stars than can be reliably monitored on video. The drawback is that the drift-scan image exposure is likely to be limited to less than a minute (depending on your chip size and focal length), so the start of the exposure must be carefully timed to catch the predicted occultation time: you only get one chance!

The following sections describe each method in more detail.

2.3.3.1 Stopwatch method

In concept, this method is most simple. You use only your telescope and a stopwatch. You locate the target star in your telescope, and select an eyepiece that gives you a comfortable view of the star. While listening to WWV, you start the watch at a known time (± 0.2 sec) about five minutes before the predicted occultation time (and write that time into your notebook).

Begin monitoring the target star in earnest about two minutes before the predicted occultation time, and plan to continue monitoring through about five minutes after the predicted time. During the four or five minutes centered on the predicted occultation time, you watch the scene in your eyepiece, with your finger on the stopwatch. When the star disappears, click a time-split on the stopwatch. When it reappears, click another time-split. Continue monitoring the star for another two minutes, in case there is a second "blink" (e.g., a satellite of the asteroid). Finally, click a final time-split at a precise WWV minute-marker (and note that time in your notebook also). That gives you four time markers, from which you can calculate the start time, ending time, and duration of the occultation, in the same way as was described in the "lunar occultations" project (refer back to Figure 2.3). Simple!

Simple, except that there are a few complexities to beware of. These are: your reaction time; your presence of mind; and the surprises of nature.

Reaction time: When you're the timekeeper at a track race, you can see the runners approach the tape, and prepare yourself to click the watch simultaneously with the lead runner crossing the finish line. Hence, skilled timekeepers can be accurate, and consistent with each other, to a few hundredths of a second. But with the asteroid occultation, you don't know when the "blink" will happen—it will take you completely by surprise, and as a result it will take you a brief moment to respond with a click on the watch. Hence, your recorded "time of disappearance" will always be a tiny bit delayed from "truth". Similarly, you don't know when the star will reappear, so there will also be a delay in your response to that surprise. The standard IOTA report form asks for your raw data, and your "estimated reaction time".

You can use a digital stopwatch to estimate the time it takes you to react to "surprise" events. It is done as follows. Tape a small piece of paper over the time display, so that you can't see the "seconds" or "tenths/hundredths" digits, but leave the ten-second digit visible. Start the watch, and make a time-split every time the "tens" digit changes. Do this for as many time-splits as your stopwatch allows. Then, examine your results. If your reaction to the digit's change was instantaneous, then your time splits would read "10.00 sec", "20.00 sec", etc. In reality, your time splits will probably show something like "10.20 sec", 20.25 sec", etc. The extra fraction of a second is your reaction time—the time it takes for your brain to recognize that the digit changed, and for your finger to click the watch's button.

The "default" estimate of reaction time is 0.2 seconds, based on a broad average of human-factors studies. However, yours may be noticeably faster or slower. Mine ranges from 0.25 sec to 0.38 sec, depending on conditions, with an average of 0.31 sec. You should probably be a little suspicious of a measured reaction-time that is much faster than 0.2 seconds, especially if you do this experiment in the warm comfort of your office, when you are wide awake. If you're monitoring an occultation that will occur in the middle of the night, when you may be cold, tired, and in unfamiliar surroundings, you may be a little slower and less "sharp". It won't hurt to conduct an on-the-spot assessment of your reaction time a short while before or after the occultation timing measurement.

Presence of mind and surprises of nature: I got a first-hand lesson on these two effects at the 1999 occultation of asteroid (375) Ursula. The occultation was predicted to occur at 4:17 a.m. local time, along a path about 200 miles north of my home. Since the target star was pretty faint, I decided that it was safest to use my "big gun" 16-inch Dobsonian telescope. My wife and I drove to Lost Hills CA, spent the evening scouting out several possible observing locations before settling on one in a large orchard off of a country lane. Then we drove on to a motel for a few hours sleep. We had the alarm set for 2 a.m. to get us up and relocated to the observing site. Yawn! I had budgeted 45 minutes to locate the target star, but luckily found it more quickly, and then spent another 10 minutes confirming that I had the stopwatch started properly. Being cold, in the dark, a little tired, and keeping an ear open for hazards (human or animal) in an unfamiliar rural location, tends to make a person a bit unreliable. The first time I checked the stopwatch, I discovered that my finger was on the wrong button ... which is why I always leave time to double-check everything. OK, finger on the correct button, WWV receiver playing gently in the background, eye staring into the eyepiece to monitor the target star, when suddenly, the star dimmed noticeably, but didn't disappear! I was so surprised that I couldn't decide whether to click the stopwatch or not. A fraction of a second later, the star completely disappeared, and I finally clicked the watch. A couple of more seconds, and the star blinked back "on", but I was unsure if this was just the first increment of a two-step reappearance, and in my confusion I waited another fraction of a second before clicking the stopwatch to mark "reappearance". It was all over in a few seconds. Afterwards, I realized what had happened, and what I should have done.

The target star was a close binary (suspected before the occultation). The asteroid first covered one component (which I should have "clicked" on), then the second component (for which I should have made a second time-split), then uncovered the first component (a third time-split), and finally uncovered the second component (a fourth time-split). If I'd known what was coming, I might have gathered a total of four event times, and in the process determined both the size of the asteroid and the separation of the components of the double star. But, when you don't know what's coming, and you're a little slow in the middle of the night, you may not have the presence of mind to respond quickly and correctly to such surprises. As it turned out, I was able to provide only a confirmation that the event occurred at my location. Accurate timing had to come from other observers.

Therein lies the primary drawback of the "stopwatch" method—there is no permanent record to review and re-evaluate. This is where the video method shines.

2.3.3.2   Video recording method

Small, lightweight video cameras and "electronic eyepieces" with surprising sensitivity are becoming quite affordable. These devices offer a way to make a permanent record of the occultation. The concept is as follows: you monitor the target star with the video camera, recording the video on either a VCR or a digital video recorder, and use the audio track to record time signals from WWV. Then, you can re-play the occultation event as often as necessary in order to get accurate timing and a description of any unusual features. Each playback can be timed using the "stopwatch" method described above, but with the advantage that the timing can be repeated to eliminate errors or mistakes. Simple!

Simple, except that there are a few special things to consider: available field of view, achievable limiting magnitude, and the ease (or difficulty) of recording the data.

The chip size of most of the appropriate inexpensive video cameras is small (typically $\frac{1}{3}$ to $\frac{1}{2}$ inch), which may lead to a very small field of view. The field of view you will achieve is a function of chip size and telescope focal length:

$$FOV = 57.3 S_c / FL \text{ (degrees)}$$

where S_c = the chip size (inches);
FL = your telescope focal length (inches);
$FL = D \times F/\#$ where D = telescope aperture (inches) and $F/\#$ = telescope focal ratio and the factor 57.3 converts the result to degrees.

If you have a short-focus telescope, then a $\frac{1}{3}$-inch chip may give a satisfactory field of view. For example, with my 6″ F/5 reflector and $\frac{1}{3}$-inch (diagonal) chip the FOV = 0.64 degree. The field of view is (in my opinion) just barely satisfactory: it's large enough that a properly aligned "Go-To" telescope will reliably put the target into the FOV, but barely large enough to provide a sufficient number of stars in the image to reliably match the field of view to the pattern on the star chart. If you can't find a distinctive pattern of three or four stars in the video image that matches a

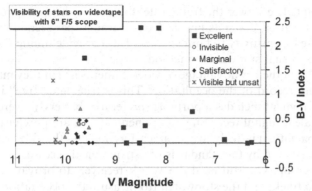

Figure 2.22. Determination of a video imager's "limiting magnitude" for asteroid occultation recording.

pattern on your star chart, then it is very hard to confirm that you're monitoring the correct star.

The same camera used with my 11″ F/6.3 SCT yields a field of view of only 10.4 arc-min. This is a very small field of view, and even with the increased light-gathering capability of the telescope (compared with the 6-inch), it is very difficult to be sure that it is pointed at the correct star. A focal reducer will be useful in such cases to increase the field of view of the video chip.

You will want to do a small project to determine the useful limiting magnitude of your video/electronic eyepiece, so that you will know which occultations are within its capability. A simple way to do this is to record a few minutes on a well-known field of view with a wide range of star brightness, and compare the visibility of the stars on the video recording with their catalog magnitudes. An example of such a study is shown in Figure 2.22. The visibility of the stars on the video tape is subjectively divided into bins ranging from "Excellent" (for the brightest stars, unmistakably steady on the video recording) through "Satisfactory" (definite and stable with only a little visible variation due to video noise), to "Marginal" (some short-duration false disappearances on the video record), to "Unsatisfactory" (either frequent noise- or scintillation-induced fluctuations or invisibility). This study showed that the video camera I've used can reliably record magnitude 9.5 to 10.0 stars with a 6″ F/5 telescope, at normal video frame rate (30 frames per second).* This restricts the system to the brighter occulted stars, but is still likely to bring three or four occultations within the range of any one observer each year.

Some of the "electronic eyepieces" and astronomical video systems allow the user to select longer exposures (slower frame rates) to increase sensitivity. This is a good

---

* Most planetarium programs report stellar magnitudes based on the Tycho and/or Guide Star Catalogs. It is well known that these magnitudes are poor by the standards of photometry (see Chapter 4), with typical errors in the range ±0.5 magnitude. This (low) level of accuracy is quite sufficient for getting a useful understanding of the realistic capability of your video imaging system for asteroid occultations.

thing, up to a point. The slower the frame rate, the poorer the timing resolution will be. As a practical matter, going as low as 5 or 10 frames per second can provide a noticeable increase in sensitivity to faint stars, and still provide timing resolution that is better than the "visual/stopwatch" method.

I glibly noted that the key advantage of video cameras is that they make it easy to create a permanent record of the occultation. This is true, and "free" if your home entertainment equipment includes a portable (preferably battery-powered) TV-VCR unit, or an equivalent digital recording component. If you are purchasing the viewing/recording capability, it is likely to cost at least as much as the video camera (albeit neither is a great expense by the standards of astronomical accessories—about $200 each). With this set-up, you still need the WWV receiver (to provide accurate time ticks on the audio track) and the stopwatch (to measure the occultation results from the recording when you play it back).

Once you've moved into astronomical video recording, you may also want to look into acquiring a time-code generator for your system [6]. This device writes accurate time ticks onto the video of each frame. This gives you the ability to play back the occultation on a frame-by-frame basis, identify the frames at which the star disappears and reappears, and thereby determine the time of each event to $\frac{1}{30}$ second.

2.3.3.3 CCD drift-scan method

More and more amateur astronomers have invested in expensive, sophisticated astro-imaging set-ups based around a CCD imager. The chip size, and hence field of view may larger than the FOV of a typical video camera. The CCD imager's cooled focal plane and ability to take long exposures enable it to record much fainter stars. Even with a half-degree or smaller FOV, a 1-minute exposure will almost certainly record enough stars that you can match the pattern to your star chart, and identify the target star for the occultation.

Of course, downloading the CCD image onto the PC that controls the imager is likely to take anywhere between 5 seconds and 30 seconds depending on your imager (more pixels means longer time to download), interface (USB is much faster than parallel-port connection), and computer. Since the frame download can take longer than the duration of a typical occultation, you can't simply take a series of short-exposure CCD images to simulate a video camera. Instead, the technique of "drift-scan" is used.

The concept of drift-scan is simple: you set the target star near the eastern edge of the field of view, turn your clock-drive off, and start an exposure. Stop the exposure just before the star reaches the western edge of the FOV. The result will be a "star-trail" image. If an occultation occurred, the trail of the target star will have a gap, indicating the time during which it "blinked" out.

An example of such a star-trail is shown in Figure 2.23. This image shows (23) Thalia occulting the 11th magnitude star TYC 4684-1624, on August 30th, 2005, UT. It was taken with an SBIG ST-8XE imager and 11" F/6.3 SCT telescope. The target star was placed at the eastern edge of the field of view. Then, about 30

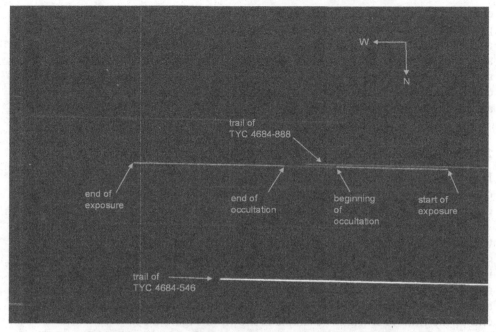

Figure 2.23. Example of a "drift scan" CCD image of an asteroid occultation.

seconds before the predicted occultation time, the telescope drive was turned off, and a 75-second exposure started.

This is a really neat way to measure the occultation of a faint star: the CCD is quite sensitive, so the star's trail is clearly defined; there's a permanent record; and most of the critical information is written automatically to the image's FITS header. The one important thing is to be sure that the exposure is short enough that both ends of the target star's trail appear on the image. Then, it's a straightforward exercise in geometry to calculate the exact time and duration of the occultation. In the case of (23) Thalia, the occultation lasted 12.7 seconds. The asteroid's angular velocity was about 16 arc-sec per hour, so this project measured the diameter of an asteroid whose size was only 0.06 arc-sec—pretty amazing for an 11-inch scope!

It is not too hard to calculate the maximum exposure duration that you can use for drift-scan imaging—that is, how long it will take for the target star to drift the full width of your image. If your field of view in the RA axis is FOV (measured in arc-min), and the target star is at Declination δ (degrees), then the time it will take for the star to drift across the field of view is approximately

$$\Delta t = \frac{\text{FOV}}{15 \cdot \cos(\delta)} \text{ minutes}$$

As a practical matter, in the half-hour before the occultation, you'll want to do a few test images, to determine how accurately you can place the star in the image, how long it takes you to initiate the imaging, and confirm the maximum permissible exposure under field conditions. That will also tell you how far in advance of the predicted occultation you want to start the image: your goal is for mid-exposure to be at the predicted occultation time. There is also a small risk that a nearby star may be at exactly the same Declination as your target star, and hence create an interfering star trail. In the example show in Figure 2.23, the star TYC 4684-888 very nearly caused this problem. Examine your pre-occultation test images carefully to see if such an interfering star trail will exist.

2.3.4   Reducing, analyzing, and reporting asteroid occultation timings

If you used the "stopwatch" method or the "video recording" method to observe the occultation, then the time reductions are done in exactly the same way as was described in the "lunar occultations" project (see Section 2.1.3).

If you used the "CCD drift scan" method, then data reduction consists of translating image position along the star trail into time. This isn't too hard, if you can dredge up memories of the geometry that you learned in high school. Use your image-processing software to examine the trail of the target star. Zoom in on the trail, and determine the pixel coordinates of the start of the trail, the point of disappearance (i.e., start of occultation), point of reappearance (end of occultation), and the end of the trail. A schematic example is shown in Figure 2.24.

Assume that the start of the exposure was T (Hr:Min:Sec, in Universal Time), and that the exposure duration was t_{exp} (in seconds). First, determine the speed of the star's drift, in pixels/second:

$$\text{speed} = S = \frac{\sqrt{(X_{end} - X_{start})^2 + (Y_{end} - Y_{start})^2}}{t_{exp}} \text{ pixels/sec}$$

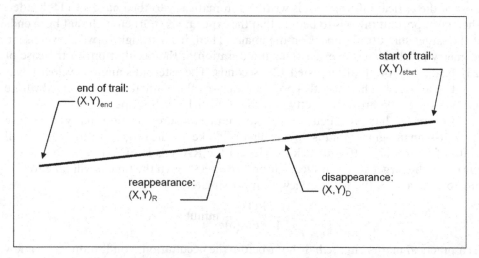

Figure 2.24. Schematic diagram for analysis of a drift-scan CCD image of an asteroid occultation.

(Note: this equation assumes that your pixels are square.) Second, determine the duration of the occultation:

$$\text{duration} = \Delta t = \frac{\sqrt{(X_R - X_D)^2 + (X_R - Y_D)^2}}{S} \text{ seconds}$$

Finally, determine the time interval from the beginning of the exposure (T_{start}) to the disappearance point:

$$(T_D - T_{\text{start}}) = \frac{\sqrt{(X_D - X_{\text{start}})^2 + (Y_D - Y_{\text{start}})^2}}{S} \text{ seconds}$$

and the time interval from the reappearance to the end of the exposure:

$$(T_R - T_{\text{end}}) = \frac{\sqrt{(X_{\text{end}} - X_R)^2 + (Y_{\text{end}} - Y_R)^2}}{S} \text{ seconds}$$

By computing both the time from start to disappearance, and the time from reappearance to end, you can use a similar technique to that described in the "lunar occultations" project, to double-check for consistency in your calculation of drift speed, occultation duration, and start/end time of the occultation.

Once you have determined the duration and time of the occultation, your data should be entered into a form that provides full explanation of your location, the circumstances of the occultation, and the timing results. All of this information should be included in your report to the collecting agency.

The central organization for collection, analysis, and dissemination of asteroid occultation observations is the International Occultation Timing Association (IOTA). The preferred method for reporting your asteroid occultation results is an Excel-based database entry form that is available at *http://www.asteroidoccultations.com* This form is illustrated in Figure 2.25. Both "positive" and "negative" reports are important, since your negative report (i.e., you didn't see a "blink") might be combined with someone else's "positive" report to define the limits of the occultation path and hence contribute to determining the size and shape of the asteroid.

Asteroid occultation observation reports should be sent by e-mail to: *http://www.reports@asteroidoccultation.com* This will put them into the hands of the IOTA's coordinators, who are responsible for analyzing, consolidating, archiving, and disseminating asteroid occultation observations.

2.4 INTERNATIONAL OCCULTATION TIMING ASSOCIATION

The center of excellence in occultation studies is the International Occultation Timing Association. The IOTA is a membership organization that provides several services for members: customized lunar and graze occultation predictions for your location, an annual package of asteroid occultation paths, and a newsletter containing reports and advice on methods and equipment. IOTA's website (*http://www.lunar-occultations.com/iota*) contains a wealth of information related to occultations,

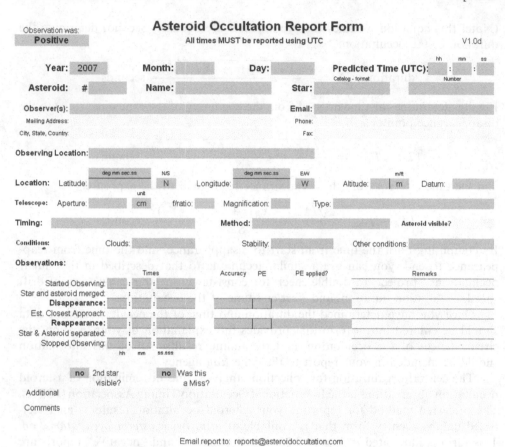

Figure 2.25. The on-line database form for reporting your asteroid occultation observations to the IOTA is available at *http://www.asteroidoccultations.com* (Used with the kind permission of the IOTA)

results, equipment, and procedures. Every occultation observer will benefit from being a member of the IOTA.

Occultation predictions are available at the IOTA website (*http://www.lunar-occultations.com/iota*), and also at *http://www.asteroidoccultations.com* The latter site's listing tends to be more extensive, since it includes occultations of fainter stars which are more difficult to observe.

2.5   ADDITIONAL READING AND RESOURCES

The most detailed description of occultation projects and procedures is the *Occultation Observer's Handbook*, by Walt Robinson and Hal Povenmire (2006). It will be a valuable reference to every occultation observer.

There is also an active occultation Yahoo® group, at *http://groups.yahoo.com/ group/IOTAoccultations*. The discussions include observing techniques, results, and plans for upcoming lunar and asteroid occultations.

Trip report: expedition to view Iphigenia (8/16/99) occultation of TYC 0587 00376—"even an unsuccessful occultation expedition can be fun!"

Dear David,

I'm beginning to think about giving up astronomy, and moving into meteorology instead! The small-scale map provided with the IOTA 1999 data package offered only a rough idea of the location of the path; and I didn't find any updated detailed predictions of the path's coordinates. I did an eyeball-estimate, averaging my best-guess as to the path's trajectory across Baja with my capacity for a one-day drive (each way), and settled on Catavina, Baja California, Mexico as the site for observing the occultation.

First, the good news: The drive from Orange County CA to Catavina was a very pleasant trip. The Hwy 1 toll road from Tijuana to Ensenada follows one of the most dramatic coastlines you could ever want to see: high hills rising directly on the east side of the road, and steep slopes ending in cliffs that drop to the clear blue Pacific Ocean, highlighted by big-sea rollers sending up fountains of spray as they crash into the cliffs. The few miles of travel through downtown Ensenada were the only complicated navigational challenge of the trip. Once you've succeeded in picking up the main road to the south of town, there's really only one paved road available to you, and if you stay on it you'll pass through farm-fields around Santo Tomas, the grape-orchards near Colonet, the rough hills and almost totally unexploited coastal plain extending the 40 miles from San Quintin to El Rosario, and the wild untamed boulders-and-arroyos landscape of the central Baja desert. According to the *AAA Guidebook*, this central desert is the only place in the world you can see the Cirio trees—straight tall poles, naked except for little tufts of spiny leaves at their crown—that look like something Steven Spielberg would invent to populate one of his planets in a galaxy far, far away. Catavina is a tiny outpost in the middle of this wilderness. The Hotel Pinta has 28 rooms (spartan, but clean), a restaurant, and a rustic Spanish-hacienda look. It also has electricity, which means it has some security and advertising lighting, so astronomical observing requires a short drive up the highway, and then off on a dirt trail to a sheltered spot in the desert. Aside from the few lights of the hotel, the nearest streetlight is probably 100 miles away, so believe me, this area is DARK at night! And it's sort of exciting, observing in a foreign desert wilderness, with no sound aside from the rustle of little critters in the bushes.

The bad news is that, despite the crystal-clear satellite photo taken the day before we left, by the evening of August 15th (i.e., about 04:00 UT 8/16/1999), high clouds were sailing in from the southeast. Ever the optimist, I set up my 16-inch F/4.5

Dobsonian anyway. My WWV receiver had no trouble pulling in a clear time-signal. The GPS receiver converged on our location (29.72802°N latitude, 114.71908°W longitude, elevation 1,674 ft.). The target star was located in a field with a distinctive pattern, and finding it was no trouble at all. I think I may have also caught a glimpse of the asteroid, a bit more than an hour prior to the predicted occultation—although it was quite a bit fainter than its predicted V_{mag} of 12.78. Then, around 06:00 UT, the target star (and every other star in the field of view) faded and disappeared behind the clouds. It peeked out coyly a few times over the next hour, but by 06:45 UT, the whole area around Aquarius, Pisces, and Pegasus was totally blocked. The only light visible in that part of the sky was the occasional flash of lightning from the distance.

Oh, well ... Once upon a time, when I was in Belize, I saw a wildlife observer's report of his encounter with a leopard: "I didn't see it, and I couldn't hear it. But I did sense its presence." In the same spirit, I can't provide a timing of the occultation, but I sense that its path must surely have passed over Catavina, else why would there have been clouds?

2.6   REFERENCES

[1]   Dunham, D.W., Dunham, J.B., and Warren, W.H. Jr., *IOTA Observer's Manual (draft version)*, IOTA, Greenbelt MD (19 March 1994).
[2]   Dunham, D.W., Nason, G., Timerson, B., and Maley, P. "Amateur-Professional Partnerships in the Observation of Occultations", in: Percy, J.R. and Wilson, J.B. (eds.), *Amateur-Professional Partnerships in Astronomy*, ASP Conference Series, Vol. 220 (2000).
[3]   Dunham, D.W., Dunham, J.B., and Warren, W.H. Jr., *IOTA Observer's Manual (draft version)*, IOTA, Greenbelt MD (19 March 1994).
[4]   Taylor, G.E., *Journal of the British Astronomical Association*, vol. 72, p. 212 (1962).
[5]   IAU-Circular 3437: "Occultation of SAO 80950 by (9) Metis."
[6]   Herchak, S. "Millisecond Video Timing for the Masses", *The Occultation Newsletter*, vol. 12, no. 2, p. 4 (April, 2005).

3

Visual variable star observing

The 19th century astronomers who laid many of the foundations of our modern understanding of the heavens did so by actually *looking* through their telescopes. No electronic sensors and computer-controlled data gathering for them! It turns out that there are still important measurements that can be made in the same way—eyeballs to eyepiece—and that amateur astronomers can contribute to science using the equipment that you probably use for every stargazing session.

3.1 PROJECT G: VISUAL OBSERVATION OF VARIABLE STARS

Poetic references to the "timeless, unchanging heavens" notwithstanding, there is a grand menagerie of variable stars* scattered throughout our galaxy. Some are now well-characterized and understood. For others, we have partial understanding, and open questions. Some are still substantially mysterious: what are they, what changes are observable as they vary in brightness, what causes their outbursts or dimming, are they involved in binary systems, etc. In order to properly and efficiently study these objects, the professional astronomer requires two types of data: long-term statistically-valid time history, and prompt alerts of unusual activity. For more than a century, amateur astronomers have provided both of these vital services to the professional community. Despite the advent of professionally-managed all-sky surveys in the past decade, the value of the amateur's contribution in these areas is undiminished.

* You will see references to "intrinsic" and "extrinsic" variable stars. Intrinsic variables are "true" variable stars—their light output changes over time. Extrinsic variables are stars whose light output is (probably) constant, but which appear to vary in brightness because of other causes. The eclipsing binary stars are the classic example of extrinsic variables.

A "statistically valid time history" of an individual variable star's activity requires that the star be kept under observation for years, decades, or centuries; that the observations be made according to well-defined, repeatable methods (so that observations from different observers can be combined without bias or error); and that the observations be collected, organized, and disseminated by a central body so that they are available to the professional astronomy community.

Keeping a variable star under observation doesn't necessarily mean that it is watched every night. The life-cycle of most stars is very long. A human lifetime is just a flash in comparison with the years that a star spends evolving through the "instability strip" of the Hertzsprung–Russell diagram, for example. The desirable frequency of observations is set by the characteristic time scale of the variable star's activity. So, a "nearly-continuous" record of observations means that the observations are spaced closely enough in time that a complete lightcurve history can be constructed, and that unusual activity is not likely to be missed. For example, Mira-type long-period variables brighten and dim in not-quite-periodic pulsations whose characteristic time scale is a year or longer from brightest to dimmest to brightest. There is no reason to check on them every hour, or every night! Once a week is quite sufficient to completely characterize their lightcurves. For many eclipsing binary stars, the situation is a little different. The timing of their eclipses is well-known and accurately periodic, so that observations can be scheduled for just those nights when the star is undergoing primary or secondary eclipse. These might be relatively frequent (e.g., Algol has a period of a little less than 3 days between eclipses) or once-in-a-decade events (e.g., ε-Aurigae's dimming as it passes behind an enigmatic extended body occurs every 27 years).* A third situation occurs with the family of stars known as "cataclysmic variables". The brightness of some of these stars may change dramatically and unpredictably on time scales of a few hours, or even a few minutes. These outbursts may be separated by weeks or years of quiescence. Therefore, these stars warrant checking (by someone, somewhere) on every night that they are visible, so that the professional community can be promptly notified of the outburst or other unusual activity [1]. The professional astronomers can then bring their large telescopes and delicate instruments to bear on the object to learn more about what is going on, and why.

Even a years-long run reporting "no unusual activity" on a particular star is valuable, because when the star's lightcurve is examined statistically, it is one thing to know for sure that "nothing happened", and something quite different (and unfortunate) to say "I don't know what happened in this time interval because no one was looking." The statistical methods that are used by professionals to analyze variable star lightcurves are best suited to well-sampled data sets. (A "well-sampled" data set is one where the data points are close enough together that there is good reason to believe that no fluctuations were missed between data points—another way of saying that the star was under sufficiently-continuous observation.) For Mira-type variables,

---

* The very special case of ε-Aurigae presents enough mysteries that monitoring of the star outside of eclipse is also needed, to understand the system's characteristics.

the long run of data (literally, a century in the case of o-Ceti) provides the basis for analyzing its pulsation modes, which contain both "frequency modulation" (gradual changes in pulsation period) and "amplitude modulation" (changes in amplitude of variation). These observational data can be cross-checked with theoretical models of stellar evolution and pulsation, to gain improved understanding of what's going on inside the star.

Some eclipsing binaries have components so close together that there is significant mass-transfer from one star to another. This effect can be identified and measured by observing gradual changes in the orbital period. The time interval between eclipses slowly changes, betraying details of the orbital evolution of the pair. Long runs of well-sampled data are required in order to detect—and have statistical confidence in—the measured changes.

Amateur astronomers began contributing variable star observations more than a hundred years ago. There have been enormous technological advances in astronomical instrumentation since then. We have also recently seen the advent of large-scale all-sky surveys funded by governments and run by professional astronomers. Given these changes, it is reasonable to wonder if there is still any value to the efforts and observations of a backyard observer staring into the eyepiece of his modest instrument. The professional astronomers involved with stellar studies will tell you "absolutely, yes: we need the amateur's contribution!" [2] In the latter part of 2005, at least five papers that used AAVSO *visual* (not CCD) data were accepted and published in peer-reviewed journals [3–7].

Regarding the major all-sky surveys, the plain truth is that they are of limited scope in time and magnitude, and most are relatively poor at photometry. They have been in operation for only a few years, and they have unproven staying power. Their future operations are subject to the continued appropriation of funding. Each of the surveys has a well-defined set of goals, and for most of them collecting data on variable stars is either secondary, or not even in the list. [An exception is the All-Sky Automated Survey (ASAS), which is specifically designed for variable star photometry.] Their variable star data—even if theoretically available—may not reach the stellar scientist in a convenient form. For example, the LONEOS survey is primarily intended to search for asteroids whose orbits make them potential threats to Earth. This is likely to take a decade or more, but it is not likely to take—or be funded—forever. Similarly, it is reasonable to expect that each survey's specific task will eventually be completed. When that happens, its data stream is likely to end, as its instruments are either mothballed, or directed to other more compelling tasks. Only the amateur astronomers will continue to collect data during, and beyond the current "era of surveys".

Even those surveys that do provide adequate photometric data are unlikely to do so with the right cadence for many variable star studies. For example, the survey may report that a particular star is varying, but follow-up observations will be needed in order to determine its lightcurve. In most cases the survey's schedule of observations will not match the timing required to prepare a complete, well-sampled lightcurve. In addition, the survey's photometry may not be on the standard system of wavelengths (filter passbands) used for decades of variable star observations. Hence, ongoing visual observations are useful to confirm that transformed survey photometry accurately overlaps with historical visual observations.

There are several practical and well-demonstrated reasons that the professional community needs and values the observations of the skilled amateur astronomer. First, there is the almost trite truth that there are many more stars, many more backyard telescopes, and many more amateur astronomers, than there are major observatories and professional astronomers. The professional stellar astronomers have a fundamental interest in maintaining long-term, complete data streams for important stars. Here, simple numbers force the professional to rely on amateur participation. Second, for many stars, the professionals cannot devote telescope time to routine monitoring of a star of interest, but they do need prompt alerts of unusual activity (e.g., an outburst of a cataclysmic variable). Third, there have recently arisen several instances where the professionals depended on amateur observations to schedule space-telescope observations of cataclysmic variables in their "quiescent" state [8]. The sensitive instruments could have been seriously harmed by the energy of the star in its "outburst" state, so amateur monitoring was used to ensure that the star was not in outburst during the scheduled observations. (Oh, if only we backyard observers had such problems—that our telescopes were too large and our equipment too sensitive!)

Finally, there are situations in which multi-spectral observations of an object of interest are made (e.g., by advanced space-based telescopes). These provide very detailed, but very short-duration characterization of the object. They must be placed into the larger context of long-term variability trends of the object, in order to get full advantage of the space-based data. "Survey" observations won't do because the space-telescope scheduling isn't (and usually can't be) coordinated with the survey's observation schedules. Professional observatories are often unable to allocate the long-observing runs that are most desirable for data correlation. So, the call goes out to the skilled amateurs, to monitor the target object either visually or via CCD. Figure 3.1 is an example of such an "alert" from AAVSO to its members, at the request of a space-based astronomy team.

I can only imagine that new astronomical technologies will make it ever more important to correlate short bursts of exotic data with longer runs of conventional (visual) data. That can only be done if there are sufficient, proficient variable star observers. And those observers will gain the requisite proficiency only by routine monitoring of variable stars. So ... don't assume that the advances in technology have made amateur visual observations obsolescent. Your visual observations are as valuable—maybe more valuable—than ever!

3.1.1   Equipment needed

You'll need three things:

- A finder chart to help you locate the variable star of interest.
- A special chart showing the set of comparison stars in the same FOV as the variable.

THE AMERICAN ASSOCIATION OF VARIABLE STAR OBSERVERS
25 Birch Street, Cambridge, MA 02138 USA
aavso@aavso.org
Tel. 617-354-0484   Fax 617-354-0665

AAVSO ALERT NOTICE 332 (January 18, 2006)

1. REQUEST TO MONITOR V426 OPH FOR XMM-NEWTON OBSERVATIONS

Dr. Darren Baskill, University of Leicester, has requested optical observations of the cataclysmic variable star V426 Oph to coincide with upcoming XMM-Newton observations.

1803+05 V426 Oph is located at RA: 18:07:51.7 and Decl.: +05:51:48. (J2000)

Both visual and CCD observations are needed. For visual observers, make an observation once per night from now until April 6, 2006. For CCD observations, please observe in B and V using the following observing schedule. Please aim for 0.01 magnitude precision. This is a relatively bright object for CCD observers, but it rises in the early morning hours so, if possible, plan ahead of time (remember, sleep is a luxury)!

Please observe during the following windows:

January 17 - February 26, 2006: At least once per night.
February 26 - March 6, 2006: Time series for as long as possible each night.
March 7 - April 6, 2006: At least once per night.

The XMM Newton observing run is scheduled for February 26 - March 6. A detailed schedule of their observations will be available in approximately 2-3 weeks.

V426 Oph is suspected to be a magnetic cataclysmic variable (polar). However, in previous X-Ray observations by the ASCA satellite the period modulation usually found in polars was absent. The XMM-Newton observations hope to either find the modulation or an explanation for their absence.

According to Dr. Baskill (used with permission):
"Our XMM-Newton observation of V426 Oph will allow us to carry out phase-resolved spectrometry; that is, we will be able to compare the Xray spectra of the numerous viewing angles that we naturally have as the two stars orbit each other. This, along with V426 Ophiuchi's relatively high inclination of 60 degrees, will allow us to search for vertical structure above and around the accretion disc. But the primary question is this: will we clearly see the periodic modulation that was hinted at in the ASCA observation? If we do, it could settle the published claims and counterclaims that this system is indeed an intermediate polar. "

Optical observations before, during and after the observing run are needed to correlate the X-Ray data with optical because "...as the gas falls through the disc, it changes colour from optical through to the X-ray."

Figure 3.1. Example of an AAVSO "Alert Notice" requesting visual observations to augment satellite-based data collection. Ongoing visual observations provide valuable correlation between short-burst detailed examination of the object, and long-term historical records. (Used with the kind permission of the American Association of Variable Star Observers)

- Your normal observing tools (telescope, eyepiece, clear dark sky, patience, a dim red flashlight, and your observing notebook).

For variable star observations, a clear sky and stable conditions are needed, but modest amounts of light pollution are not a serious impediment to making excellent brightness estimates.

3.1.1.1 Telescope considerations

Virtually any sort of observing equipment can be effectively used for variable star monitoring. The nature of your equipment will play a role in selecting the stars that you put into your observing program. With my 7×50 binoculars, I can follow stars to about magnitude 7.5. With my $16''$ Dobsonian, stars down to about 14th magnitude are available to me. There are plenty of stars in both magnitude ranges needing regular observation, so you need not be concerned that your equipment is too modest for valuable work. In fact, there are some situations where your "modest" telescope will be a better choice than your giant light bucket.

The usual array of mechanical and optical qualities that you look for in any telescope will be valuable for variable star observing: well-figured and well-collimated optics, good baffling, and a steady mount with smooth motions. A clock-drive is definitely nice. The goal is to put yourself in a situation where you can concentrate on perceiving the star's brightness, rather than on operating the telescope. Reflector, refractor, or catadioptric 'scopes all have their unique merits and weak points. I assume that you have at least one telescope that you're comfortable with, and that you've tweaked it into its best possible performance. That is the one you should use.

Two telescope performance parameters that are of special concern for variable star observations are off-axis aberrations, and vignetting. You probably know that it is common for star images to appear sharpest at the center of the field of view and a bit "softer" at the far-edge of the field. This is the symptom of off-axis aberrations. Most telescope designs are affected by such aberrations as coma, field curvature, etc. The result is that stars are "pinpoints" at the center of the field but become slightly distorted blurs as they approach the edge of the field. This blurring tends to make the star appear fainter. The same total energy is contained in the "pinpoint" and the "blur circle", but subjectively the "blur circle" will appear dimmer. This obviously corrupts your brightness measurements if one comparison star is near the edge of the FOV and the target is near the center of the field (or vice versa).

Vignetting is a bit more subtle, and merits a special experiment so that you'll understand the vignetting situation in your instrument. Most telescope designs have some degree of vignetting. The center of the FOV is "fully illuminated" by the primary objective, but off-axis points are less than 100% illuminated. With reflector telescopes, the secondary mirror and the focuser draw-tube are usually the sources of vignetting. With Schmidt–Cassegrain 'scopes the internal baffles and the star-diagonal are typical culprits. Good-quality refractors tend to be free of vignetting, but do not take this as a blanket assurance—use it as incentive to do the experiment that will show you if you should limit your field of view when observing variable stars. The principle, and the experiment, are described in the next section.

When you are making your estimates of the brightness of the variable star relative to the comp star, both should be placed close enough to the center of the FOV that neither is significantly affected by vignetting or off-axis aberrations. In some situations, that may mean that you have to put first one star at the center of the field, "remember" its brightness, and then move the other star to near the center of the field, in order to compare them. Obviously, that "remembering" step imposes on your perceptive ability, and probably gives somewhat worse accuracy compared with the situation where both stars can be simultaneously close enough to the center. Still, it's usually a better practice to have a somewhat higher random error, rather than a consistent bias in your observations.

3.1.1.2 Determination of vignetting

There has been long discussion among aficionados of the Dobsonian "light bucket" telescopes on the merits and drawbacks of large vs. small secondary mirrors. This issue is illustrated in Figure 3.2. With a large secondary mirror, you can design the telescope so that even the edge of the FOV is fully illuminated (i.e., essentially zero vignetting). If two stars are exactly the same brightness, then they will appear the same even if one is near the center of the field, and the other is near the edge of the field of view. But that large secondary mirror creates a large central obscuration, reducing the total amount of light that comes into the image, and increasing the effect

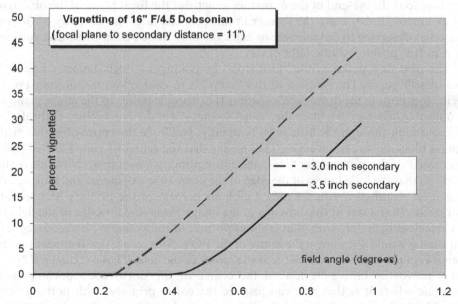

Figure 3.2. Vignetting increases, and the fully-illuminated field decreases, when a fast Newtonian telescope uses a small secondary mirror.

of diffraction. The "softening" of the image due to this diffraction is particularly evident if you use high magnification and attempt to observe fine detail. Some observers find the softening so unattractive that they want to reduce the central obscuration and its adverse effects.

A small secondary mirror creates a smaller central obscuration, which means less light-loss and less-pronounced diffraction effect. Images near the center of the field of view appear "crisper" than in a large-obscuration instrument (all other things being equal). The penalty that you pay for using a small secondary mirror is vignetting—the illumination of the field is less than 100% at modest field angles. If two stars are of exactly equal brightness, the one that is closer to the center of the field will appear to be the brighter. Many amateur astronomers—particularly deep-sky observers—find that this is a good tradeoff. You get a crisper image near the center of the field (where you'll put the object you're looking at), and suffer a barely-perceptible fall-off in brightness near the edge of the field (where you're not really concentrating your gaze, anyway). It is, however, a more troubling tradeoff for variable star observers. For variable star observing, a fully-illuminated field of view is virtually a requirement; and since many telescopes (not just Dobsonians) will show some degree of vignetting, it is important to do an experiment to determine the fully-illuminated field of your instrument.

The experiment to determine if your 'scope has noticeable vignetting is pretty easy [9]. It will take one evening and some time the next day. Pick an evening when the Moon is bright in the sky. Point your telescope at the Moon, remove the eyepiece, and cover the draw-tube's opening with a small piece of vellum or tracing paper. Focus so that you get a sharp image of the Moon on the tracing paper. What you've done is placed the eye-end of the draw-tube exactly at the focal plane of the objective lens or mirror. Either leave the focuser in this position until the next day or (if you have other observing to do) measure or mark the draw-tube's position so that you can put it in this position in the light of day.

The next day, make a little "field stop" by poking a $\frac{1}{16}$-inch diameter hole in a piece of stiff paper. The purpose of this "stop" is to enable you to put your eye at specific locations in the draw-tube's opening (i.e., specific points in the image plane of the objective lens/mirror). Place the "stop" so that it's in the exact center of the draw-tube, and look through the little hole. Normally, you'll see the entire objective, with nothing blocking any of its edges. That means that the center of your FOV is "fully illuminated"—the center-point "sees" the entire primary lens/mirror. Now, move the "stop" so that the little hole is at the edge of the draw-tube's opening, and look again. Do not be surprised if you see only a gibbous portion of the primary lens/mirror. That means that a star at this far-edge of the image plane (the far-edge of the field of view) receives light only from that gibbous portion of the primary—it's receiving less light than it would if it were at the center of the FOV. As a result, it will appear fainter when it is at the edge than when it is at the center of the field. That's vignetting! Now, slowly move the little hole toward the center of the draw-tube's opening, and determine where it is that you can just see the entire primary, with nothing "cut out". That's the limit of your "fully illuminated field". Measure how far from the center of the draw-tube that limit is. Call that distance X_{FI} (inches).

The limit of your fully-illuminated field can now be calculated. It is given by:

$$\theta_{FI} = 57.3 X_{FI}/\text{EFL} \text{ (degrees)}$$

where EFL is the focal length of the primary lens/mirror (in inches). Since you know (or can measure) the field of view of your eyepieces, this enables you to decide how close to the center of the FOV you must put the variable and comp stars to ensure that their brightness isn't corrupted by vignetting.

3.1.2 The importance of standardized observing methods

The time span over which data on any given variable star should be collected can be very long—often exceeding the lifetime of the individual astronomer. Even on a given night, there is great value in having geographically dispersed observers. That way, if one is "weathered-out", another may still achieve a good observation. Both of these considerations imply that results from many observers must be combined into a single data stream in order to achieve a long-duration, well-sampled lightcurve record. This in turn requires that the data must be collected and reported in a consistent way to eliminate bias between observers. For example, suppose that for some reason I usually report brightness that is a half-magnitude brighter than you report when we observe the same star at the same time. What then is the professional astronomer to make of it when he sees that last night you saw the star at 8.0 mag and tonight I say it is 7.5 mag? Did it really brighten, or is this just another case of observer bias? So, you can see that we need to figure out how to be as sure as possible that, if we are looking at the same star at the same time, we will report the same brightness. Only then will the eventual user of the data have confidence that the observations are reporting real changes in the star, rather than observer-to-observer inconsistencies. Observer training and experience obviously contribute to reducing bias and accidental errors. However, consistency in everyone's observing methods is just as important. This is what the industrial engineer would call "process control"—we must all do essentially the same thing, in the same way, using the same reference stars, in order to minimize inconsistencies between our measurements. The widely-accepted standard operating procedures for visual measurement of variable stars are described in the next section.

3.1.3 Making the observations

Making variable star observations sounds simple on paper: "this star is brighter than that one . . . estimate how much brighter . . . write it down . . . done!" However, during your first few attempts at the eyepiece, it is likely to seem to be nearly impossible: "where is the target star? . . . is that one brighter, or this one? . . . I'm not sure . . . the red star seems brighter now than it did a moment ago . . . I'm confused!"

Don't let a troublesome beginning session discourage you. Do you remember the first time you looked through a friend's telescope, and she described all the wonderful detail she saw in M-51—swirling spiral arms, bright nucleus, the streamer that connects to the companion galaxy, and those two foreground stars that she once mistook for supernovae—but the only thing you could perceive was a fuzzy blob?

Or the first time you tried to observe the markings on Mars, but saw only a muddy orange dot? Observing is a learned skill. The more you practice, the more skilled you become, and the more accurate and detailed your observations will be. If you stick with variable star observing, you'll quickly gain the necessary skill, as your eye and your mind become proficient in the technique.

Say, for example, that your first target for the night is YZ Andromedae, whose finder chart, matched to your telescope FOV, is shown in Figure 3.3 (see opposite page). You locate the field of view, noting the characteristic pattern of stars—such as the ragged pentagon of bright stars that's vaguely reminiscent of the constellation Auriga—to confirm your location, and identify the variable star. You also check the chart to confirm that you can identify the "comparison stars", whose brightness (magnitudes) are indicated on the chart. Since the comp stars are all quite close to the variable in your 1-degree FOV, you put the variable at the center of the field, and carefully consider the relative brightness of the stars. Find the two comp stars that are closest to the target in brightness—one just a little brighter than the target variable and the other just a little fainter. Suppose that you decide that this night, YZ And is definitely brighter than the M-13.5 comp star, a little bit brighter than the M-13.0 comp star, not quite as bright as the M-12.3 comp star, and that YZ And is definitely fainter than the M-11.9 comp star. You record this brightness sequence in your notebook:

$$
\begin{array}{l}
\uparrow \quad \text{M} \quad 11.9 \\
\qquad \text{M} \quad 12.3 \\
\qquad \leftarrow \text{YZ And} \\
\qquad \text{M} \quad 13.0 \\
\qquad \text{M} \quad 13.5
\end{array}
$$

increasing brightness

Then you more carefully consider the position of YZ And in this list. Is it closer in brightness to the M-12.3 star, or the M-13.0 star? If you decide that it is exactly halfway in brightness between the two, then you record it as YZ And = M-12.7. If you can mentally divide the brightness difference between M-12.3 and M-13.0 into four equal steps, decide which "step" the variable falls in to. If YZ And is just one step fainter than the M-12.3 comp star, then your estimate is $12.3 + (0.7/4) = 12.475 \approx 12.5$.

You record this information in your notebook, and you've completed one observation. Clear your mind for a moment, and perhaps double-check your estimate. Then you're ready to move on to the evening's next target.

This quick run-through of the procedure probably leaves a few questions and has glossed over some details. I'll come back to the details in the next two sections. The obvious questions include:

- Where did you get the chart? How do you read it?
- Why were those particular comparison stars selected? Do all observers use them?
- What if the variable is brighter than the brightest comp star? What if it's fainter than the faintest comp star (or, what if I can't see it at all)?

- How did you decide that YZ Andromedae was tonight's target?
- What do I do with the observations I've recorded?

Let's take these questions in order.

3.1.3.1   AAVSO star charts

The chart shown in Figure 3.3 is from the AAVSO and is typical of the standard AAVSO charts. Charts are available in a variety of scales, ranging from binocular field of view to very narrow CCD field-of-view charts. This one is a D-scale chart. It spans a 1-degree field of view, typical of the FOV of my 6-inch F/5 Newtonian telescope at 30× (using a 26-mm eyepiece). The target variable star is indicated with a special symbol. The comp stars are labeled with their magnitudes, but the decimal points are omitted, to avoid extraneous dots on the chart. The star labeled "105" is

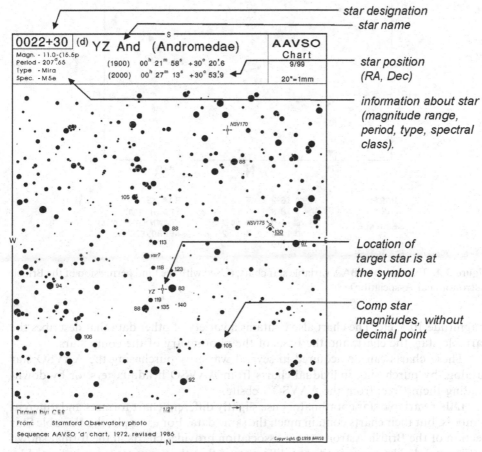

Figure 3.3. Sample AAVSO star chart, and explanation of the information it provides. (Used with the kind permission of the American Association of Variable Star Observers)

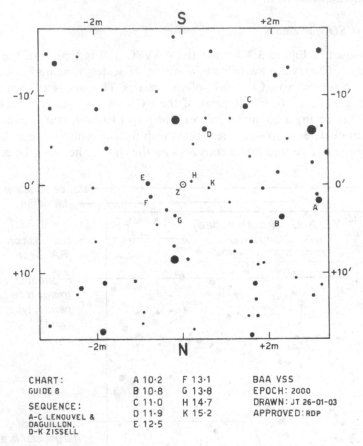

Figure 3.4. Example of a BAA variable star chart. (Used with the kind permission of the British Astronomical Association)

magnitude 10.5, etc. The chart also contains a variety of other data that describes the variable star, the chart, and the basis of the photometry of the comp stars.

These charts can be acquired in several ways: by purchasing the AAVSO star catalog, by purchasing individual charts from AAVSO headquarters, or by downloading them (free) from the AAVSO website.

Other variable star coordinators use slightly different chart formats and different symbols, but their charts contain much the same data. For example, the Variable Star Section of the British Astronomical Association provides charts in the format shown in Figure 3.4. The most significant difference is that the comp stars are indicated by letters, with their actual magnitudes in a table at the bottom of the chart.

3.1.3.2   Comparison stars

The selection of "comp stars" is a little art in itself. The ones on the standard charts have been selected to cover the whole range of the variable star's magnitude change (so that there's always at least one comp star brighter than the variable's expected maximum, and one star fainter than the variable's expected minimum). They are reasonably close to the variable so that both the comparisons and the target can be seen in the same FOV. None of the comps are close double-stars (so that the observer isn't confused by which member of the pair is intended). Ideally, they are approximately the same color as the variable so that the observer isn't forced to disentangle the perceived effects of different brightness and different color (as tricky as it can be to decide if star A is brighter than star B, it's really confusing to try to decide if this blue star is brighter or fainter than that orange star!) Unfortunately, many types of variable star tend to be pretty red, so this goal must often be compromised. The most critical feature is that the photometry (brightness in two or three spectral bands, usually the standard Johnson–Cousins B, V, and R bands) has been accurately determined and checked for reasonable assurance that the comp stars are not themselves variable.

The visual variable star observer won't get involved in measuring or selecting the comp stars. (A few CCD photometrists might take on the challenge of creating photometric sequences, as described in Chapter 4). What is critical is that you *use* the defined comp stars as your magnitude references and that you *record* which comp stars formed the basis of your magnitude estimate for the target variable star. If everyone uses the predefined comp stars, then there is no source of error or bias because of "my" comp stars being different from "yours". If all observers, over the world and over the years use the same set of comp stars, then all of our observations can be reliably combined into a single long-term lightcurve history. Recording the identity of the specific comp stars used in each magnitude estimate will help confirm the validity of the estimate (people do, after all, transpose numbers or make illegible marks in their notebooks from time to time), and may be useful if it turns out that the photometry of a comp star is adjusted at some future time (e.g., suppose we discover that one of the comp stars is a very-long-period variable, or that there was some problem with the underlying photometry of that star).

3.1.3.3   Variations outside the range of the comp stars

It will occasionally happen that the variable may appear brighter than the brightest comp star. If the variable were to appear brighter than the brightest (mag 8.3) comp star on the chart, then you would record it as ">8.3", meaning "brighter than mag 8.3". Similarly, if it is fainter than the faintest (mag 14.0) comp star, then you record "<14.0", meaning "fainter than mag 14.0".

You will also run into situations where the variable star is invisible. For example, with my 6″ Newtonian, viewing from my suburban backyard with my middle-aged eyes, I can realistically expect to see down to about magnitude 12.5. So, not only might the variable be invisible, but the mag 14.0 comp star is also invisible, as is the mag 13.5 comp star. Suppose that the faintest comp star I can see is the mag 12.3 star, but that YZ And itself is invisible. I would then record YZ And as "<12.3", meaning that it is fainter than 12.3 magnitude. Perhaps someone with a larger telescope or darker/clearer sky will be able to actually bracket its brightness. In that case, the data analysts at AAVSO who collate our results will understand exactly what happened, and will see that our observations are, in fact, consistent.

3.1.3.4 Selecting your first targets

When you begin observing variable stars, you are embarking on a learning process, and it is appropriate to select as your first targets stars that are relatively "friendly". First, select a couple of constellations that are rising at astronomical twilight. Picking your target variables in these regions will give you the opportunity to follow them through the entire season. Second, identify a few variable stars in these constellations that are well within the magnitude range of the telescope you intend to use (e.g., no fainter than 12th magnitude if you're using a 6-inch telescope). Highlight those that are well-observed by visual observers (by checking the AAVSO lightcurve generator). High-amplitude, medium-period intrinsic variables make good friendly targets. By measuring their brightness once per night every few nights, you'll gather a large array of data in the course of a few months, and can compare your results with those of other observers.

You may find that some of your more experienced astronomical friends recommend against selecting an eclipsing binary target during your "training" period. The reason is that there will be little or no variation in the star when you're observing it out of eclipse. If you do select an eclipsing variable as one of your initial targets, you'll want to find a night that offers a deep eclipse, and devote the entire night to estimating the brightness of that star, at intervals of 20 minutes or so. If all goes well, you'll get a complete lightcurve in one night, and will have given your variable star muscles a great workout.

Two good sources of friendly variable stars are *Sky & Telescope* magazine's "variable stars of the month" column, and AAVSO's "Variable Star of the Season" article on their website.

3.1.4 Human factors considerations

Because visual variable star observations are very much a "personal" matter, you'll want to be aware of several aspects of human factors related to comparing brightness. First, as is true of any meticulous task, your general comfort (or discomfort) will affect the accuracy and repeatability of your comparisons. If you're comfortably seated, warm, and secure, then you can focus all of your attention on the quality of your observations. If you're shivering cold, legs cramped from standing in a contorted position, then you will be hurried, distracted, and generally a less-accurate observer. And (as you probably know from experience) a tired astronomer in the wee hours of the morning is far more prone to making really dumb mistakes than he or she was earlier in the evening! Such things as meticulously measuring the brightness of a random field star instead of the target variable, or transposing entries in your notebook, are hardly rare. Some of the duplicative entries in your logbook (such as entering both the name and the designation of the target star) are designed to help you unravel such goof-ups. When you do make such a mistake, just smile at yourself and try again—you're in good company!

As was discussed in the previous section, you'll want to place the target and comp stars near the center of your field of view to avoid the confounding effects of vignetting and off-axis aberrations. This discipline also tends to ensure that the stars' images will be at approximately the same position on the retina of your eye. Yes, your eye's sensitivity to brightness (and brightness differences) probably does vary across your field of view, so it's best to place the stars at about the same position in your visual field. Most of the time, it's recommended to use "direct" vision (rather than averted vision) when making brightness comparisons. If the variable happens to be near your limit of perception, and you're more comfortable using averted vision with it, then be sure that you apply averted vision to the comp stars also.

If the target and/or comp stars are near the limit of your perception—your "limiting magnitude"—then you may be able to improve your view by increasing the magnification. Increasing magnification tends to reduce the effect of sky background, relative to the stellar "point" of light, and may improve your limiting magnitude by a few tenths of a magnitude. If the stars are faint, this might be just the improvement you need to have more confidence in your measurements.

The human eye is most sensitive to brightness differences when the objects are not too bright. Stars that are within two or three magnitudes of your "limiting magnitude" present you with a situation that is best for human perceptive accuracy [10]. If your target star is very much brighter (e.g., an 8th magnitude target star observed with a 6-inch telescope), then you may get better accuracy if you use a smaller telescope, or stop down your aperture to 2 inches, so that the stars are closer to the "limiting magnitude", where your perceptive abilities are more refined.

It turns out that the way the human eye–brain combination processes brightness comparisons is affected by the orientation of the two objects (the "Ceraski effect"). You are probably slightly more accurate when comparing the brightness of two points that are oriented side-by-side, than you are if they are oriented one above the other. So, if you have the option, tip your head or rotate your star diagonal so that the two comp stars and the variable are in an approximately horizontal row. Similarly, if one star is at the center of your field of view, and the other is off to the side, it is probable that you have a personal tendency to consistently rate one of the two positions as "brighter" (even if the stars are, in actuality, identical brightness). Each person probably has a different predilection in this regard, but it appears to be a real effect. For this reason, it is recommended that you bring the variable to the center, then bring the nearest-brighter comp star to the center, and also the nearest-fainter comp star to the center of your FOV, to confirm your assessment of their relative brightness.

Color can be a real confounding problem. If the target variable star is noticeably different in color from the comp star, then you have to differentiate between "color" difference vs. "brightness" difference—an unsatisfying thing to try to do. One trick that has been successful is to slightly defocus the star images. This makes them dimmer, and your eye is less sensitive to color differences when the light is dim, so that you're left with more of a pure "brightness" comparison to make.

Another color effect that is commonly encountered is the "Purkinje effect" [11]. This has two dimensions. First, and most commonly, the apparent brightness of very red stars seems to gradually increase the longer you stare at them. The solution to this is that if your target star is noticeably red (as some types of variable stars are), make your brightness-assessment of it in short, quick glances rather than long, steady stares. The second aspect of the Purkinje effect is that your sensitivity to red vs. white stars is not equally proportional. Suppose that you are presented with two stars of significantly different color: R (a red star) and W (a white star). If these two stars, R and W, are perceived to be equally bright in your telescope, then doubling their brightness (e.g., by using a larger telescope) will likely be perceived as $2R \neq 2W$. If your target is quite red, and the comp stars white or blue, then you'll want to arrange your observational set-up so that they are near your limiting magnitude, to minimize your eye's sensitivity to color difference, and hence to this confounding effect [12].

In addition to these human factors considerations, there is a matter of "discipline" and "forced objectivity" that you'll want to learn. Let's say you're monitoring a rapidly-varying star (such as a short-period eclipsing binary, in which the star fades by up to a full magnitude in just a few hours). You make your first, careful, comparison and decide that your target is 10.5 mag. When you make your next measurement—say, 15 minutes later—you must consciously determine not to be influenced by the previous measurement. Your second measurement must be an objective, independent comparison of target and comp star brightness, during which you convince yourself to "forget" what the previous measurements were. This is a tricky discipline, but one that is important in order to ensure that your recorded magnitude estimates are unbiased. You will probably find it useful to step away from the eyepiece for a few minutes, to let the memory of the previous observation "fade away". It may even be useful to go to another field (perhaps measuring a different star), and then return to the first target with a fresh eye, and enforced ignorance of the previous measurement.

By training yourself carefully, and following the standard observing procedures, you should strive to estimate the variable star's brightness to an internal consistency of ±0.1 magnitude. It is quite realistic to expect that your estimates will be within ±0.25 magnitude of other observer's estimates.

3.1.5   Recording your observations

I emphasized earlier the importance of maintaining complete, objective records of your observations. I'll challenge you again on that subject in Appendix C. For variable stars, the following parameters must be recorded in order for your observations to be understood and combined with those of other observers [13]:

- Date and time of observation: the accuracy required depends on the rapidity of variation of the target star, but as a general rule time accurate to 1 minute is good practice. Higher accuracy may be warranted for stars whose variations are quite rapid.

- Instrument description (aperture, magnification, telescope type).
- Identity of the "comp stars" that bracketed the variable star's brightness.
- Name and designation of the target variable star and its brightness. Record both the name and the designation—i.e., YZ And and 0022+30—as protection against illegibility or accidental errors in your notes.
- Identity of the chart used (e.g., the scale and revision date, in the case of AAVSO charts).
- Miscellaneous notes such as sky conditions, weather conditions, and other effects that may be useful to know if you come back to re-examine these notes in the future. If for some reason you have less-than-normal confidence in your magnitude estimate, describe why (e.g., "bitterly cold night, eyes watering, uncertain measurement".)

This information should be recorded, while you're conducting your observing session, in your permanent notebook. It is bad practice to take "rough notes" on scrap paper at the telescope, and then attempt to improve, expand, or embellish them the next day. Doing so just adds the risk that your recollections are different than what your on-the-spot observations were. The purpose of your notebook is to record objectively and precisely what you saw in the eyepiece. The only way to be sure that your notebook represents an accurate record of what you *actually* saw is to record those observations "on the spot".

3.1.6 Reporting your observations

After you've learned the techniques, applied the standard observing methods, and collected some data points on one or more variable stars in your notebook, it is essential that they be collated with those of other observers and made available for the professional astronomers. A data point that lies hidden in your notebook, seen and used by no one, might just as well not exist! There are several well-respected organizations whose mission includes the collecting, checking, collating, and disseminating of variable star observations. The premier organization is the American Association of Variable Star Observers (AAVSO).

Before you can submit observations to AAVSO, you must request and receive an "observer code". This identifies you, and enables AAVSO to link all of your submitted observations to your credit. Any variable star observer can request an observer code; you do not have to be a member of AAVSO (although if you find that variable star observing is a regular part of your astronomical life, you will benefit from being a member). The request form can be filled in on the AAVSO website, where you'll enter your name, address, e-mail, and some information about your observing equipment. Your observer code will be assigned and sent to you within a few days. You then use your code to identify yourself whenever you submit observations.

The most convenient method for submitting your observations to the AAVSO is their on-line "WebObs" system. The contents and format for the entries are shown in Figure 3.5. You can follow this example as a template for your own data entry. The variable star is identified either by its "Designation" (0022+30) or its "Name" (YZ And). This particular observation was made on January 12, 2006, at 08:15 UT, so that is the reported "Date". If you prefer, you can convert the time to Julian Day and use JD in your report. (See Appendix A if you are not familiar with UT or JD.) The interpolated brightness of the star is entered as 12.5 (rounded to a single decimal place for visual observations). If the star was invisible, enter the faintest comp star that you could see in the "Mag" cell, and check the "Fainter Than" button. Since the mag 12.3 and mag 13.0 comparison stars were used as the basis for estimating the target's brightness, they are entered in the indicated field. The comp stars were selected from the chart dated 9/99, so that is used as the chart identification. The "Comment Codes" field contains a drop-down menu of the most common comments (e.g., "Bright Sky", "Haze", "Moon Interfered", etc.) This was a visual observation, so from the "Band Field" drop-down menu, select "Visual". The other fields ("Photo-metric Uncertainty", "Airmass", and "Transformed") are only needed for CCD observers. Leave them blank for visual observations.

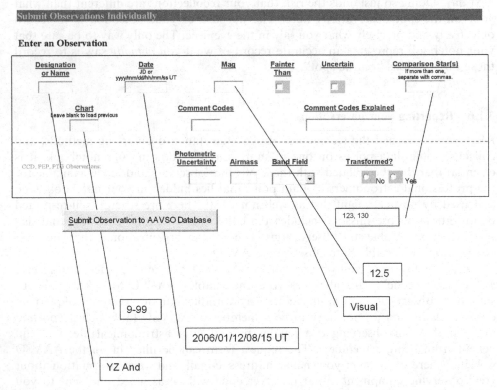

Figure 3.5. AAVSO web observation-submittal form. (Used with the kind permission of the American Association of Variable Star Observers)

That completes the data entry for this observation. Hit the "Submit Observation" button, and your results will be sent to the AAVSO server and incorporated into their database. Within a day or less, it will be available to any researchers who request data on this star.

The BAA's Variable Star Section accepts observations via e-mail, and processes them in a similar fashion, to make them available to scientists worldwide. For details, see *http://www.britastro.org/vss*

3.2 COORDINATING ORGANIZATIONS: AAVSO, BAA-VSS, AND AFOEV

The central coordinating organization for variable star observations in the USA is the American Association of Variable Star Observers (AAVSO). Every active variable star observer in North America should be a member of this organization. Their publications will keep you up-to-date on techniques, variable star research results, and specific stars that are in need of current observation. The association's website (*http://www.aavso.org*) contains a wealth of information for both the observer and the researcher. For the beginning observer, their *Manual for Visual Observing of Variable Stars* is available for free download. Of particular interest to the astronomer who has made variable star observations are the lightcurve generator (which shows you the current database of observations for each star), the WebObs application (that allows you to enter your observations into the database), and the VSX system (that gives you descriptive summary data on virtually all known variable stars.)

In Great Britain, the Variable Star Section of the British Astronomical Association (BAA) maintains an active variable star observing program, a large and respected database of observations, and star charts that are similar to those of the AAVSO. For British amateur astronomers, the BAA-VSS's mentoring program is a wonderful way of meeting other variable star observers, and receiving personal training and advice for the pursuit of your variable star research observations. The VSS can be reached on the web at *http://www.britastro.org/vss*

The French Association of Variable Star Observers (Association Française des Observateurs d'Etoiles Variables—AFOEV) is the corresponding organization of the European variable star community. This organization can be found on the web at *http://cdsweb.u-strasbg.fr/afoev*, in both French and English-language versions. A unique feature of the AFOEV database is that it incorporates the data from a great many smaller European variable star associations.

3.3 REFERENCES

[1] Menali, H., "A date with U Gem—An amateur's humble contribution to science", in *Eyepiece Views #313*, AAVSO (March, 2006).
[2] Templeton, M., "Astrophysics from Visual Observations", in *Eyepiece Views #313*, AAVSO (March, 2006).

[3] Rudnitskii, G.M. and Pashchenko, M.I., "Long Term Monitoring of the Long-Period Variable Y-Cassiopeiae in the 1.35-cm water vapor line", *Astronomy Letters* (Pleiades Publishing, Inc.), vol. 31, no. 11, p. 760 (2005).

[4] Templeton, M.R., Leaman, R., Szkody, P., Henden, A., Cook, L., Starkey, D., Oksanen, A., Koppelman, M., Boyd, D., Nelson, P.R. *et al.*, "The Recently Discovered Dwarf Nova System ASAS J002511+1217.2: A New WZ Sagittae Star", *Publications of the Astronomical Society of the Pacific*, vol. 118, no. 840, p. 236 (2006).

[5] Long, K.S., Froning, C.S., Knigge, C., Blair, W.P., Kallman, T.R. and Ko, Yuan-Kuen, "Far-Ultraviolet Spectroscopy of the Dwarf Novae SS Cygni and WX Hydri in Quiescence", *The Astrophysical Journal*, vol. 630, issue 1, p. 511 (2005).

[6] Templeton, M.R., Mattei, J.A., and Willson, L.A. "Secular Evolution in Mira Variable Pulsations", *The Astronomical Journal*, vol. 130, issue 2, p. 776 (2005).

[7] Wheatley, P.J. and Mauche, C.W. "Chandra X-ray Observations of WZ Sge in Superoutburst", in Hameury, J.-M. (ed.), *The Astrophysics of Cataclysmic Variables and Related Objects*, Proceedings of ASP Conference, vol. 330, p. 257, Astronomical Society of the Pacific (San Francisco, 2005).

[8] Henden, A.A., "From the Director", *AAVSO Newsletter*, no. 32 (December, 2005).

[9] North, G., *Observing Variable Stars, Novae, and Supernovae*, Cambridge Univ. Press (2004).

[10] North, G., *Observing Variable Stars, Novae, and Supernovae*, Cambridge Univ. Press (2004).

[11] Southall, J.P.C. (translator), *Helmholtz's Treatise on Physiological Optics, Vol II: The Sensations of Vision*, The Optical Society of America (1924). Electronic edition (2000) © University of Pennsylvania at URL: *http://psych.upenn.edu/backuslab/helmholtz*

[12] Grouiller, H., "The Purkinje Phenomenon and the Minima of Long Period Variables", *The Observatory*, vol. 59, p. 86 (1936).

[13] Beck, S. J. (editor), *Manual for Visual Observing of Variable Stars*, American Association of Variable Star Observers, 2005.

4

CCD photometry

4.1 INTRODUCTION

CCD imagers have revolutionized amateur astrophotography. They have also given amateur astronomers the ability to conduct a variety of projects that entail the accurate measurement of the brightness of celestial objects. That is the essence of the science of photometry: accurate characterization of the quantity of light that is coming from an object. In this introductory section, I'll cover the barest summary of what you'll want to know about light, the atmosphere, and the CCD. Photometry is a broad subject, and one where amateur astronomers are actively collaborating with professionals on a wide variety of projects. It is also one of those subjects where "the more you know, the more you realize you need to learn." The references at the end of this chapter are well worth reading (and owning for reference) as your knowledge and skill increases and you discover yet more things that you don't know.

The value of photometric studies that are within the range of the advanced amateur astronomer is as broad as the sky itself. There are many types of variable stars that need regular monitoring. For some, the goal is to alert professional astronomers when something unusual occurs (e.g., an outburst). For others the goal is to characterize the orbits, sizes, and temperatures of the stars in a binary system. Closer to home, photometry of asteroids can provide the data that astronomers need to determine the rotation rate, shape, and pole orientation of these objects. As a real challenge, precise photometry can confirm the existence of extra-solar planets and help determine their mass.

When you look at them with your eyeball to the eyepiece, these objects—single stars, close binaries, asteroids—all appear to be tiny points of light. Careful study of that light, and an understanding of the things that happen to the light as it makes its way to your CCD imager, can yield valuable scientific information.

The light that we receive from a celestial object—say, a star—can be described by several parameters. Is it bright, or dim? What color is it? Is it constant (invariant), or does it change with time? Of course, as scientists we want to precisely define, and quantify, the answers to these questions.

Bright or dim is a way of describing how much light we receive from the star. The amount of light can be measured in two ways: as an energy flux (watts/m^2), or as a flux of photons (photons/m^2/sec). The energy flux from a typical star is a very small number. A 1st-magnitude star delivers about 1.14×10^{-8} watts/m^2 to your eye. That flux represents an enormous number of photons (about 5×10^9 photons/m^2/sec). Neither of these numbers is easy to measure directly, and they're not even very easy to talk about. Hence, it is most common for astronomers to define a reference object and then describe other objects by comparison with the reference object—for example, "the target star is twice as bright as the reference." Further, objects in the sky present an enormous range of brightness: the Sun is nearly a million times as bright as the full Moon, which is about 25,000 times as bright as a 1st-magnitude star, which is about 160,000 times as bright as Pluto. Therefore, it is most convenient to use a logarithmic scale to describe relative brightness. Suppose that star #1 delivers intensity I_1 (watts/cm^2) to the aperture of your telescope, and that star #2 delivers I_2 watts/cm^2. The "magnitude difference" between two stars is related to their intensities by:

$$\Delta m = (m_1 - m_2) = -2.5 \log(I_1/I_2)$$

Let's decide that star #2 will be the "standard star" with which we'll compare all other stars. If star #1 is twice as bright as star #2, then we have

$$\Delta m = (m_1 - m_2) = -2.5 \log(2) = -0.75$$

and we say that "star #1 is 0.75 magnitude brighter than star #2."

As another example, suppose that star #3 is only one-tenth as bright as star #2. Then,

$$\Delta m = (m_3 - m_2) = -2.5 \log(0.1) = 2.5$$

and we say that "star #3 is 2.5 magnitudes fainter than star #2."

Note that we have described the brightness of star #1 and star #3 relative to the "reference object", without assigning a particular magnitude value to the reference object. The selection of the reference object is arbitrary, but obviously the math becomes particularly simple if we decide that the reference object will be defined to be zero-magnitude. Although there are a variety of special caveats, almost all photometric systems use the star Vega as the fundamental standard reference star whose magnitude is defined to be $m = 0$ in all spectral bands [1].

Color is a way of defining the spectral characteristics of the light that we receive from an object. Light is an electromagnetic wave, and as such it can be described by its frequency or, equivalently, its wavelength. The human eye is sensitive to light with wavelengths in the range from approximately 0.5 μm at the blue end to 0.65 μm at the red end. Different sensors are sensitive to different ranges of wavelength. Typical silicon CCD imagers are sensitive to light from about 0.35 to 0.9 μm. The typical star generates light over a far broader range of wavelengths, ranging from the far-ultraviolet (0.2 μm) to the thermal infrared (10 μm) and beyond into radio emissions

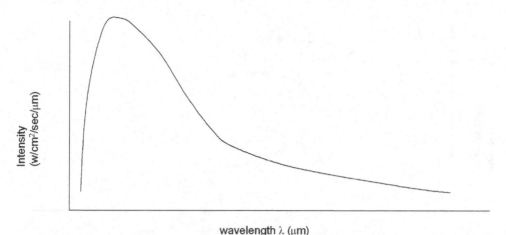

Figure 4.1. Idealized plot of the spectral output of a celestial object.

of very long wavelength. If we plot the intensity of an object's light output as a function of wavelength, we get a spectral plot similar to Figure 4.1.

The total amount of light is the integral of $I(\lambda)$—that is, the area under the curve. Thus, as illustrated in Figure 4.2, a brighter star gives us more light than a faint star:

Or, as shown in Figure 4.3, two stars that are of the same total brightness might have different spectral curves. One star puts out more red light, and the other puts out more blue light.

The light we receive from an object may be different than the light that was sent. The spectral plot lays the groundwork for understanding the distinction between the light that was sent by the star vs. the light that we see and measure. For example,

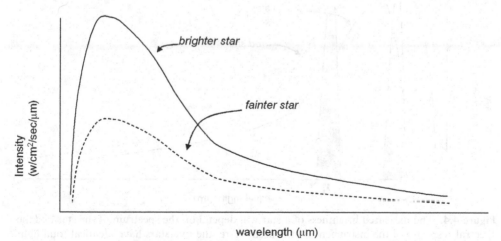

Figure 4.2. Stars of different brightness may have identical spectral characteristics (i.e., the same color).

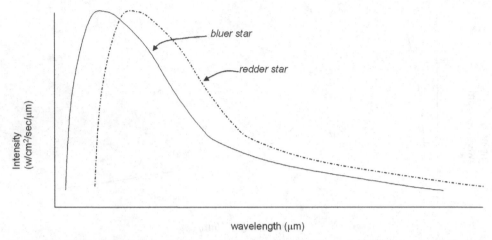

wavelength (μm)

Figure 4.3. Celestial objects may have identical total brightness, but different spectral characteristics (i.e., different colors).

suppose that two different astronomers have different sensors. This situation is illustrated in Figure 4.4. The astronomer whose sensor detects light at wavelength λ_b will say that the blue star is the brightest, but the astronomer whose sensor detects light at wavelength λ_r will say that the red star is the brighter one.

Our real sensors have a sensitivity curve that describes their response to light of different wavelengths (colors). A typical CCD imager's sensitivity curve is shown in

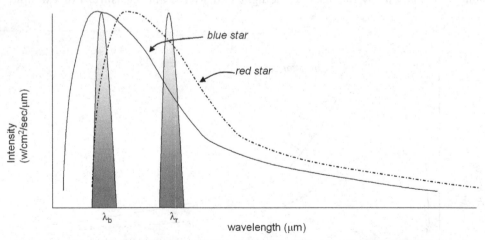

wavelength (μm)

Figure 4.4. The measured brightness of a star will depend on the spectrum of the star *and* the spectral response of the instrument that is used. Here, the two stars have identical total light output, but the instrument that responds to short wavelengths sees the "blue" star as brighter, while the instrument with long-wavelength response sees the "red" star as brighter.

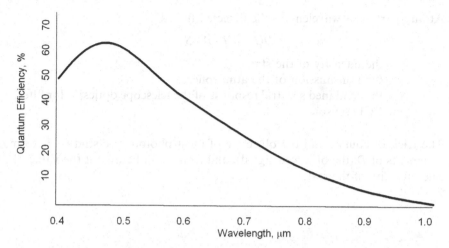

Figure 4.5. Typical spectral response of front-illuminated CCD imager.

Figure 4.5. Obviously, the shape of this curve will affect the measurement that we make of different stars.

The sensor's response isn't the only thing that alters the nature of the light that we detect from the star. The optics of the telescope pass some wavelengths better than others (e.g., glass doesn't transmit far-infrared at all, and aluminum doesn't reflect ultraviolet very well). The imaging system may include color filters that pass certain wavelengths and block others. The atmosphere absorbs and scatters some starlight before it reaches our telescope. Each of these effects has a characteristic spectral curve, and the net result is that we only detect the light that passes through all of them. A typical situation is illustrated qualitatively in Figure 4.6.

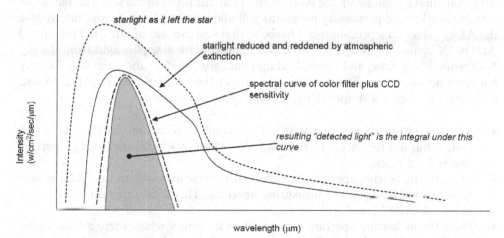

Figure 4.6. The starlight that you measure is only that portion of the star's light that reaches, and is detected by, your CCD imager.

At any particular wavelength, the detected flux is

$$D(\lambda) = I \cdot A \cdot S$$

where I = the intensity of the star;
   A = the transmission of the atmosphere;
   S = the combined spectral response of the telescope optics, color filter, and
      CCD sensor.

The trick of course, and the objective of most photometric studies, is to make measurements of D (the observed signal), and from them figure out the value of I— the true intensity of the star.

4.2   GENERAL PRINCIPLES OF APERTURE PHOTOMETRY

In order to put this into a practical context, we'll go through an example that will explain the procedure for CCD photometry. Then we'll deal with some important details related to image processing, calibration, and measurement.

Figure 4.7 is an image of the field of view that contains the cataclysmic variable star "V378 Peg" (except of course when you take your own image, it won't include the arrow to identify the star of interest!). As you probably know, the CCD image is made up of pixels, and each pixel is assigned an "ADU" value that expresses the amount of light that it received during the exposure. The inset shows an expanded view of the target star, in which the individual pixels can be seen. The inset also shows the ADU values of some typical pixels.

The total amount of light in the star's image is a measure of its brightness. You determine that by adding up the ADU count of all the pixels in the star's image. Most astronomical image-processing programs will allow you to zoom in and interrogate the ADU values on a pixel-by-pixel basis, so that you can add up the total integrated ADUs. Of course, any single pixel receives light from the star plus light from the sky background (sky glow, and unresolved dim background stars and galaxies). The sky background must be subtracted in order to get the "starlight only" ADU count. Conceptually, that's a simple thing, consisting of four steps:

- Select a measuring aperture that is large enough to encompass the whole star image, but not too large. In the example of Figure 4.7, the measuring aperture size is 5×5 pixels.
- Place the measuring aperture over the star's image and add up the ADU counts from all pixels inside the measuring aperture. This total represents the "starlight + skylight" count.
- Place the measuring aperture someplace on the image where there are no visible stars, and add up the ADU counts from all pixels inside the measuring aperture. This gives the "sky light" count.

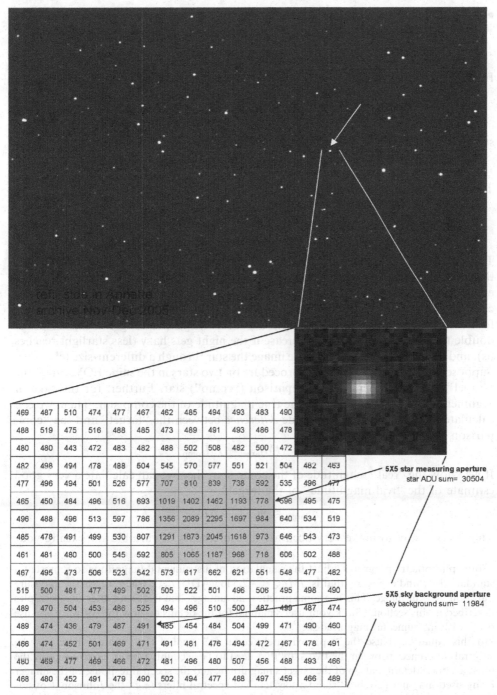

Figure 4.7. Typical CCD image containing a variable star (V378 Peg); and (inset) the ADU values of the star image and sky background pixels.

- Subtract the "sky light" count from the "starlight + skylight" count, to get the "starlight only" ADU count: starlight only = "starlight + sky light" minus "sky light".

In the example of Figure 4.7,

$$\text{starlight} + \text{skylight} = 30{,}504 \text{ ADU}$$

$$\text{skylight} = 11{,}984 \text{ ADU}$$

subtract these two, and we get

$$\text{starlight only} = 18{,}520 \text{ ADU}.$$

Then, turn this into a magnitude:

$$IM_{\text{star}} = -2.5 \cdot \log(18{,}520)$$

$$= -10.67$$

This is called the instrumental magnitude (IM), because it represents the brightness of the star as measured by your instrument, with your spectral characteristics, and at the exposure that you used for your image. The instrumental magnitude is peculiar to the circumstances of that particular image: the "starlight only" ADU count that we just found will increase if we use a longer exposure* (e.g., doubling the exposure will double the ADU count), it will decrease if the night gets hazy (less starlight reaches us), and of course it will change if we image the star through a different-size telescope. Suppose, however, that we do this procedure on two stars in the same FOV: one is the "target", and the other is the comparison ("comp") star. Further, for the sake of argument, suppose that we know the true magnitude of the comp star. Then, we can calculate the instrumental magnitude *difference* between the "target" and the "comparison star":

$$\Delta IM = (IM_{\text{target}} - IM_{\text{comp}})$$

If we know the "real" magnitude of the comp star, then we can calculate a very good estimate of the "real magnitude" of the target†:

$$m_{\text{target}} \approx m_{\text{comp}} + \Delta IM$$

This gives a way to make very accurate measurements of variable stars using your

---

*Some photometry programs adjust the calculation of instrumental magnitude to represent a standard 1-second exposure: that is, $IM_{1\text{sec}} = -2.5 \log(\text{ADU}/t_{\text{exp}})$, where t_{exp} is the exposure duration of your image. This has some practical advantages, but it doesn't alter the principles described in this section. Check your user's manual for the details on how the software you use deals with instrumental magnitudes.

†In this equation, I use the "approximately equal to" symbol ($\approx$) because there may be spectral differences between the two stars that need to be accounted for in order to make the most accurate determination of the target's magnitude; and the spectral response of the imager being used may not match a "standard system" of spectral filters (e.g., the Johnson–Cousins UBVRI system). These spectral adjustments are accomplished through the use of "transformation coefficients", as described in Section 4.7.3.

CCD imager. *Very* accurate: brightness changes of a few hundredths of a magnitude can be detected in this way, using amateur-class equipment.

4.2.1 Selection of comp stars

In Chapter 3, I described the considerations that go into the selection of "good" comp stars for visual variable star observations, and the reasons that it is best if all observers use the same comp stars. That discussion is also applicable to CCD variable star observing. If practical, select comp stars and check stars that are similar in brightness and color to the target variable star. Use the same comp stars as were used by other observers studying the same object (e.g., use the stars identified on the AAVSO or BAA star charts, if you are monitoring a known variable star). It is also good practice to record the identity of the stars selected as "comps" so that if there is any question about their stability, they can be re-investigated, and your results can be adjusted accordingly.

4.2.2 Software packages for photometry

Doing the four steps (select aperture size, measure "star + sky", measure "sky-only", and subtract to get "star-only") for two or three stars on each image, and then repeating for several images, is a tedious and error-prone process if you do it manually. Happily, all of the most popular astronomical image-processing software programs include routines that automate most of these steps. CCDSoft (Software Bisque), AstroArt (MSB Software), MaximDL (Diffraction-Limited), MPO Canopus (BDW Publishing), IRAF (distributed free by National Optical Astronomy Observatories) all provide powerful photometry routines that automate the four steps, and most of the other functions that are discussed in this chapter. The user's manual for the software that you use for image acquisition and processing will explain the capabilities and procedures for using its photometry tools. Some packages will enable you to define the target and comp stars, and bang, out comes the target star delta-magnitude. Most will allow you to define two or more "comp stars" (e.g., a "comp star" and a "check star"). The "check star" is used to confirm that the comp star isn't varying in brightness. Some of these programs can be instructed to analyze a whole series of images, and create a light curve almost unattended.

Hence, you probably won't go through the meticulous manual detail of the four steps for your projects (except perhaps once or twice, to confirm that you understand what your software is doing behind the scenes). It is useful to understand the fundamental principles, however, so that you'll understand the reasons behind several of the image-processing and calibration steps, as well as the pitfalls that can come your way when you're doing a photometric research project.

Some software packages will link your image to a star chart database and attempt to give you the actual magnitude of the target (instead of its delta-instrumental magnitude relative to the comp star). Unfortunately, all of the popular star catalogs used in the planetarium programs have serious photometric problems [2]. The *Guide Star Catalog*, which forms the basis for most planetarium programs' stellar database, cannot be relied upon for photometry. Errors of ±0.5 magnitude are not unusual in the GSC. The V-magnitudes reported in the *Tycho Catalog* (also a popular basis for planetarium programs) are better, but still are good to only about ±0.1 magnitude. (There is a pretty simple formula that will transform Tycho magnitudes to the standard V-band with much better accuracy [3, 4], but that isn't included in the popular planetarium programs.) The *UCAC2 Catalog*—the recommended catalog for astrometry—is also not a very good photometric reference (±0.2 magnitude). This isn't a failing of the planetarium programs, nor of their star catalogs. These catalogs were developed specifically for astrometry, and they are very, very good at that. They were never intended to be used for accurate photometry.

Therefore, for most photometric projects, you will refer to a well-determined photometric database (such as the AAVSO star charts) for the "true" magnitude of your comp stars. You will then determine the brightness of your target from the delta-magnitude between your target and the comp stars.

If for some reason you don't have well-characterized comp stars in the FOV of your target, it is still possible to determine accurate photometry on the standard B, V, R color system. I'll describe how in Section 4.7. The procedure is well within the capability of a patient amateur astronomer, but it does entail some additional work compared with differential photometry. Happily, it is not necessary for most amateur photometric projects.

4.2.3 Absolute vs. differential photometry

The discussion above implicitly assumed that the "comp star" was in the same image as the target star, and that we were happy to monitor the brightness of the target *relative to the comp star*. In some cases, we might not know the "true" magnitude of the comp star, but that's OK since the variation of the target is completely described by its light curve relative to the comparison. Since we determined the magnitude difference between target and comp stars, this method is called "differential photometry".

Differential photometry is a surprisingly robust method of monitoring the brightness of a target. The delta-magnitude that you determine for the target is unaffected by changing the exposure duration, and almost unaffected by changing atmospheric conditions. For example, if your comp star is in the same FOV as the target, and your image FOV is pretty small (a fraction of a degree is typical), then you can be pretty confident that any haze or wispy cloud that affects the target star will also affect the comp star. Similarly, if your imager's shutter isn't precisely accurate, so that one image may have a slightly longer or shorter exposure than the next image, that doesn't matter because both target and comp stars were exposed for the same duration on any one image, and $(ADU_{target}/ADU_{comp})$ isn't affected by changing exposure durations. This means that the delta-mag is unaffected by modest amounts of haze or changing atmospheric extinction. (There are some important caveats to this, discussed in Section 4.6.2.)

For this reason, almost all amateur research projects rely on differential photometry. This method is quite effective for studying variable stars, determining asteroid rotation periods, and detecting extra-solar planets.

"Absolute photometry"—to determine the brightness of a target on the standard stellar magnitude scale, without having well-characterized comp stars in the FOV—is quite a bit more complicated because of the many additional factors that must be considered. If you need to image the target star in one frame, and then move the telescope through 20 or 30 degrees to get to the "standard" comp star, you need to be sure that the imager's exposure is exactly the same on both images,* and you need to determine the atmospheric extinction in both directions so that it can be accounted for, and you need a way to be confident that the atmospheric conditions didn't change between one image and the next, etc. These can all be handled, and absolute photometry is well within the capability of the advanced amateur astronomer, but beware that it does require a quantum step in effort compared with differential photometry. The basics of "absolute photometry" will be described in Section 4.7.

For most variable stars, your differential photometry can be based on comparison stars whose photometry has already been determined. For example, the AAVSO star charts include comparison stars whose V-magnitude and color indices have been accurately determined. They are, in effect, secondary standards. By using a V-filter when you make your images, and using the AAVSO-recommended comp stars, you are closely anchored to the standard system, even though you are doing differential photometry.

All of this background is probably more than you really need to know in order to conduct the most common amateur CCD photometry project—variable star measurements—but hopefully it will help you understand the rationale behind the procedures that you will use for variable stars, and other projects.

4.3 PROJECT H: CCD PHOTOMETRY OF VARIABLE STARS

The *General Catalog of Variable Stars* (GCVS) contains over 38,000 variable stars, of various types. A few of these warrant ongoing study at any particular time, but for most of them the professional astronomers can only justify allocating telescope time when a special, peculiar event occurs (such as an outburst of a cataclysmic variable). The role of the amateur research community in this case is to monitor the stars, and announce when an event occurs. For other stars, the science information lies in the existence of a long continuous record of their activity (sometimes for many decades). For example, in a close, massive binary system, the slow change in orbital period can

---

*Some software, such as MPO-PhotoRed will perform the calculations correctly even if the two exposures are not identical. In order for the calculations to be valid, the imager's *reported* exposure duration must be equal to its *actual* duration. If very short exposures are used (e.g., less than a second), then you are placing demands on shutter speed and consistency that should be checked by experiment with your equipment. In addition, for very short exposures, you may find that the effective exposure duration varies across the field of view.

give evidence of mass transfer between the stars. Here, the amateur research community is the best resource for long-term monitoring of these objects.

Compared with visual monitoring of variable stars, the CCD imager is capable of reaching fainter stars, and achieving significantly better accuracy. It also has the advantage of bypassing the "human factors" considerations that variable star observers are subject to.

4.3.1   Equipment needed

CCD measurement of the brightness of known variable stars requires all of the usual equipment for a CCD imaging session, plus a few special pieces of data and software unique to the photometry projects. The normal equipment required for any night of CCD imaging includes:

- Telescope.
- CCD imager.
- Personal computer (to run the imager and collect the images on its hard drive). The PC may also operate the telescope and other instruments;
- Software for CCD control and operation. (The specific software will depend on the make of your CCD imager, and on your personal preferences; some of the most popular programs include CCDSoft, MaximDL, and AstroArt, but there are many other fine products available.)

For CCD photometry of variable stars, you also need the following "special" photometric equipment:

- Photometric V-band filter.
- AAVSO or BAA star chart for the target star.
- Software for differential photometry reduction of your data (CCDSoft, MaximDL, and AstroArt offer this capability as does specialized software such as MPO Canopus and IRAF).
- WWV receiver or other source of accurate time to set your PC's clock.

The telescope should be polar aligned and equatorially driven. There is no reason to prefer any particular type of telescope for this project, and almost any monochromatic CCD imager can be used for variable star photometry. Ideally, the CCD's pixel scale should be reasonably matched to the telescope, such that the full-width-at-half-maximum (FWHM) of a star's image equals one to two pixels. The idea here is that you don't want to have pixels so large (in angular size) that a star's light is hitting only a portion of a pixel, nor so small that you spread the star's light over a great many pixels. For most amateur set-ups, this implies striving for a pixel size in the range of 1 arc-sec to a few arc-sec. This "ideal" guideline can be violated without ill effect, up to pixel sizes of perhaps 10 arc-sec. If the pixels are larger than this, you must begin to worry a bit about a single pixel encompassing more than one star (which obviously will throw your photometry off). If your pixels are on the large side, you'll want to study a star chart to confirm that your selected target and comp stars aren't corrupted by near neighbors.

The only piece of equipment that is not "standard" for CCD imagers is the photometric V-band filter. Its purpose is to ensure that the spectral response of your system is a reasonable match to the Johnson V-band, which in turn is very similar to the response of the human eye. Neither an unfiltered CCD imager, nor any of the standard color filters used by CCD imagers for color astro-imaging (e.g., the R-G-B or C-Y-M filter sets) will provide the proper spectral response. You will have to buy a Johnson–Cousins V-band filter. They are available from most of the manufacturers of commercial filter wheels, or directly from filter manufacturers such as Custom Scientific and AstroDon. It is a modest investment—about $150 for a 1.25-inch filter, or a few hundred dollars for a 2-inch filter.

The accuracy of time required for most variable star observations is modest: Setting your PC to within a few seconds is usually acceptable. However, just in case short-period fluctuations or critical eclipse timing turn out to be important, I usually keep the PC's time accurate to better than ±1 second as a routine discipline.

4.3.2 Conducting the observations

If you haven't previously observed the night's target, then your preparation for the observing session begins a few hours before sunset. You need to acquire the standard star chart for the target and acquaint yourself with the chart, the object, and the comparison stars. The most convenient method for getting the star charts is to download them from the AAVSO website at *www.aavso.org/charts* There, you can search for charts by star identity, by coordinates (RA, Dec), and by constellation. For most stars, the charts come in different scales (ranging from the relatively wide field of view C- or D-scale charts, to the smaller field of view F-scale CCD charts). Select the CCD chart if it is available and confirm that its scale is appropriate for your imaging setup. If you have an unusually short focal length telescope and unusually wide field of view imager, then the CCD chart might conceivably be too small, in which case you can use one of the larger-scale charts.

Figure 4.8 illustrates how to find the target field of view, and your target in it, by finding the pattern of stars in your CCD image. On the image, I've added an outline of the area covered by the AAVSO chart (this is a CCD F-scale chart) to help you identify the variable star and the comp stars. For the record, this image was taken with my 11-inch Celestron NexStar, operating at F/6.3, and an SBIG ST-8XE imager, through a V-filter. It is a 3-minute guided exposure. It has been reduced by dark-subtraction and flat-fielding using Software Bisque's CCDSoft. Note that all of this, aside from the V-filter, is typical entry- or medium-level CCD imaging equipment.

Once night falls and your target star is reasonably high in the sky, you're ready to locate the target in your imager. As a general rule of thumb, it is a good idea to have the target at least 30 degrees above the horizon. That will minimize most of the atmospheric effects that can corrupt your photometry. There will, of course, be situations where you have no choice but to observe at lower elevation angles (such

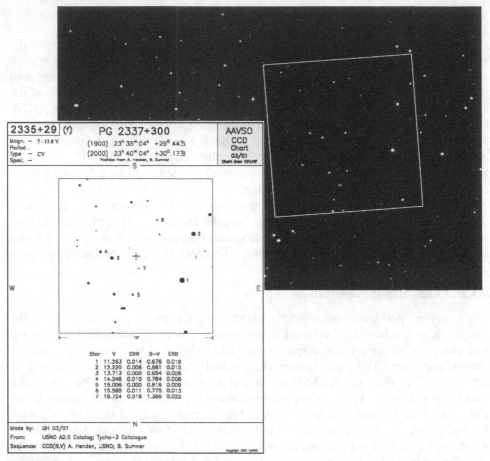

Figure 4.8. Matching an AAVSO chart to a CCD image. (Used with the kind permission of the AAVSO)

as a star just emerging from behind the Sun, visible only low in the sky just before twilight, or a star that is at a low southern declination that never gets high in the sky for northern-hemisphere observers).

Take your photometric images as you would any other CCD images. Save them in FITS format (preferably), or native uncompressed imager format (if FITS is not an option). Do not use a compressed image format such as JPEG; image compression severely garbles the photometric data. It is also good practice to save in "image-only" mode, not with "auto-dark" or "auto-flat". Set the imaging software to record the image time into the FITS header. This makes your record-keeping simpler. If for some reason it isn't possible to have your CCD control software write the image time directly to the FITS header, you'll need to record the time and exposure duration for each image in your notebook. Exposure duration should be long enough that both the target and the comp stars have good signal-to-noise ratio, but not so long that any are saturated. For most targets, and most set-ups, exposures in the range 1–5 minutes will be appropriate.

Even though a single image contains all of the information needed to determine the target star's brightness, it is usually good practice to take at least three images, sequentially, using identical imaging parameters. That gives you two meritorious types of redundancy. First, if by chance either the target star or comp star are corrupted by a cosmic ray hit on one image, odds are that they will be fine on the other two. Second, with two or three images, you will be able to assess their consistency, so that you can toss out an image that gives discordant data, and you can improve your accuracy by averaging the results from "good" images.

At this stage, it is also a good idea to examine each star's image to determine the peak-pixel ADUs, and (if your software provides it) the signal-to-noise ratio. The peak pixel ADU of each star allows you to confirm that the image is not saturated. If your brightest comp star is saturated, just delete it from your analysis. If your target star is saturated, then you'll need to re-take the images, using a shorter exposure. If your target signal-to-noise ratio is low (e.g., less than 30), then your photometric accuracy will not be as good as it could be, and you may want to consider re-taking the images using a longer exposure. If you are not familiar with the concepts of sensor linearity and saturation, refer to Sections 4.5.2 and 4.5.3.

In addition to your science images, you need "dark frames" and "flat frames" that match the conditions of your science images. The dark frames should be taken at the same exposure and same chip temperature as your science frames. The flat frames can be twilight flats, "T-shirt" flats, or "lightbox" flats (whichever is most convenient), and they can be taken either immediately before or after your photometry session. One purpose of flat-fielding is to compensate for the effects of "dust donuts" in the image. Therefore, it is essential that the flat frames be taken through the same V-filter as the science images, and that they be taken before the CCD orientation in the telescope is changed from what it was during the science frames. Otherwise, the "dust donuts" in your science frames won't line up with those on the flat frames. If your image-processing software provides for use of "dark flats"—dark frames of the same exposure as your flat frames—take those also. If you are an experienced astro-imager, then you probably already understand the purpose and use of darks and flats, and have developed methods that you are comfortable with. If this subject is relatively new to you, then I offer a few suggestions about darks and flats in Section 4.5.

4.3.3 Reducing and analyzing your observations

Image reduction for photometry is identical to the "best practice" for any astro-image reduction: apply dark-frame subtraction and flat-field compensation to your image. I usually take a dozen darks and flats, and median combine them to eliminate noise spikes and cosmic ray hits. When image reduction is complete, I archive both the "raw" (un-reduced) science images and the "reduced" science images, along with the associated darks and flats, on non-volatile media such as CD or DVD. The rationale behind this is that if something so unusual as to be implausible is found in the reduced science images, then it can be investigated in the raw images and/or the darks and flats. It's rare, but not unheard of, for some sort of problem in one of the compensation images to corrupt the "reduced" science frame. Having the raw science frame gives you the opportunity to re-reduce the images if necessary.

The reduced science frames are then examined with your photometry software. You will establish a "measuring aperture" that is used to collect all of the light from your target star and (separately) from the comp stars, and a "sky aperture" that is used to collect light from the sky background. The shape and relative locations of these apertures will be to some degree dependent on our choice of photometric software. Some programs use square measuring apertures (matching the square array of pixels in your image), while others use round measuring apertures (matching the typically round star images). Either is acceptable. Some programs measure the "sky" in a ring surrounding the "target" aperture; others allow you to position the target and sky apertures independently. Again, either approach is acceptable.

For the measuring aperture, you want to select a size that is large enough that it comfortably encompasses the entire star image but doesn't include too much surrounding sky. It should definitely not be so large that any visible background field stars are included in it. If you had some guiding errors, or modest change in focus between images, make the measuring aperture large enough to comfortably encompass the worst stellar image, with some room to spare. The resulting size of the measuring aperture will depend on your equipment, and on seeing conditions. As an example, my set-up has 1 arc-sec pixels. Autoguiding errors plus atmospheric-blur usually results in stellar images whose width is about 2–3 arc-sec (full-width at half-maximum of the point spread function—FWHM). I usually use a measuring aperture in the range of 13 to 19 pixels diameter (i.e., about 5 to 7 times the FWHM). That is large enough that I'm almost guaranteed to be collecting all of the starlight. The purist will point out that such a large measuring aperture reduces the signal-to-noise ratio compared with an "optimally-sized" measuring aperture. As a practical matter the penalty is small, and using a relatively large measuring aperture makes the results immune to image-to-image changes in focus or guiding.*

For the "sky" aperture, you want to have approximately the same number of pixels as the "target" aperture (say, within a factor of 2 or 3). Ideally, the "sky" measuring aperture will see only sky, with no background stars within the measuring aperture. Some photometry programs include algorithms that can reject the effect caused by one or two not-too-bright stars in the "sky aperture".

Your photometry software will probably automatically take care of the arithmetic involved in summing the target ADUs over the measuring aperture, summing the ADUs over the sky aperture, scaling each in proportion to the number of pixels in

---

*If you are working a particularly faint target, then the SNR penalty of using a relatively large measuring aperture may not be acceptable. In that case, you may want to use a measuring aperture as small as two to four times the FWHM, to maximize the signal-to-noise ratio. If you use a small measuring aperture, then you are placing more stringent requirements on your telescope. It must maintain focus better, and guide more accurately, than would be allowed if you used a relatively large measuring aperture.

comp star #	signal (integrated ADU)	instrumental mag = -2.5 log(ADU)	Vmag (AAVSO chart)
1	2933.59	-8.6685	11.263
2	1144.46	-7.6465	12.220
3	285.63	-6.1395	13.713
4	177.66	-5.624	14.248
5	85.59	-4.831	15.006
6	52.89	-4.3085	15.585
7	17.12	-3.084	16.724
var	222.13	-5.8665	

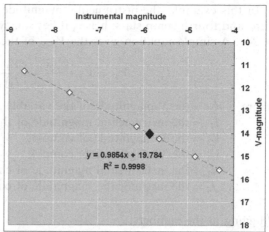

Figure 4.9. Differential photometry analysis of a target star.

each aperture, and subtracting "sky" from "target + sky" to yield the target ADUs. It may also calculate the instrumental magnitude of each object that you place in the measuring aperture, according to:

$$\text{Instrumental magnitude} = -2.5 \cdot \log(\text{summed target ADUs})$$

Using your photometry software, you select each of the comp stars and your target star, in turn, and determine their instrumental magnitudes, and, then, graph the results to make a plot of IM vs. V-magnitude for the comp stars. An example of a typical result (using the image of PG 2337+300 from Figure 4.8) is shown below as Figure 4.9.

The discipline of graphing the result is not required by AAVSO, but it is a very good way to self-check your observations and the calculations. The comp stars should all fall cleanly on a straight line when plotted as instrumental magnitude vs. V-magnitude. The minor random deviations of the points from the best-fit line give you an indication of the accuracy of your instrumental magnitudes. In this case all points are within ±0.03 magnitude of the best-fit line, so the data looks pretty good. If one of the stars falls noticeably off the line, then there is something awry with that star's measurement. The problem might be a hot or cold pixel, or a cosmic ray may have corrupted the image of that star, or the star may be an unrecognized variable, or it may have been the victim of some other sort of random error. If all of the other stars fall on the straight line, then you can be confident in simply ignoring the "problem child", and performing your calculation of the variable star's brightness based on other comp stars. If more than one comp star falls significantly off the line, or if the whole group is noticeably scattered, then something more serious has gone wrong and this image should not be used for creating the variable star observation.

In this example, the instrumental magnitude of PG 2337+300 falls between the third and fourth comp stars: hence, it lies somewhere between mag 13.713 and mag 14.248. A simple equation gives the V_{mag} of the variable star:

$$V_V = V_3 + \left[\frac{(V_4 - V_3)}{(I_4 - I_3)}\right](I_V - I_3)$$

where V_V = the V-magnitude of the variable star;
         I_V = the instrumental magnitude of the variable star;
         V_3 = the V-magnitude of comp-3;
         V_4 = the V-magnitude of comp-4;
         I_3 = the instrumental magnitude of comp-3;
         I_4 = the instrumental magnitude of comp-4.

Plugging in the numbers:

$$V_V = 13.713 + [(14.248 - 13.713)/(-5.61485 + 6.1395)] \cdot (-5.8665 + 6.1395)$$

$$= 14.01$$

So, the variable star was $V_{mag} = 14.01$ on this night. As a check of the arithmetic, this value is plotted on the best-fit line in the graph (Figure 4.9) to be sure that it does, indeed, fall on the line. If you don't like algebra, you could have found the same value for the variable star's V_{mag} by using a graphing technique.

It is a wise discipline to create a few images for each observation and do this analysis independently for each of them. Doing so provides another self-check on your results. For slowly-varying stars, you expect essentially no change between images taken a few minutes apart. For rapidly-varying stars, you may expect to see a consistent trend of falling or rising magnitude over the short interval between several images.

4.3.4   Submitting your observations

The central organization whose mission is to gather, analyze, and disseminate variable star observations is the AAVSO. Although it has a North American focus, it is open to all variable star observers, worldwide. Before you can submit observations to the AAVSO, you must request and receive an "observer code" to identify yourself and link all of your submitted observations to your credit. Any variable star observer can request an observer code; you do not have to be a member of AAVSO (although if you find that variable star observing is a regular part of your astronomical life, you will benefit from being a member). The request form can be filled in on the AAVSO website, where you'll enter your name, address, e-mail, and some information about your observing equipment. Your observer code will be assigned and sent to you within a few days. You then use your code to identify yourself whenever you submit observations.

The most convenient way to submit your observations to the AAVSO is their "WebObs" submittal system. This is available through the AAVSO website, for all

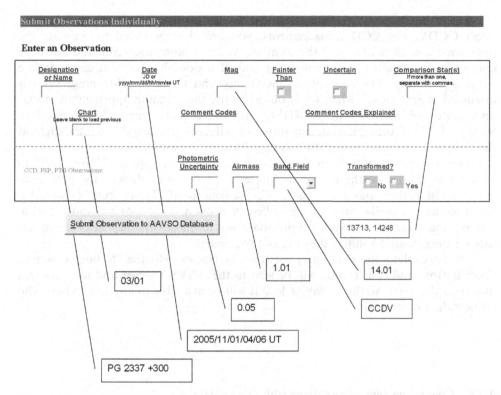

Figure 4.10. AAVSO WebObs observation-submittal form. (Used with the kind permission of the American Association of Variable Star Observers)

registered users. The contents and format for CCD entries is described in Figure 4.10. You can follow this example as a template for your own data entry. The variable star is identified either by its "Name" (PG 2337+300) or its "Designation" (2335+29). I use Universal Time on my observatory computer. This particular observation was made on November 1, 2005 at 04:06 UT, so that is how the "Date" is reported. If you prefer, you can convert the time to Julian Day and use JD in your report. (See Appendix A if you are not familiar with UT or JD.) The interpolated brightness of the star is entered as $V = 14.01$ (CCD observations should be entered to the nearest hundredth of a magnitude). If the star was not visible on your image, then you would enter the brightness of the faintest comp star that you could detect, and check the "Fainter Than" box. Since the mag 13.713 and mag 14.248 comparison stars were used as the basis for calculating the target's brightness in this example, they are entered in the indicated field. You can enter, or suppress, the decimal points when entering the comp star identities. The comp stars were selected from the chart dated 03/01, so that is used as the chart identification.

The "Comment Codes" field opens a drop-down list of the common comments (such as "Hazy", "Moon Interfering", etc.).

This was a V-band CCD measurement, so in the "Band Field" drop-down list, select CCDV. For CCD measurements, you are also requested to estimate the photometric uncertainty, and the airmass. Your photometric reduction software may calculate these two parameters for you. If it doesn't, then you can calculate them quite easily. The concept of uncertainty and CCD signal-to-noise ratio is discussed in Section 4.5. Most CCD observers use the common approximation that uncertainty $= 1/\text{SNR} = 1/(g \cdot \text{ADU})^{1/2}$. The concept of airmass is discussed in Section 4.7.2. If your planetarium program tells you the target's zenith angle at the time of the observation, then you can calculate the airmass $= X = \sec(\theta)$.

The observation entry form also asks whether you have applied "Transforms" to your determination of the target's brightness. If you have followed the procedure described here, then your answer is "No". The concept of "Transforms" (also called "transformation coefficients") is described in Section 4.7.3. For the most critical observations, transforms may be appropriate, but for most variable star observations using a photometric V-filter, they are not required.

That completes the data entry for this observation. Hit the "Submit Observation" button, and your results will be sent to the AAVSO server and incorporated into their database. Within a day or less, it will be available to any researchers who request data on this star.

4.3.5   Comparing your observations with other's data

After you've made some measurements of a variable star, you may be curious as to how your observations fit in with other's measurements of "your" star. This information is also easily available on the AAVSO website, as immediately-downloadable "unvalidated" data. "Unvalidated" data has not been subjected to AAVSO's checking and analysis process, and should not be used for formal reports or publication. It represents the sum of all observer reports, as received. When an astronomer requests AAVSO data for use in a research project, the AAVSO conducts a quality-evaluation on the raw data to be sure that unreliable observations aren't included in published papers. Figure 4.11 shows all of the data that AAVSO received during the six months around our example observation of PG 2337+300. (Astronomical nomenclature is frustratingly replete with duplicate names for each object. This star is also known as V378 Peg, which is how it is labeled on Figure 4.11.) The data point representing our example is shown as a filled circle. As you can see, it is consistent with the other CCD observations, to within the internal consistency of my data (±0.03 mag).

This is a genuine tidbit of new scientific data, and gathering it took less than an hour (considering observing time, image reduction and analysis, and magnitude calculation). As you can see, it isn't difficult and it isn't time-consuming. But it is valuable! So, do add a few variable star fields (and a photometric V-filter) to your observing plans!

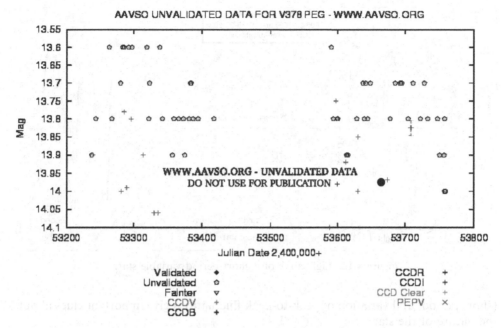

Figure 4.11. Example plot of "unvalidated" AAVSO data accessible at their website. (Used with the kind permission of the American Association of Variable Star Observers)

4.3.6 Rapidly-changing (short-period) variable stars

In the example above, we determined the brightness of the target star at a single point in time. If we (and other astronomers) check this star periodically for many nights, or many years, we can examine the data to evaluate the way that the star's brightness changes over time. The result is called a "lightcurve". Many variable stars have lightcurves with periodicity that is measured in months, years, or decades, which is why a long series of observations (usually from many different observers) is required to characterize their activity.

However, there are some variable stars whose brightness changes on time scales of only a few hours. For these stars, one observer can fill in a complete cycle of the lightcurve in a single night. The observing, data reduction, and analysis procedure is exactly the same as for any other variable star, with one exception. Instead of a few images, you take a continuous series of images all through the night.

After reducing your data (as described in Section 4.3.3) you can plot the star's magnitude vs. time to create a complete picture of its light fluctuations. Figure 4.12 shows an example of such a lightcurve. This is a high-amplitude δ-Scuti type star, whose characteristic period is so short that it goes through more than one cycle in a single night.

The lightcurve shows that this particular star brightens and dims with a period of about 2.6 hours, and that its brightness variation isn't uniform. Some maxima are brighter than others, and some minima are fainter than others. These two features

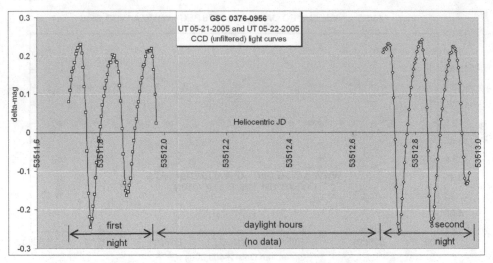

Figure 4.12. Lightcurve of a short-period variable star.

(short period, and variation of peak-to-peak fluctuation) give important clues about the nature of the star.

The measurement and study of variable stars is just one of the avenues that amateur photometry can take. You can also contribute to science by shifting your focus from the distant stars to the small bodies of the Solar System. That's our next project.

4.4   PROJECT I: DETERMINING ASTEROID LIGHTCURVES

The purpose of this project is to determine lightcurves of asteroids. For many asteroids, there is no known lightcurve, or one whose characteristics are poorly determined. Amateur astronomers are leading the effort to generate lightcurves that are "secure results" for these objects, often providing better data than is achieved by professional observatories. The amateur's advantage in this area (compared with professionals) is that it is easy for the amateur to devote several consecutive nights of telescope time to the study of a single asteroid, enabling the development of a lightcurve that is more complete, and more densely sampled, than the typical "professional" lightcurve. This complete, densely-sampled lightcurve is needed to be sure that the inferred rotation period is correct and to provide high-accuracy data for asteroid shape determination.

An accurate lightcurve of an asteroid provides some fascinating information. Consider the lightcurve of (125) Liberatrix shown in Figure 4.13. The lightcurve period of 3.966 hours is the rotation period of the asteroid. The peak-to-peak amplitude of 0.3 magnitudes tells us something about the shape of the asteroid. The projected area of the asteroid changes by about 31% as it rotates. This result

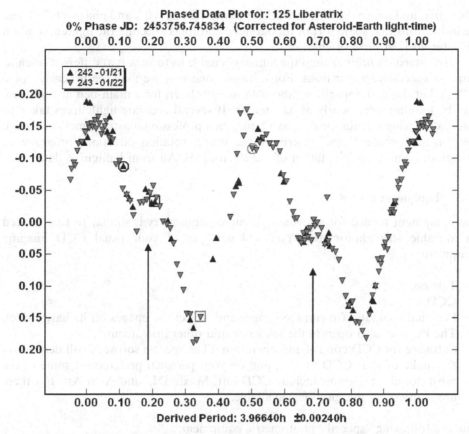

Figure 4.13. Typical asteroid lightcurve determined by CCD differential photometry.

is calculated as follows: assume that there is no reflectivity or color change across the asteroid's surface, so that the brightness variation is solely the result of changing projected area. Thus,

$$0.3 \, \text{mag} = -2.5 \log(A_{\text{max}}/A_{\text{min}})$$

and therefore

$$(A_{\text{max}}/A_{\text{min}}) = 1.31$$

The assumptions of constant reflectivity and constant color over the asteroid's surface are generally valid, based on a variety of observational and theoretical considerations. However, the asteroid's surface is not smooth. There are hills, valleys, and craters across its surface. You can probably imagine that the shadows of these topographic features will affect the lightcurve. If a mountain casts a shadow across the asteroid's surface, then the lightcurve will dip. When the shadow is eliminated (either because the asteroid has rotated so that the Sun is "straight overhead" as viewed from the mountain peak, or because the mountain has rotated out of our view), then the lightcurve will rise slightly. Look carefully at the lightcurve of

Liberatrix, and notice the little inflection points at phase $= 0.2$ and phase $= 0.7$. These are indications that some sort of topographic feature is affecting the lightcurve when the asteroid is at these orientations.

The asteroid's orbit around the Sun may enable us to view it at different orientations on successive apparitions. For example, one year we may see it nearly "pole on", and at the next apparition (about 18 months later, for a main-belt asteroid) we may be looking more nearly at its equator. If several accurate lightcurves taken at different viewing orientations are available, then professional astronomers can solve the "inverse problem" and determine the shape, rotation direction (prograde or retrograde), and pole orientation of the asteroid [5]. All from lightcurve data!

4.4.1   Equipment needed

The equipment needed for asteroid lightcurve photometry is similar to that needed for variable star photometry. You will need all of your usual CCD imaging equipment:

- Telescope.
- CCD imager.
- Personal computer (to run the imager and collect the images on its hard drive). The PC may also operate the telescope and other instruments.
- Software for CCD control and operation. (The specific software will depend on the make of your CCD imager, and on your personal preferences; some of the most popular programs include CCDSoft, MaximDL, and AstroArt, but there are many other fine products available.)

Plus the following "special" photometric equipment:

- Planetarium program with up-to-date asteroid orbital elements (to locate your target asteroid).
- Software for differential photometry reduction of your data (CCDSoft, MaximDL, and AstroArt offer this capability, as does specialized software such as MPO Canopus and IRAF).
- WWV receiver, GPS, or other source of accurate time to set your PC's clock.
- GPS or other means of accurately determining your observing location (latitude, longitude, and elevation).
- Spreadsheet program (or specialized software such as MPO Canopus or Peranso) for analyzing and plotting the lightcurve.

The telescope requirements are the same as for any CCD imaging session. Matching of the CCD to the telescope FOV should follow the same guidelines as were described above for variable star photometry.

In order to point your telescope to the correct position and find your target asteroid in the field of view, your planetarium program must contain up-to-date orbital elements for the target asteroid. Most of the popular planetarium programs provide simple methods for downloading orbital elements from the Minor Planet Center (MPC), so this is not a great challenge. For numbered asteroids (whose orbits are well-known), you can safely use an orbital element database pretty much forever. For newly-discovered or un-numbered asteroids, whose orbits may be less certain and hence may be revised periodically, you will probably want to update your planetarium program a week or so before you plan to begin your observations. (As a practical matter, most "un-numbered" asteroids are so faint that they are beyond most amateur's reach, for accurate photometry.) For near-Earth asteroids during close-approach to Earth, their motion across the sky can be so rapid that you'll need to update your orbital elements every day.

The accuracy of time required for asteroid observations is somewhat more stringent than for most variable star observations because of the more-rapid light variations that you can expect from an asteroid. For most asteroid lightcurve photometry, time accuracy of a few seconds is acceptable. However, it is wise to strive for more accurate time synchronization, so that your asteroid photometry images can also be used for astrometry. Setting your PC to within ±0.5 second of WWV is good practice. For really fast-moving near-Earth asteroids, you may need to be within ±0.1 seconds, in order for the time error to not be a major contributor to astrometric accuracy.

Note that—unlike variable star photometry—I do not insist on using a V-band spectral filter for asteroid observations, and that there is no mention of specific star charts to select predefined "comp stars". The reason for this liberality will be discussed below.

4.4.2 Making the observations

Measuring an asteroid's lightcurve uses techniques that are very similar to those used for variable stars. In general, everyone uses differential photometry to track the brightness of the asteroid relative to one or more comp stars. Pretty much everything that was discussed under "variable stars" is also appropriate for asteroid photometry: select an exposure duration that will give you a good signal-to-noise ratio without saturating any star images; save your raw images in FITS format; do not apply auto-darks or auto-flats; and do not use any sort of image compression.

There are a few special things about asteroids that will affect your observations and your evaluation of the results. First, asteroids move! Whereas you could count on your variable star being in the same place every night, that's not true with asteroids. You'll need a decent ephemeris just to find the asteroid, but with modern planetarium programs that isn't a challenge. All of the popular programs can be set to plot asteroid positions, and their internal databases can be updated from the Minor Planet Center. You tell the program what time you'll be observing, and it can show you the coordinates and speed of the object. My technique is to take a short exposure (10–15 seconds), examine the pattern of bright stars in the target field, compare them with the planetarium star chart, and determine which one is the asteroid. If the asteroid is too faint to show up in the short-exposure image, the star pattern will show you where it is located. Then, I adjust the telescope pointing to put the asteroid where I want it in the FOV. Those of you who trust your "Go-To" mounts might just dial in the asteroid's coordinates; but following President Reagan, I believe that we should trust, but verify!

The subject of placing the asteroid in the field of view will occasionally give you fits, if the asteroid is moving rapidly, your field of view is small, and you have an integral autoguider (rather than a separate guide-scope). Generally, you want to place the asteroid so that its motion during the course of the night will keep it within a single FOV. That way, you can use the same set of comp stars all night. If your target is a fast-mover (as a near-Earth object will be), then you may find that you need to adjust your FOV periodically through the night to follow the asteroid. In that case, it's advisable to arrange that the "new" FOV have a healthy overlap with the "previous" FOV. That way, the two FOVs can have at least one comp star in common during data reduction. You may also need to make compromises to all of this in order to get a good guide star into your autoguider frame.

Note that I've referred to an "all night series" of images. That's the second way in which asteroid photometry differs from variable star photometry. With most variable stars, you're the member of a team building a grand cathedral, and each of your observations is one irreplaceable brick in the edifice. Because the star's light may change very slowly (taking months or years for a lightcurve to emerge), it may be a long, long time before you see the whole structure take shape, as each worker adds his or her observations, night by night, and year by year. Photometry of an asteroid, on the other hand, is more like making a small cottage. You may be able to complete it in a few nights, and you can probably do it all by yourself. The thrill of plotting a complete lightcurve using your own data after just a few sessions on an asteroid can be downright addictive! In any case, it changes the pace of your project compared with variable star photometry. One or two isolated data points per night don't do much to fill in an asteroid lightcurve. A typical asteroid rotates in about 4 to 12 hours (there are exceptions, of course), and over the course of just an hour the asteroid's brightness may change significantly. Hence, think in terms of making a continuous series of images, one after another, all night.

The downside of the required pace for asteroid photometry is that such a session can get pretty boring. It cries out for automation! The upside is that this is just about the easiest project to automate: set your telescope on the target, initiate autoguiding, and tell your image acquisition software to keep taking and storing images until you tell it to stop. Then you can go do something else (like sleep) while the telescope and imager take care of business. Just before dawn (or at the time the asteroid's elevation gets too low), you turn everything off and you're ready to reduce your data. Neat!

Some observers attempt to gather data for two or more lightcurves per night by taking an image of asteroid #1, then moving to asteroid #2 and taking an image, then moving back to asteroid #1, etc. This may increase your productivity, but it does carry a risk. Since you will have fewer data points per asteroid, and those data points are separated more widely in time, you increase the risk of missing some feature in the lightcurve. In the worst case, you may even find that your data isn't dense enough to establish a secure result for the lightcurve period. I know people who have been very successful with the "multiple asteroids in one night" approach, but I'm conservative enough that I'd prefer to get a solid result on one target per night, with relatively little effort, rather than shoot for two or more at greater effort and with greater risk of coming away with inadequate data. Hence, my practice is to set up on an asteroid, instruct my telescope to follow it all night, and instruct the CCD imager to take a continuous series of images, with only enough waiting time between images for the autoguider to settle down.

Because the asteroid moves, you probably won't be using the same comp stars on succeeding nights. Each session will be anchored to its own comp stars, and, of course, the comp stars that you use are probably not the same ones that were used by other observers, if anyone else happens to be studying "your" asteroid.

Many asteroid photometrists use unfiltered (rather than standard V-band) images. This is done for several reasons. Most obviously, using a spectral filter reduces the signal that you get from the asteroid. Since many asteroid targets are faint, that's unattractive. You may be using unguided exposures (either because you don't have an autoguider, or because there's no convenient guide star), which—depending on the quality of your mount—probably limits you to 1 or 2-minute exposures. That again puts a premium on using all of the available light. If the target asteroid is a fast-moving near-Earth object (NEO), you may be forced to use very short exposures so that it doesn't trail on your image (NEO tracking rates of 10 arc-sec per minute aren't unheard of). Finally, you will need to achieve high signal-to-noise ratio in order to have good photometric accuracy (as discussed in Section 4.5), which encourages you to use all of the light that's available, rather than restricting yourself to the narrow spectral band of a V-filter.

Considering the warnings that I gave you about using the agreed-upon comp stars and standard-spectral-color filter for variable star photometry, you may ask, "Why is it not terribly risky to neglect that advice when doing asteroid photometry?" The answer is three-fold. First, for the project of determining an asteroid's lightcurve, these disciplines are usually safely neglected because of the more-rapid pace of asteroid photometric data compared with variable star data. Since you will most likely be collecting the entire data set for a complete lightcurve, you have nearly eliminated the need to correlate raw data from two or more observers. Second, the asteroid's rotational characteristics are contained within the differential photom-etry—it isn't necessary to know its magnitude on the standard system in order to glean the desired information from the lightcurve. Finally, since asteroids don't show color changes as they rotate, and they generally fall within a narrow range of colors, there is less risk of spectral effects confounding your differential photometry. Never theless, for some asteroids and some purposes, it *is* necessary to maintain the discipline of standard comp stars and filtered photometry. In Section 4.7, I'll discuss the question of when and why filtered photometry and standard comp stars are sometimes needed.

4.4.3   Reducing and analyzing your observations

Having made a night-long series of images, you're ready for image reduction and data analysis. Image reduction should follow the same procedure as I described in Section 4.3.3 for variable star photometry: save your raw images (including darks and flats) on a non-volatile medium; do the routine CCD image reduction of flat-fielding and dark-subtraction (and bias subtraction, if you're using "scaled" dark frames); and do not do any kind of image enhancement algorithm. With your reduced images, you're ready for data analysis. This has three steps: finding and following the asteroid, measuring its brightness, and plotting the results.

To start, you'll need to find the asteroid in each image so that you can place your photometric measuring aperture directly on it in each image. The standard way to do this is to "blink" several images in your sequence. The "blinking" algorithm that comes with most astronomical CCD image-processing software is the digital-age equivalent of the mechanical "blink comparator" that decades of 20th century astronomers used to search their images (on glass plates!) for moving objects. The idea is that two images taken at different times are aligned, and then the screen rapidly shifts to display first one, then the second, then back to the first, etc. If the images are finely aligned, what you'll see on your screen is the stars unchanging (or changing or shifting only very slightly) as the image "blinks". The asteroid, on the other hand, will bounce back and forth as the screen shifts from image to image. A little care in watching, and a few notes on scratch paper, should be sufficient for you to identify the asteroid on any of your images as it moves across the FOV. My habit is to select one of the first images of the night, an image about mid-way through the session, and one of the final images, and blink the three in sequence. That helps me determine the asteroid's path across the stars. It's also a chance to search the field for any other moving or changing objects (e.g., an undiscovered asteroid). It hasn't happened to me yet, and the odds are very much against such a discovery (see Chapter 6), but I figure that I'll never know if I don't look.

Now that you know the location and path of the asteroid, you are in a position to place your photometry aperture onto the comp stars and the asteroid, in frame after frame. If you're doing this manually, it can be something of a chore. A continuous series of 2-minute exposures for a 6-hour observing session, you'll have 180 images to reduce! For my first asteroid lightcurve, I used a rudimentary CCD imaging program that allowed me to put the measuring aperture over one comp star to display the integrated ADUs. I typed this value into an Excel spreadsheet, then moved the measuring aperture to the asteroid, and entered its ADU into the spreadsheet. Then I loaded the next image, and repeated the process ... until all 180 images were reduced. This laborious routine had several meritorious consequences: I learned a lot about the various flaws that you will find from time to time in your images, I avoided the cost of buying special-purpose software until after I had tried my hand at photometry, and (happily) I got a very nice lightcurve from an asteroid that fortuitously happened to have a large amplitude (over 0.5 magnitude peak-to-peak). This effort showed me that photometry was within my grasp, and made my subsequent investments in specialized photometry software a lot more palatable! (It also was the initiating force that eventually led to the purchase of a better CCD, a filter wheel, a set of photometric filters, a new telescope, and a backyard observatory; but that's another story.)

Some widely-available and modestly-priced software will make the photometric reductions much easier. TheSky has an asteroid lightcurve routine that will automatically move the measuring aperture to follow the asteroid, and generate the lightcurve. MPO Canopus also does that, as well as providing several features that make it relatively easy to merge several nights' data and interpret the lightcurve after it's been measured. If your software doesn't offer to follow a moving object with the photometry aperture, than you'll have to do it by hand, manually placing the aperture over the object in each image. I've done it both ways, and they give equivalent results, but if a few projects convince you that asteroid photometry is "your thing", then you'll definitely want to invest in a software package such as MPO Canopus that simplifies the lightcurve reduction process.

Plotting the differential lightcurve from a single night is pretty straightforward—any spreadsheet can do it as well as the more sophisticated photometry packages. With luck, your first night will show a noticeable variation in the asteroid's brightness relative to the comp stars, and may even give a tantalizing hint of a periodic cycle. Most likely, in order to observe the complete cycle of brightness variation, you'll need to gather two or three more nights of data. It is preferable to make your observations on consecutive nights, but that isn't a hard and fast requirement. If you miss a night or two between observing runs, chances are that you'll still be able to construct a complete lightcurve and determine the asteroid's rotation period.

Merging data from several nights and combining them together to determine the rotation period and complete lightcurve shape can be done with a spreadsheet, but it is a bit of a challenge. Here, special-purpose software such as MPO Canopus shines. The nature of the problem can be best explained by an example. Figure 4.14 shows the data from two nights' observation of asteroid 755 Quintilla. This is a plot of "target minus comp" where "comp" in this case is the average instrumental magnitude of five comp stars.

There are a few items to note on this graph. First, there's a huge gap containing no data. That's the daytime between these two adjacent nights. Second, the x-axis reports time in terms of "Julian Days", to simplify aligning observations across long periods of time. (See Appendix A for an explanation of JD.) Third, the time is "corrected for asteroid–Earth light time" so that the time scale represents the time that the light left the asteroid, rather than the time that the light arrived at the Earth. (See Appendix A for a discussion of this topic, also.) Fourth, the two nights did not have any comp stars in common, because the asteroid moved through a distance greater than my FOV between the two nights.

The use of different comp stars for the two nights explains the vertical offset between the two nights' curves. The standard asteroid photometrist's method of dealing with this is simple: select one night as the reference curve, and adjust the other nights up or down by a "delta-comp" that brings their curves into line. "Delta-comp" is a purely arbitrary vertical offset. Each night has its own "delta-comp", and

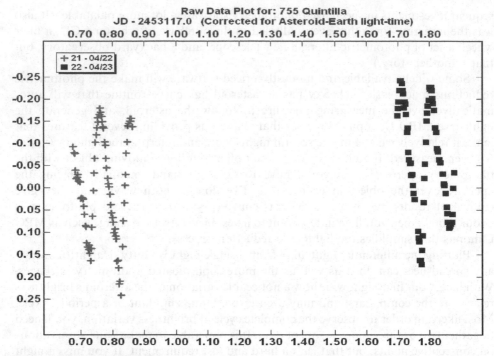

Figure 4.14. Raw differential photometry from two nights' observation of an asteroid. The apparent brightness difference between the two nights is an artifact of using different comp stars on the two nights (because the asteroid had moved through a distance larger than the imager's FOV).

you use a "try and check" approach to get the curves lined up in the vertical axis. It sounds a little sloppy, but it almost always works.

Figure 4.15 shows the two nights' curves lined up in magnitude. Each data point in the second night was adjusted by:

$$\text{adjusted mag} = (\text{raw mag}) + (\text{delta comp})$$

and in this example, delta comp ≈ 0.06 magnitude.

Careful study of the shape of the curve gives you some clues about how to time-align the two nights. The idea is that the asteroid is rotating, sort of like a poorly thrown American football (or a potato). When we view it "point on", it is faint. When we view it "side on", it is brightest. So we expect that the lightcurve will go up and down, in time with the asteroid's rotation. Since a football has two "points" and two "side on" orientations, we expect the lightcurve to be "double-humped". That is, one complete rotation of the asteroid normally gives two "peaks" and two "bottoms" in the lightcurve. Usually, these "peaks" and "bottoms" are not exactly the same magnitude. The asteroid's deviation from a perfect triaxial ellipsoid shape will make one "peak" brighter than the other, and one "bottom" fainter than the other.

In order to get the two nights time-aligned in terms of the asteroid's rotation

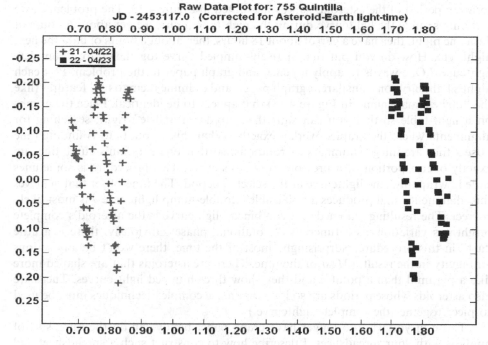

Figure 4.15. The raw photometry from Figure 4.14, with the second night adjusted by "delta-comp" = 0.06 mag so that it lines up with the first night.

period, let's take a close look at that "brightest maximum" that appears on both nights. Reading off the graph, the times of the peak are approximately:

$$JD$$
$$\text{night 1 maximum} = 0.76$$
$$\text{night 2 maximum} = 1.71$$

Now, apply some geometrical thinking. If the asteroid is, in fact, in the same orientation at JD = 0.76 and at JD = 1.71, then it must have gone through an integer-number of complete rotations in the time interval = 1.71 − 0.76 = 0.95 days × 24 hr = 22.8 hr. That is, its rotation period must be one of:

$$P = 22.8/N \text{ hours, } \quad \text{where } N = 1, 2, 3, \text{etc.}$$

So, the rotation period might be $P = 22.8$ hours, or 11.4 hours, or 7.6 hours, or 5.7 hours, or 4.6 hours.

You can usually sort out which period estimate is correct by (a) getting another night or two of data, (b) applying a qualitative test of "reasonableness" to the combined lightcurve shape, and (c) using "Occam's razor" to select the simplest solution.

Here's the idea: If you could observe the asteroid continuously through several rotations, you would expect the lightcurve to repeat, with a cyclic period equal to the rotation period of the asteroid, as was illustrated in Figure 4.13. The problem is, we can only sample the lightcurve during the nighttime hours: we get about 6 hours of data one night, then have a gap of about 18 hours, then about 6 hours of data the next night, etc. How do you put that sparsely-sampled curve together into a complete lightcurve? One way is to apply a pencil and graph paper to the problem. Plot each night's lightcurve on transparent graph paper, and examine them to find features (like the "highest maximum" in Figure 4.15) that appear to be identical. Place the graphs on a light-table so that you can shift them up/down and left/right, searching for alignments where the graphs overlap exactly. What this amounts to is finding time offsets that are integer numbers of rotations, so that one rotation's curve lines up exactly over a portion of a previous rotation's curve. This procedure is sometimes called "wrapping" the lightcurve at the selected period. The time-offset that achieves this alignment—and produces a "plausible" double-hump lightcurve—is most likely correct. The resulting "aligned and combined" lightcurve is the asteroid's complete brightness variation, as a function of rotational phase. Obviously, there is a little "art" in this procedure. Surprisingly, most of the time, there won't be any serious ambiguity in the result. (*Most* of the time. There are asteroids that are shaped more like a pyramid than a potato, and they show three-humped light curves. There are also asteroids whose periods are so long that more complex techniques must be used to piece together the complete lightcurve.)

Those of you who no longer have graph paper handy can do the same sort of analysis with your spreadsheet. I describe how to construct such a spreadsheet and use it for period analysis in the following section. If you're not experienced with spreadsheet analysis, or don't plan to use it, feel free to skip ahead.

4.4.3.1   *A spreadsheet approach to lightcurve period determination*

What we'll do in this section is create an Excel spreadsheet that will enable you to determine the period of a lightcurve. If you created your differential photometry with TheSky, then you can export your data into Excel and use this spreadsheet to "wrap" the data from several sessions into a single lightcurve. This spreadsheet approach may be useful for your first few projects. It is a bit time-consuming, and it demands that you have some experience with spreadsheet formulas and graphs, but with it you can create a nice lightcurve. After you've used it a few times, and assuming that you get the "asteroid lightcurve" bug, you'll recognize the value of special-purpose data analysis software such as MPO Canopus or Peranso.

The general idea is that your data points form a series of samples of the lightcurve function $f(t)$. Since we assume that this is a periodic function, with period $= P$, it should obey the rule that

$$f(t) = f(t - NP)$$

where $t =$ time of observation (in the same units as P—either hours or days);

$P =$ period of rotation (in hours or days, whichever is most convenient for your analysis);

$N =$ an integer (1, 2, 3 ...).

	A	B	C	D	E	F	G	H	I	J	K	L
1	Excel Worksheet for determining asteroid lightcurve period											
2	P_{est}=	5.68	(hours)	Session 1	Session 2		Session 3					
3					delta-comp$_2$=	0.5	delta-comp$_3$=	-0.12				
4	Observational data				N_2=	4	N_3=	8	Data for Plot			
5	J.D.	t, hrs	Obj-Comp	Obj-Comp	t_2-N_2*P	session 2	t_3-N_3*P	session 3	time (hrs)	sess 1	sess 2	sess 3
6	2453117.6777	0	-0.063	-0.063					0	-0.063		
7	2453117.6800	0.0554	-0.049	-0.049					0.0554	-0.049		
8	2453117.6823	0.1114	-0.047	-0.047					0.1114	-0.047		
9	2453117.6846	0.1668	-0.009	-0.009					0.1668	-0.009		
10	2453117.6869	0.2218	0.01	0.01					0.2218	0.01		
11	2453117.6893	0.2779	0.008	0.008					0.2779	0.008		
12	2453117.6916	0.3336	0	0					0.3336	0		
13	2453117.6939		-0.015	0.015						0.015		
50	2453118...	24.4606	0.437		1.7406	-0.063			1.7406		-0.063	
51	2453118.6992	24.5165	0.42		1.7965	-0.08			1.7965		-0.08	
52	2453118.7016	24.5738	0.405		1.8538	-0.095			1.8538		-0.095	
53	2453118.7040	24.6317	0.379		1.9117	-0.121			1.9117		-0.121	
54	2453118.7064	24.6886	0.357		1.9686	-0.143			1.9686		-0.143	
55	2453118.7088	24.7464	0.357		2.0264	-0.143			2.0264		-0.143	
56	2453118.7112	24.8033	0.364		2.0833	-0.136			2.0833		-0.136	
57	2453118.7136	24.8614	0.38		2.1414	-0.12			2.1414		-0.12	
58	2453118.7160	24.9187	0.387		2.1987	-0.113			2.1987		-0.113	
59	2453118.7183	24.9751	0.418		2.2551	-0.082			2.2551		-0.082	
60	2453118.7207	25.0327	0.405		2.3127	-0.095			2.3127		-0.095	
82	2453119.7027	48.6000	0.005				3.1600	0.125	3.1600			0.125
83	2453119.7056	48.6708	0.02				3.2308	0.14	3.2308			0.14
84	2453119.7086	48.7409	0.049				3.3009	0.169	3.3009			0.169
85	2453119.7330	49.3270	-0.054				3.8870	0.066	3.8870			0.066
86	2453119.7361	49.4011	-0.116				3.9611	0.004	3.9611			0.004
87	2453119.7391	49.4746	-0.152				4.0346	-0.032	4.0346			-0.032

(Left margin labels: data from Session 1, data from Session 2, data from Session 3)

Figure 4.16. Overall layout of a spreadsheet for lightcurve period determination.

Enter all of your data into your spreadsheet as a table of observation time (t) and differential magnitude [$f(t)$]. Select an (arbitrary) value of P—based on your best guess from a geometrical analysis such as described above. Then for night #2 and subsequent, derotate the time of each observation by $t \to t - NP$, using whatever value of N will make the data overlap in time with night #1. Plot the result, adjusting the level of night #2's data by a delta-comp if needed. Iterate through estimates of P and N until you find the "best-fit" estimate of the period, where the data from multiple nights overlaps nicely, and the curve is a plausible rotational lightcurve.

The overall structure of the spreadsheet is illustrated in Figure 4.16. The spreadsheet is built and used in seven steps. Steps 1 through 5 are devoted to creating the spreadsheet, and entering and formatting your data. Steps 6 and 7 are where the actual data analysis happens. You'll have to go through all seven steps for your first project, but once the spreadsheet has been created, you can enter new data and analyze it without having to repeat steps 1 through 6.

Step 1: Set up the headings, enter your raw data, and calculate the time of each observation   As shown in Figure 4.17, your raw data is entered in columns A and C. Column A contains the Julian Date (JD) of each observation.* Column C contains the differential photometry (object minus comp star) in magnitudes. Column B

---

*If your observations span more than a few nights, it is wise to include light-time correction in the observations time, so that the "observation time" represents the time that the light left the asteroid, rather than the time that it arrived at Earth. (See Appendix A for a discussion of "light-time" corrections, and of Julian Days.)

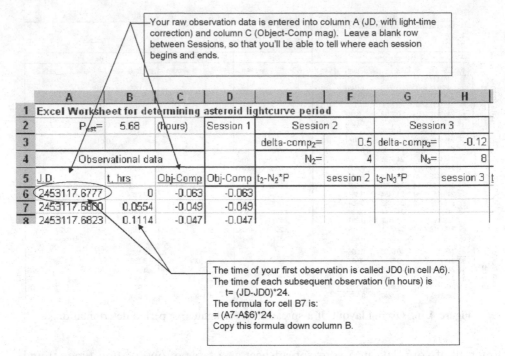

Your raw observation data is entered into column A (JD, with light-time correction) and column C (Object-Comp mag). Leave a blank row between Sessions, so that you'll be able to tell where each session begins and ends.

The time of your first observation is called JD0 (in cell A6).
The time of each subsequent observation (in hours) is
  t= (JD-JD0)*24.
The formula for cell B7 is:
  = (A7-A$6)*24.
Copy this formula down column B.

Figure 4.17. Step 1 of creating the spreadsheet for lightcurve period determination.

contains the formula that translates JD into "hours since the first data point." This is primarily a mathematical convenience, since it's easier to work with small numbers, and it's common (although not mandatory) to report lightcurve periods in "hours" rather than "days".

Step 2: Set Column D to contain the Tgt-Comp data from Session 1   Enter the equation that copies Session 1 data into column D, as shown in Figure 4.18.

Step 3: Calculate the "wrapped" time of Session 2 observations, and the associated "adjusted" Obj-Comp   As shown in Figure 4.19, column E is where you enter the formula that "derotates" Session 2 data by an integer number of rotations of the asteroid. Cell F4 contains the number of rotations through which to perform the derotation (N_2). Note that when you are doing the analysis in Step 7, you must only enter integer values for N_2 into cell F4. Column F contains the formula that adjusts the differential magnitude up or down by delta-comp, to align sessions that used different comp stars. Cell F3 contains the delta-comp by which the differential photometry will be adjusted; delta-comp can be any number, and may be positive or negative.

In cell B2, create a named variable "*P*", and enter any arbitrary value. This is your estimated period (in hours).

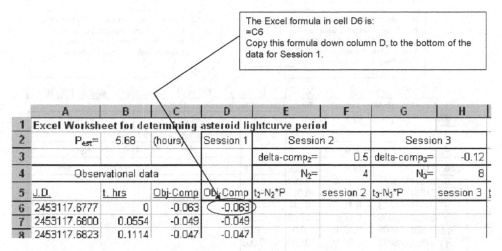

Figure 4.18. Step 2 of creating the spreadsheet for lightcurve period analysis.

If you have more than two nights of data (i.e., more than two sessions) on this asteroid, repeat Step 3 for each session (using columns G and H, etc.) In the example that is shown here, there are three sessions. If you have more sessions, just continue this process of adding columns, and establishing an N and delta-comp to use for each session.

Step 4: Create columns I, J, K, L for plotting the data   The purpose of this step is to copy the data that has been time-wrapped and adjusted by delta-comp into four adjacent columns, so that you can easily create a graph. Refer to Figure 4.20.

Step 5: Create a graph of the data in columns I, J, K, and L   In this step, you select the entire data set in columns I, J, K, and L, and create a "scatter"-type graph, as shown in Figure 4.21. With delta-comp and N set to zero for all sessions, this is a graph of your raw data.

Your spreadsheet is now complete, and you're ready to use it to analyze your data, to determine the lightcurve shape and period of this asteroid.

Step 6: Using delta-comp to align the sessions vertically   The first step in data analysis is to apply a "delta-comp" to each session, moving its differential photometry up and down so that all sessions line up vertically with Session 1. This delta-comp compensates for the fact that each session used different comp stars. Enter a value for delta-comp$_2$ (cell F3) that moves Session 2 data up or down to align it vertically with Session 1 data. Do the same with Session 3, by entering a value for delta-comp$_3$ (cell H3), that aligns it vertically with Session 1 data. The concept, and the result of aligning the sessions, is illustrated in Figure 4.22.

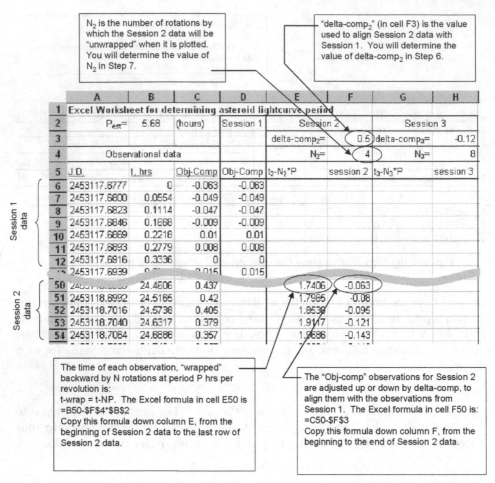

Figure 4.19. Entering formulas to derotate the observations, and apply delta-comp to align multiple sessions that used different comp stars.

Step 7: Finding the period by iterative approximations Now that all sessions are approximately aligned vertically, you can begin estimating the period of the light-curve. Working with just Sessions 1 and 2, determine a plausible approximate period. As explained in the previous section, the starting point for guessing the period is to identify a point on the partial lightcurve of Session 1 that also appears in Session 2. Sometimes, a noteworthy wiggle in the curve may appear in both sessions. In most cases, a reasonable starting point is to identify the "brightest maximum" or the "faintest minimum", and assume that the asteroid has spun through an integer number of rotations in the time between these two points. For this example, the faintest minimum of Session 2 occurs almost exactly 22.8 hours after the faintest minimum of Session 1. If these two points do, indeed, represent the same rotational orientation of the asteroid, then it must have gone through one or more complete

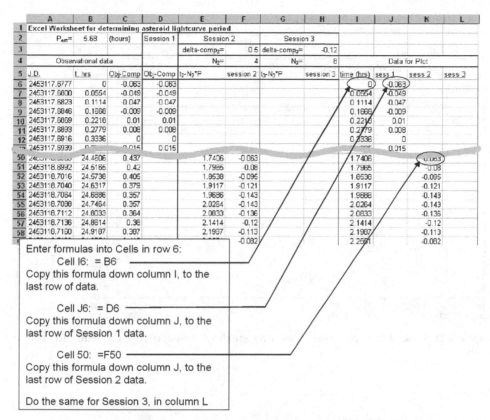

Figure 4.20. Preparing to graph the adjusted data.

rotations during this period of time. That is, the period of the lightcurve must be approximately $P = 22.8/N$ (where N is an integer). We know that P isn't as short as 2.28 hours because if it were, then we would have seen more than one complete rotation on a single night. So, we'll start by examining the shortest plausible period, $P = 22.8/5 \approx 4.6$ hr. Enter 4.6 in cell B2.

If the period really is 4.6 hours, then during the 24 hours between Session 1 and Session 2, the asteroid will have gone through $N_2 = \text{INT}(24/4.6) = 5$ revolutions ("INT" is the integer function). So, enter 5 into cell F4. This tells the spreadsheet to "derotate" Session 2 through five complete revolutions of the asteroid (at the assumed period of 4.6 hr). Remember, the value of N_2 (and of N_3) must be an integer. As shown in Figure 4.23, using $N_2 = 5$ makes Session 2 nearly overlap with Session 1. That is a sign that we're on the right track.

Session 3 occurred two nights after Session 1, so in the intervening 48 hours the asteroid would have rotated $N_3 = \text{INT}(48/4.6) = 10$ times. Therefore, enter 10 into cell H4, to derotate the data from Session 3 by ten complete rotations of the asteroid. This brings Session 3 into pretty good alignment with Sessions 1 and 2, as illustrated in Figure 4.24, which is encouraging!

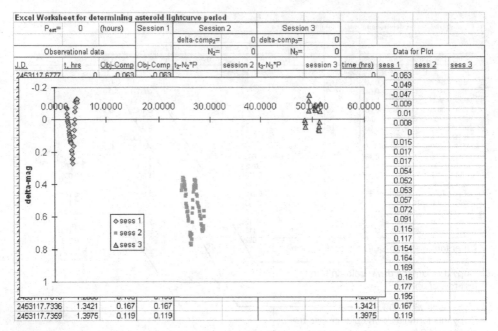

Figure 4.21. Completed spreadsheet, showing graph of un-adjusted data, ready for period analysis.

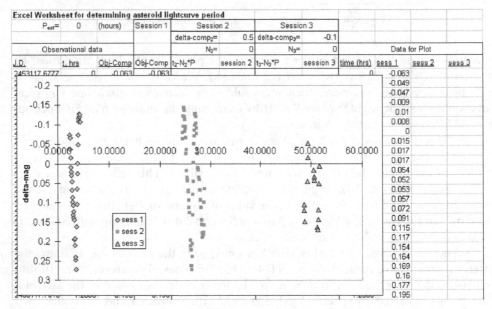

Figure 4.22. Step 6—using delta-comp to align each session vertically.

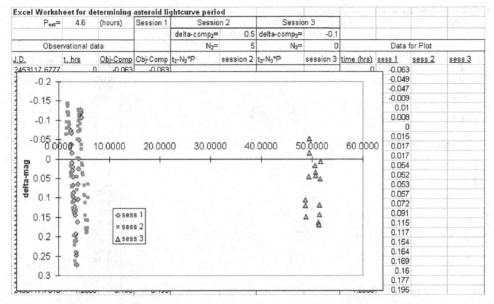

Figure 4.23. Derotating Session 2 to bring it into time-alignment with Session 1.

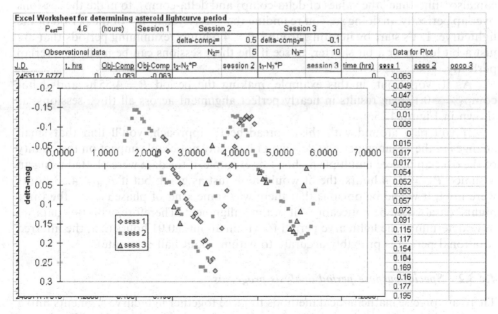

Figure 4.24. An "almost good" overlap of lightcurves from three sessions.

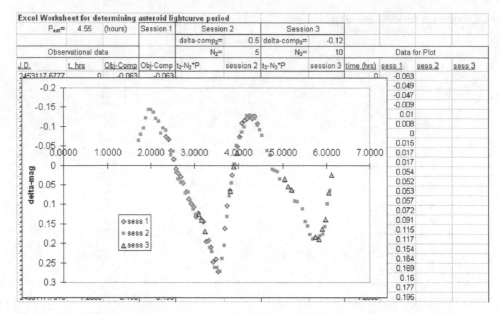

Figure 4.25. Wrapping the sessions to a period of $P = 4.55$ hours gives a very clean lightcurve.

However, the fit clearly isn't perfect, which suggests that either (a) our estimated period is a little off, or (b) we need to examine other possible periods. Try adjusting the assumed value of P_{est}, to make the three sessions line up in time. If necessary, you can also "fine-tune" the values of delta-comp$_2$ and delta-comp$_3$, to make the sessions overlap perfectly. In doing so, you're finding the "best-eyeball-fit" of the period of the lightcurve. Let's start by iterating around $P = 4.6$ hours, examining periods that are just a bit longer or a bit shorter, to see if the three sessions can be made to overlap perfectly.

As it works out, in this example, making the period $P = 4.55$ hr and delta-comp$_3 = -0.12$ mag results in nearly perfect alignment across all three sessions, as shown in Figure 4.25.

If you play around with this "spreadsheet" approach, you'll find that small changes in the assumed period (P_{est}) won't have any noticeable impact on the quality of the overlap between multiple nights. For example, if the data were plotted using an assumed $P_{est} = 4.54$ hours, the fit would look just as good; but if $P_{est} = 4.53$ hours were used, it would be obvious that there was some kind of phase error—the three nights' data would be noticeably out of time-alignment. Therefore, with this data we have determined the lightcurve period to within about ± 0.01 hours (i.e., the inferred rotational period is probably accurate to within about half a minute).

4.4.3.2   Special-purpose period analysis programs

There are precise mathematical methods to piece together lightcurve segments into a complete lightcurve. One popular algorithm is based on a special application of

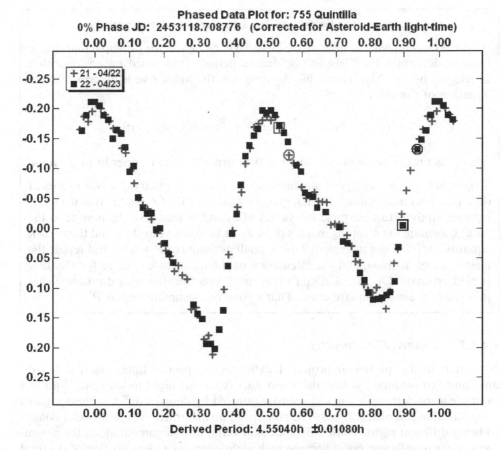

Figure 4.26. Two nights' asteroid lightcurve data, "wrapped" to the best-fit rotation period.

Fourier analysis. The two readily-available software packages that I'm aware of that implement this Fourier algorithm are Peranso by Tonny Vanmunster, and MPO Canopus by Brian Warner. What this algorithm does is simplify the process by automatically searching through all possible periods to find the best time-alignment. The result of doing that on 755 Quintilla is shown in Figure 4.26. The best-fit period is 4.55 hours (pretty close to our original very-approximate estimate of 4.6 hours!)

The advantage of programs such as Peranso or MPO Canopus is that they eliminate the need to re-format the spreadsheet for each new data set, they do all of the necessary calculations for you, they provide good estimates of the accuracy of the fit, and they each provide a wealth of other special features that are beyond the ability of a spreadsheet to offer. So, if you find yourself doing more than a few asteroid lightcurve projects, you will almost certainly want to invest in one or both of these fine programs.

Fourier analysis for determining an asteroid's rotation period

If you're familiar with Fourier analysis, you will recognize the principal that can be used to determine the "best-fit" lightcurve period. It is based on an algorithm developed by Dr. Alan Harris [6]. Assume that the lightcurve is described by an equation of the form

$$M(t, P) = a_0 + \sum_{i=1}^{n}[a_i \sin(2\pi t/P) + b_i \cos(2\pi t/P)]$$

where t is time, P is the period, and n is the "order" of the Fourier fit to the data.

Take that equation, and your measured differential photometry (i.e., your series of data points of magnitude vs. time), pick a starting value of P, and perform a least-squares analysis to determine the values of a_i and b_i that give the best fit to the data. Calculate the resulting mean-square error between your data and the best-fit equation $M(t)$. Then increment P by a small amount (to $P + \Delta P$), and repeat the least-squares analysis, and the calculation of the mean-square error for this new period estimate. Do that a zillion times, until you find the period estimate that minimizes the mean-square error. That's your best-estimate period P^*.

4.4.3.3 "Unfiltered" photometry

Note that in this particular project, I didn't use a spectral filter, and I didn't do anything extraordinary to link the comp stars from one night to the next. With this example as background, you can see why asteroid lightcurves differ in these regards from variable star observations. The concept of an arbitrarily-selected "delta-comp" to bring different nights (with different comp stars) into alignment along the magnitude axis is usually successful because each night contains a large portion of the total light curve, and because the data is dense enough that you can visually recognize the portions that "overlap" between two or more nights.

Hence, aside from the challenge of placing your photometric measuring aperture over an object that is in a slightly different position in each image, asteroid photometry is in some ways less complex, and less demanding on your equipment, than is variable star photometry.

4.4.4 Reporting your results

The type of result shown in Figure 4.26—the rotation period and lightcurve for a single asteroid—is a valuable (and publishable) contribution to Solar System science. If it is the first lightcurve for this object, then astronomers can add it to their statistical studies of asteroid rotation rates. If lightcurves for this object have already been determined at previous apparitions, then they may be able to combine the lightcurves to calculate the three-dimensional shape of the asteroid and determine the direction of its rotation and its pole orientation.

Asteroid lightcurve results are shared with the planetary science community by reporting them in the *Minor Planet Bulletin*, published quarterly by the Minor Planets Section of the Association of Lunar and Planetary Observers (ALPO). This is a peer-reviewed journal, so publication of your results not only serves the planetary science community, but may also serve to add a prestigious item to your *curriculum vitae*. The typical article in the *Minor Planet Bulletin* ranges from less than one page to a few pages long, and covers one to a half-dozen asteroid lightcurves. You can download recent copies of the *Minor Planet Bulletin* at no charge from *http://www.minorplanet observer.com/mpb.default.htm* This will enable you to see the type of reports that other astronomers (many of them amateurs) are making, and hopefully will encourage you to try your hand at asteroid lightcurve photometry. The same website can direct you to the Instructions for Authors when you're ready to submit your own observations.

How important are amateur astronomer's asteroid lightcurve observations? There are more than 100,000 known asteroids. As of February 2005, lightcurves have been reported for only about 2,400 of them. Of those, only about 1,100 are considered to be "secure results" (i.e., full lightcurve and no ambiguity in the period determination). Pole orientations have been determined for fewer than 120 asteroids. So, your data is desperately needed to better understand the population of asteroids! You should definitely try at least one or two asteroid lightcurve projects, if only for the "chops". Who knows, you may find that you enjoy showing people the graphs made from your own data as much as you enjoy showing off the fruits of your astro-imaging. In that case, you will have joined the small community of active asteroid photometrists.

4.4.5   The challenge of long-period asteroids

While most asteroid rotation periods are in the range 4–12 hours, some asteroids are very slow rotators, with periods longer than 24 hours. These present a peculiar challenge to your data reduction, and to your observing procedure. Imagine what would happen if you were observing an asteroid whose rotation period was about 50 hours. On the first night, you'd get a partial lightcurve, but no clear evidence of periodicity. On the second night, there would be no obvious "overlap" between the first and second night's data (since the asteroid hasn't completed even half of a rotation yet). The third night might also give you a sample of the lightcurve that doesn't overlap with the first or second nights. Absent a clear feature on which to apply the delta-comp, it's impossible to confidently combine data from several nights.

One solution to this problem is to arrange to "bridge" your comp stars from night to night. Make your observations as described above on night # 1. On night # 2, the asteroid will probably have moved so far that it is no longer in the same FOV as the comp stars that you used on night # 1. You'll have to select new comp stars in the asteroid's FOV on night # 2. You can link the two sets of comp stars by taking a few special images on night # 2. When the asteroid is near culmination (so that the atmospheric effects are minimized), return to the FOV that was used on night # 1, and take three images of that FOV. Then return to your target asteroid and continue your lightcurve imaging for the balance of the night. This gives you images of the "night #1" comp stars, taken at practically the same time, and same atmospheric conditions, as your "night #2" comp stars. You can use them to determine the delta-comp between night #1 and night #2.

The math is pretty simple. Call $A(t)$ the brightness of the asteroid, as a function of time. On the first night, using comp star C1, you used differential photometry to created a graph of $[A(t) - C1]$. On the second night, using comp star C2, you made a graph of $[A(t) - C2]$. On the second night, you also measured the instrumental magnitudes of C1 and C2 at essentially the same time and at identical airmass. You can use these simultaneous measurements of the instrumental magnitudes of the comp stars to calculate delta-comp $= (C1 - C2)$.

Now, recognize that

$$A - C1 = [A - C2] - (C1 - C2)$$

$$= A - C2 - \text{delta-comp}$$

That is, $[A(t) - C2 - \text{delta-comp}]$ equals the differential magnitude that you *would have measured* on the second night if you could have used C1 as the comp star on that night. Be careful that you watch the signs when you're doing the algebra: delta-comp can be a positive or a negative number.

Use the derived delta-comp to align the magnitude data from night #2 with that of night #1. Follow the same procedure for subsequent nights—always return to the "night #1" FOV for a few images near the time of culmination so that you can determine the delta-comp that will act as the bridge back to the first night.

In this way, you can piece together an accurate lightcurve of a slow-rotating, long-period asteroid. This is a particularly valuable analysis, because relatively few long-period asteroids have good published periods and lightcurves. I suspect that this is not so much because they are unusually rare, but rather because people get discouraged. If you haven't seen a recognizable lightcurve after a few nights of observing, you may be inclined to give up and move on to another target. Don't give up! The procedural bias against slow-rotators and small-amplitude lightcurves needs to be corrected. If you find yourself on one of these frustrating objects, stick with it! Good lightcurve information on the slow-rotators is especially valuable, so that planetary scientists can improve their understanding of the statistics of asteroid rotation periods.

4.4.6 Choosing your target

With so many asteroids calling for photometric attention, how do you select your target for the night? Your planetarium program will help you identify asteroids that are rising in early evening (so that you can observe them all night), and that are bright enough to offer a good SNR (which might be anywhere between 11th magnitude and 15th magnitude, depending on the size of your telescope). That will probably still leave you with a long list of potential targets. There are several ways to narrow that list down.

First, review the most recent issue of the *Minor Planet Bulletin*. Each issue provides a list of lightcurve opportunities and shape/spin modeling opportunities. The "Lightcurve Opportunities" list suggests several dozen asteroids that are nicely placed for viewing in the following three months, and that have no known lightcurve, or relatively uncertain ones. These may be prime targets. The "Shape/Spin Modeling Opportunities" are asteroids for which good lightcurves are available from several apparitions. Lightcurves from just a few more apparitions may be sufficient to complete the calculation of their three-dimensional shape and the orientation of their spin vectors. These are also valuable targets.

Several classes of asteroids are of particular interest [7]. A larger sample and improved statistics are needed for asteroids whose orbits cross the inner planets. Small, very rapidly-rotating asteroids are poorly understood, and more lightcurves are needed. Many of these are near-Earth asteroids, hence tricky to observe because of their fast motion across the sky. Amateur photometry is used to augment radar observations of these objects, providing valuable correlation between the two modes of observing [8]. Lightcurves taken at very large solar phase angles (i.e., either very far from the ecliptic, or very far from opposition) are particularly valuable in shape-modeling projects. It is particularly useful to collect lightcurve data on an asteroid within a few nights before or after it occults a star (see Chapter 2). A complete lightcurve at the same epoch as the occultation provides a valuable adjunct to the occultation observer's shape and size measurements [9].

4.4.7 Collaborations: the CALL website

A single observer working alone can often determine the lightcurve of an asteroid quite effectively in just a few nights. However, there are situations where the lone observer simply can't succeed. In these situations a collaboration between multiple observers separated by many time zones can be the key to success. One such case is the very-slowly rotating asteroid. Imagine that your target has a period of 30 or more hours. If you are the only observer recording its lightcurve, you may gather 6 or 7 hours on one night—less than $\frac{1}{4}$ of a complete rotation. If you devote a great many nights to this asteroid, linking the comp stars along the way, you can eventually figure out how to piece together those disconnected 6-hour segments into a complete light-curve. Doing so will be challenging because of the number of nights required and it is liable to leave you with a nagging lack of confidence in your result, because you'll never be completely certain that you found the one-and-only way to piece the nights together into a plausible lightcurve and rotational period. Wouldn't it be nice if you could arrange to have multiple observers around the globe, to keep the asteroid under nearly-continuous observation?

I once participated in a project that included an observer in Japan, another in Switzerland, and me in southern California. Imagine how that improved the completeness of our data. The observer in Switzerland gathered a full night's data. Then, as the Sun was rising in Switzerland, it was beginning to set in California, and I started gathering data only a few hours after the Swiss observer concluded his session. Then, as dawn was breaking in California, it was evening twilight in Japan, and the observer there was getting ready to start gathering his night's data. Finally, at dawn in Japan, it was late afternoon in Switzerland and almost time for that observer to gather his second night's data. With this spacing of observers, we could keep the asteroid under almost continuous observation around the clock. With just a few nights of observing at each site, a complete lightcurve could be gathered for even a very slow rotator. (If we had added observers in the Middle East, eastern North America, and Hawaii, we could have had a world-girdling 24-hour telescope network!)

The other situation in which collaboration between widely spaced observers is virtually mandatory is that of an asteroid whose period is a multiple or sub-multiple of 24 hours. Suppose, for example, that you're trying to determine the lightcurve and rotation period of an asteroid whose (as yet unknown) period is almost exactly 12 hours. On your first night, you gather a few hours of photometry. The next evening, you gather more data, but what you see is virtually identical to what you got on the first night, because in the intervening 24 hours, the asteroid has made exactly two complete rotations, and it is back in the exact same orientation that it was in the night before. Another night, another two rotations of the asteroid, and you're still seeing the same portion of the lightcurve. Phooey! Suppose, however, that you had a friend who lived four or five time zones away. She would be having the same problem (seeing the same fractional lightcurve night after night), but because her night starts and ends four or five hours before or after yours does, she is gathering a *different* portion of the lightcurve than you are. By combining your results, you can achieve more-complete coverage of the asteroid's rotation. If you think through how this will work, you'll recognize that it isn't critical that you and your partner observe on the same nights. Even if your observations are separated by a week or so, you'll probably be able to merge your data sets to determine the asteroid's rotation period and its complete lightcurve.

All well and good, but most of us don't have friends scattered around the globe. How do your identify interested collaborators? That's where the Collaborative Asteroid Lightcurve Link (CALL) comes into play. CALL is a resource for organizing multiple observers in campaigns to monitor selected asteroids. This web resource is hosted by the Palmer Divide Observatory, at *http://www.minorplanetobserver.com/ astlc/default.htm* There are several ways to use it. If you are conducting a lightcurve study, you can notify everyone about the asteroid that you're working, and request collaborators. Or, you can check to see if someone else has already started working one of the asteroids on your candidate list. An e-mail offer to collaborate will usually be welcomed. Finally, you may see cases where an observer is specifically requesting assistance with an asteroid. If it's within your magnitude range, this is a wonderful opportunity to contribute to science and make a long-distance friendship. Using this resource, you can join widespread observing teams, and become acquainted with amateur and professional astronomers whom you never would have met otherwise. The normal etiquette is that after your collaboration successfully determines a good lightcurve, you agree on who will write up the report for the *Minor Planet Bulletin*, and all contributors are listed as co-authors.

4.5 SIGNAL, NOISE, AND PHOTOMETRIC ACCURACY

In the previous two sections, you have seen how amateur photometry can provide valuable scientific data, using techniques that are very similar to those used for normal astrophotography. The accuracy of CCD photometry can be remarkable—measurement errors of only a few hundredths of a magnitude are well within amateur capability. However, if you study your results carefully, you'll notice a slight randomness in your results. If you make several measurements of the relative brightness of two stars, following the procedures for variable star measurements, your results will not be precisely identical on each try. Star #1 may be 1.10 mag brighter on the first measurement; but 1.09 mag brighter on the second measurement, and 1.12 mag brighter on the third measurement. Is one of the stars really fluctuating quickly at the 0.01-mag level? Or is there something in your measurement method or equipment that is causing this fluctuation? How do you decide which it is?

Regarding asteroid lightcurve photometry, I mentioned that the main lightcurve cycle describes the asteroid's rotation, and that there are wiggles that provide subtle information about the shape of the asteroid. Look closely at Figure 4.26. There are point-to-point fluctuations of about ±0.02 magnitude. Are those real—does the asteroid's brightness really change by tiny amounts in just a few minutes—or are they the signature of some sort of measurement errors? In this case, I'm quite confident that these little wiggles in the asteroid's lightcurve are just "noise" in my data. I don't think that there's a significant discovery hidden in them.

Here's a trickier situation. When a relatively nearby star passes directly in front of a more distant star, the gravitational effect of the close star can bend the light from the distant star, in effect amplifying the star's apparent brightness. The effect is called "gravitational lensing". Theory predicts that the expected signature of such a lensing event is a gradual rise and fall of brightness, by a magnitude or so, over a period of a few weeks. The predicted lightcurve is smooth and symmetrical. These events are not common, but now that several professional photometric surveys are searching for them, quite a few have been detected. Figure 4.27 shows a particularly remarkable such detection [10]. The lightcurve displays the expected smooth rise and fall of brightness. The intensity amplification is equivalent to 1.2 magnitudes in this case.

What makes this curve remarkable is the "wiggle" on the falling part of the lightcurve. That wiggle (of about 0.15 magnitude) is claimed to be the signature of a planet in orbit around the lensing star. This curve is the first reported detection of an extra-solar planet based on gravitational microlensing. The credibility of this remarkable discovery hinges on the confidence that the researchers have in their photometry, that the wiggle in the curve is real, and not an artifact of some sort of measurement error. That credibility rests on two foundations: (a) analysis of the inherent accuracy of the photometric data, and (b) the confirmation offered by the fact that the wiggle was observed at more than one observatory.

This section explains how the "signal-to-noise" ratio of your image affects the accuracy of your photometry, and ways that you can improve that accuracy. If you've done a few variable star measurements or asteroid lightcurves, it will help put some of your procedures on a firmer foundation. You will need some photometry

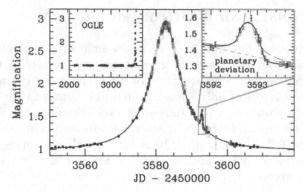

Figure 4.27. The first reported photometric detection of gravitational microlensing by an extra-solar planet. (Reprinted by permission from Macmillan Publishers Ltd.: *Nature*, J.-P. Beaulieu *et al.* "Discovery of a cool planet of 5.5 Earth masses through gravitational microlensing", vol. 439, pp. 437–440, © 2006)

experience, and a familiarity with the challenge of achieving extreme accuracy in photometry, in order to take on the next project that I'll describe (extra-solar planet transit detection).

4.5.1 "Signal" defined

When a photon enters your optical system and hits the photosensitive surface of your CCD imager, it lands on one of the pixels. There is some probability that the photon's effect on the silicon chip will knock free an electron. This probability is called the "quantum efficiency". For modern CCDs at their wavelength of peak sensitivity, the quantum efficiency is about 60%. That means that for every 100 photons that come in, about 60 photoelectrons are freed. The electrons that are collected during the exposure are "read out", detected by the CCD's electronics, digitized, and reported as "ADUs" (analog–digital units). On any given pixel, during the exposure, photoelec-trons are generated by at least two sources: celestial objects (stars, galaxies, etc.), and "sky background" (which includes both true sky-glow from atmospheric emission and light pollution, and a cosmic mist of distant, unresolvable sources). In addition to these photoelectrons, the pixel also collects "thermal electrons" that are sponta-neously liberated by the energy of heat (which in turn is controlled by the temperature of the CCD's chip). So, the accumulated charge on a pixel is*

$$N = N_{\text{star}} + N_{\text{sky}} + N_{\text{thermal}}$$

In most photometry projects, the objective is to isolate just those photoelectrons that arose from the target. That is why you do two things during your photometric

---

*You also occasionally see cosmic ray hits on some pixels, and the inevitable satellite and airplane trails. We'll ignore images on which any of those interfering sources compromise the target, comp star, or sky background ADU counts.

reduction. First, you subtract a "dark frame" that contains nothing but thermal electrons. Second, you subtract an estimate of the sky background ADU count. Then,

$$N_{star} = N - N_{sky} - N_{thermal}$$

The sum of the "target" output of all of the pixels within your measuring aperture is the photometric signal. You will occasionally run across a bit of sloppiness in terminology: sometimes we loosely refer to the sum of the ADUs as the "signal"; sometimes what is meant is the sum of the photoelectrons. They are related by the "gain", g, in that

$$\# \text{ electrons} = g \cdot \text{ADUs}$$

For clarity, I will define the signal in terms of the number of photoelectrons; but of course your imager's output is always expressed in ADUs. For some purposes, the distinction isn't critical. However, signal-to-noise ratio calculations must always be done in terms of photoelectrons, not ADUs.

4.5.2 Gain of CCD sensors

If you check the specifications on your CCD imager, one of the parameters will be the gain, in electrons per ADU. For example, the specified gain of my SBIG ST-8XE is 2.3 electrons per ADU. If a pixel's output is reported as 1,000 ADUs, that means that 2,300 electrons were collected on that pixel during the exposure. (Assuming a quantum efficiency of 60%, that is equivalent to $2,300/0.6 = 3,833$ photons.)

The concept of "gain" is based on the idea of linearity of the CCD: that more photons create more photoelectrons which yield more ADUs, in a strict linear relationship. Since increasing the exposure increases the number of photons that are collected, you expect that a plot of signal vs. exposure should be a straight line, as in Figure 4.18. As long as the target and comp stars are within the linear range, then signal (ADUs) are proportional to photoelectrons:

$$(\# \text{ electrons}) = g \cdot (\text{ADU count})$$

If the signal becomes too large, then it is no longer proportional to the number of photoelectrons. This effect marks the limit of the "linear range" of the sensor.

4.5.3 Linearity of CCD sensors

The photometric principles described above are based on the assumption that the response of the CCD imager is precisely proportional to the amount of light it receives. If the amount of light coming into the CCD imager is doubled, then the number of photoelectrons and the ADU count will double. Modern CCDs are remarkably linear, but they do have limits! Each pixel can only hold so many electrons (its "full well capacity"). From your CCD imaging experience, you know that it is possible to saturate the sensor. If the pixel is "full", and more light comes in, then more electrons are generated, but the pixel can't hold them. At that point "more light" does not equal "more photoelectrons" on a single pixel. Instead, the excess charge spills out and bleeds into adjacent pixels, causing "blooming spikes" and other artifacts.

Accurate photometry requires that you operate in the linear range of the sensor. There are (broadly speaking) two types of CCD chips available to the amateur: those with "anti-blooming" gates, and those that are "non-anti-blooming". The "anti-blooming" gates (ABG) draw off some charge before the pixel saturates, so that your astro-images are less susceptible to bleeding around bright stars. This is a concern for photometry because the process of "drawing off" charge makes the sensor non-linear above a certain charge level. Non-anti-blooming sensors (NABG) tend to be linear almost up to their full-well capacity (and, since they don't include the anti-blooming gates, they also tend to be more sensitive), but of course once the wells are full, even NABG sensors become non-linear.

Accurate photometry can be done with either type of sensor. Whichever you have, in the spirit of "know what your equipment can, and cannot do", you will want to make a study of the linearity limit of your system. I have used both ABG and NABG CCDs for photometry. The key in either case is to determine the linear range of the sensor (i.e., the maximum number of ADUs allowed), and then be sure that the exposure of your science frames is set so that both the target and the comp stars stay within that linear range.

The procedure for checking the linear range of your system is simple. Select a convenient FOV, and make a series of exposures of different lengths (e.g., 15-sec, 30-sec, 1-minute, 2-minute, and 4-minute exposures). Select a few convenient stars, and plot the ADU of the *peak* pixel of each star's image vs. exposure length. Figure 4.28 shows an example of such a plot. This example is for a NABG sensor, so the linearity limit is very near the full-well capacity. Up to a peak signal of about 50,000 ADU, the signal is proportional to the amount of light that came in. An ABG sensor's non-linearity will begin at lower ADU level (usually about 50% of the full-well capacity), and will show a more gradual onset, compared with the NABG sensor. As long as your images stay below the onset of non-linearity, either type of imager will work just fine.

All of the photometric project data analyses assume that you are operating in the linear range of your sensor. For most situations, your challenge is to collect more photons, to acquire sufficient signal. However, there are projects (particularly the extra-solar planet transit search, in which the target stars tend to be quite bright) where too much signal can be a concern. Therefore, it is important to know the ADU count at which your system reaches the knee of its curve—the limit of its linearity.

Since photometry depends critically on the linearity of the sensor's response, it is important to be sure that your image processing doesn't inject non-linearity into the data. This means that some of the image processing that you may be used to doing for astrophotos is forbidden in your photometry projects. The normal flat-field, dark-frame, and bias-frame corrections are linear processes, so continue to use them. It is also acceptable to add images together (to increase the signal from a faint target), because addition is a linear process. It is even acceptable to "shift and add" to increase the signal from a moving target.

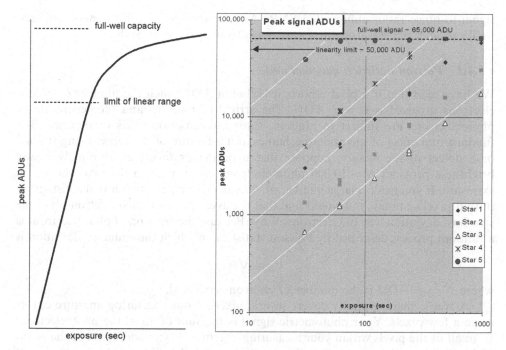

Figure 4.28. CCD imagers are "linear"—ADUs are proportional to photons in—up to the limit of the linear range. (a) Concept of the linearity limit. (b) Linearity check of a NABG CCD imager.

However, almost all "sharpen" algorithms, unsharp-masking, histogram stretching, image compression (e.g., to JPEG), and the like are non-linear processes. Do not do any of them to your photometry images!

Good flat-fielding is probably *more* important for photometry than it is for conventional astro-imaging. Imagine if your system has some vignetting—almost inevitable in most set-ups—and your target is in the center of the field, but the comp star is near the vignetted edge of the field. If you don't properly flat-field the image, then the target will appear systematically "too bright". Worse, if you're studying a moving target, and you don't flat-field properly, then the target brightness will appear to fall and rise as it moves past a "dust donut".

4.5.4   Noise sources and types

For a variety of reasons, there is some inevitable randomness in your measurement of the signal. If you measure the same signal several times, you'll rarely get the same answer twice. The culprit is "noise", in a variety of guises. The following sections describe the noise sources that most amateur photometrists will encounter, and the characteristics of each noise source. In general, the accuracy of your photometry is driven by the ratio of signal to noise. This ratio (signal divided by noise) is called the signal-to-noise ratio, abbreviated SNR. After covering the various noise sources, we'll examine the few straightforward ways to increase your SNR.

4.5.4.1 Poisson statistics: quantum noise

We have seen that if a pixel reports its value in ADU, then the number of photo-electrons involved is $N = g \cdot \text{ADU}$. The arrival of photons and the generation of photoelectrons are subject to random, statistical variation. This randomness is a fundamental effect of quantum mechanics, not a feature of the target being studied or a defect of the imager. Suppose that a particular target of absolutely steady brightness provides $N = 10,000$ photoelectrons on average to the pixel during an exposure. If you made a large number of "identical" images, and plotted a histogram of that pixel's output, you would find that the average was 10,000 electrons, and the standard deviation was 100 electrons. Why? Because the creation of photoelectrons is a random process described by Poisson statistics, in which the standard deviation is

$$\sigma_N = \sqrt{N}$$

where $N = g \cdot \text{ADU}$ is the number of electrons collected.

A star's image typically covers several pixels. Your measuring aperture covers quite a few pixels. Your photometric signal is the sum of all of the photoelectrons from all of the pixels within your measuring aperture. If you add up all of the pixels within your measuring aperture, this principle still holds: the standard deviation of the (summed) signal is equal to the square root of the (summed) signal.

If your situation is dominated by Poisson noise, then the signal-to-noise ratio will be

$$\text{SNR}_{\text{Poisson}} = N/\sqrt{N} = \sqrt{g \cdot \text{ADU}}$$

Since N is proportional to the exposure duration, we can also write

$$N = n\Delta t$$

where $n =$ the photoelectron flux (in electrons/sec of exposure);
            $\Delta t =$ exposure duration (seconds);

so that

$$\text{SNR} = \sqrt{n \cdot \Delta t}$$

and we see that you can increase the SNR by increasing the exposure. In this situation (which is the most common CCD imaging situation), the SNR is proportional to the square root of exposure duration.

4.5.4.2 Thermal noise ("dark current noise")

Thermal energy liberates electrons in the silicon chip of the CCD. The rate of thermal electron generation depends on the temperature of the chip. Typically, the rate of thermal electron generation roughly doubles for every 5-deg increase in chip temperature. This stream of thermal electrons is sometimes referred to as "dark current".

Long-exposure astronomical CCD imagers are generally cooled to reduce the number of thermal electrons.

The number of thermal electrons generated in a single pixel during an exposure of Δt seconds is $N_{\text{thermal}} = r(T)\,\Delta t$, where $r(T)$ is the thermal electron flux at chip temperature T. When you subtract a dark-frame from your science image, you are subtracting the effect of the thermal electrons. Unfortunately, that subtraction can't be perfect because there is a Poisson-noise effect in the thermal electron count. The *average* number of thermal electrons on a given pixel is N_{thermal}, but the *actual* number of thermal electrons on that pixel in a given image will vary randomly, with a standard deviation of $\sigma_{\text{thermal}} = \sqrt{N_{\text{thermal}}} = \sqrt{r(T)\cdot\Delta t}$. This residual variation, after subtracting the average dark frame, is thermal noise.

In general, modern amateur CCDs have low enough dark current that this noise source is negligible compared with the Poisson noise in the sky background and target signal.

By the way, the fact that the thermal electron flux is strongly temperature-dependent is one reason that it is valuable to control the chip's temperature—you want the dark frames to contain the same number of thermal electrons as the science frames, so that when you dark-subtract, what's left is the best estimate of the signal. If your dark frames are taken at a different temperature from your science images, then after dark-subtraction what you have is "signal plus delta-dark flux". If your chip isn't temperature-controlled, you also run the risk that as the chip's temperature changes during the night, your signal will be corrupted by incomplete subtraction of thermal noise.

4.5.4.3 Pixel-to-pixel non-uniformity

Even in the best CCD chip, the individual pixels do not have absolutely identical performance. Some pixels are a little bit more sensitive than others. A few are spectacularly sensitive ("hot pixels"), and likely to be saturated just by thermal electrons in an exposure of a few minutes. Some pixels may be almost completely unresponsive to either signal or thermal electrons ("cold" or "dead" pixels).

Dark-subtraction and flat-fielding may only partially correct the signal estimate on "hot" or "dead" pixels because there is a good chance that they violate the requirement of linearity. Therefore, if you have the choice, it is a good idea to place both your target object and your comp stars away from the "dead" and "hot" pixels of your imager.

4.5.4.4 Guiding/Tracking errors transformed into photometric noise: the importance of flat-fielding

The existence of "hot pixels", with abnormally high gain, and "cold pixels" that are nearly unresponsive to both photoelectron and thermal electron generation, can transform guiding/tracking errors into photometric noise. Consider the following situation: The target variable star is located in a "clean" area of the chip, but the comp-star image falls on a "hot pixel". If the gain of the "hot pixel" were, say, 10 times that of a normal pixel, then the comp star would appear noticeably brighter than it would have if it were placed over "normal" pixels. Recall Figure 4.9, in which we plotted comp-star standard V-mag vs. ADUs. This "hot/cold pixel" effect is one reason that one of the comp stars might fall off of the expected line. If that happened, you'd just exclude the aberrant star and continue with your variable star magnitude analysis.

Now, suppose that you are making a lightcurve of a short-period variable (such as the one described in Figure 4.12). The "hot/cold-pixel" effect would be nearly irrelevant if guiding was perfect and the target wasn't moving. But what if your guiding is poor? Each image will place the comp star on slightly different pixels. One image might have the comp star directly centered on the "hot pixel", so that the signal is significantly over-stated in that image. On the next image, the comp star may have moved a couple of pixels (due to tracking/guiding error), so that it is relatively unaffected by the "hot pixel". It will appear as if the variable star got brighter in the second image (relative to the comp star, which is assumed to be constant).

You'll probably notice if one of your comp stars, or your target, is so seriously affected by a "hot" or "cold" pixel that the individual pixel is near saturation, or nearly "black" (zero signal). For the majority of "normal" pixels, the effect still exists, but it is tiny.

A similar sort of effect can be caused by dust donuts, if you haven't flat-fielded your images. Imagine that your target happened to be positioned right near the edge of the image of a "dust donut". If a tracking/guiding error moves the target slightly deeper into the dust donut, it will appear as if the target has faded. If the target moves slightly away from the dust donut, it will appear as if the target has brightened.

In this way, the presence of pixel-sensitivity variations and/or flat-field errors, in combination with tracking errors that cause the target and comp stars to wander on the focal plane, can create a random, false fluctuation in the differential photometry.

Flat-fielding is your first line of defense against this noise source. Either "twilight flats" or "lightbox" flats can be used to achieve uniformity to within about 1% to 2% across the image. Twilight flats must be taken during that fleeting interval of time when the sky is dark enough that you don't saturate the sensor, but light enough that you don't take long exposures that show stars. However, the occasional star aside, with proper procedures you can be pretty confident that the twilight sky is in fact "flat". The problem of background stars is easily dealt with by taking multiple flat frames, and changing the telescope pointing slightly between frames. Doing a "median-combine" to generate the master flat frame will eliminate the stars.

A lightbox provides greater flexibility, since you can take flats at your convenience. However, there are plenty of ways that a lightbox can give a false sense of security (e.g., how do you know that your lightbox is in fact presenting uniform illumination to your telescope?) If you do use a lightbox (as I do), my recommendation is that you occasionally take a set of "twilight flats", followed by "lightbox" flats under identical conditions (i.e., nothing moved in the optical train between them). Dividing the "lightbox" flat by the "twilight flat" will give you a good indication of the residual "non-flatness" of your lightbox. A 1% or so flatness should be achievable, with careful lightbox design.

Many astro-imagers rotate the CCD imager as they search for guide stars and pleasing image composition. If you do this, remember that you need to take new flat frames every time that anything in the optical train is moved, so that the dust donuts and vignetting on your flat frames are properly registered to the same defects in your science images. If you have a filter wheel, you need to take separate flats for each filter position (each filter has its own unique pattern of dust).

The filter wheel presents a challenge: every time you cycle your filter wheel, you've moved an optical element. Most commercial filter wheels are very good, but they aren't absolutely perfect at filter wheel position repeatability. Hence, the R-filter position for your flat frame is ever so slightly different from the R-filter position after you've cycled through the filters a few times during a sequence of science frames. If the filter's position is non-repeatable by just a thousandth of an inch, it is moving dust particles as much as a few pixels (0.001 inch = approximately 30 microns, and most amateur CCD imagers have pixel sizes in the range 9 microns to 35 microns). That means that the corresponding dust donuts have also moved a few pixels. The filter wheel non-repeatability may also cause the vignetting to change ever so slightly. For most situations, it is not practical to take new flats every time you rotate the filter wheel. The best that can be done is to take one set of flats for each filter, and accept that there will be slight imperfections in your photometry as a result.

You should run a simple experiment to determine how "slight" or gross the residual imperfections are. Make a series of flats, rotate the filter wheel a couple of times, and then make another set of flats. By dividing "flat #1" against "flat #2", the residual imperfections are highlighted. With my set-up (SBIG CFW-8 filter wheel), I see residual imperfections of about 1% (peak) that appear to be caused by filter wheel non-repeatability. For most projects, this is of no consequence; other noise sources are far larger. However, for the most critical differential photometry project described below—extra-solar planet transits—it is wise to plan your strategy so that you make flat-fields and then do not touch the filter wheel for the rest of the night.

With care in your routine of dark-frame and flat-fielding, you can reduce these noise sources to levels that are acceptably small for most projects—photometric accuracy and consistency of better than 3% (about 0.03 magnitude) should be achievable with good practice.

4.5.4.5 *Target motion transformed into photometric noise*

For variable star photometry, autoguiding will keep the target and comp stars "anchored" to the pixel array, minimizing the effects described in the previous section. For asteroid photometry, you have a moving target. In the course of a night, the asteroid is likely to move across a sizable fraction of your field of view. Your photometric aperture must, of course, follow the target as it moves across the field. This gives rise to two noise sources. First is the effect described in the previous section, in which the target's motion is transformed into brightness fluctuations by the action of sensitivity variations and imperfect flat-fielding. Good flat-fielding can largely mitigate this noise source. Second is the fact that the "sky background" is itself not precisely uniform.

Break the "sky background" into two portions. First are the stars and galaxies that you can recognize in the image. If the asteroid happens to pass so close to one of these that your measuring aperture will collect a noticeable amount of light from this clutter (e.g., a star), then you'll see it happen, and can simply toss out the affected frames. The effect of this clutter will be obvious in your data, as well. If your asteroid appears to brighten just as it passes a field star, then you can be pretty sure that your "asteroid signal" was contaminated by additive "star" signal.

Second are background sources that you can't recognize in the image. Even if you don't see any star clutter, it's probably there, but at small amplitude. This portion of the sky background includes a pattern of photons coming from imperceptibly faint stars, undetectable galaxies, etc. That subtle signal gets into your measuring aperture, which is in effect scanning across this pattern as it moves to follow the asteroid. There's really nothing you can do to eliminate this effect, but it can be recognized, and "averaged out", by dedicating sufficient time to each asteroid that your light-curve covers two or more complete rotations. Little ($\sim$0.03 mag or smaller) inconsistencies from rotation to rotation are most likely due to this background clutter, rather than to real topography on the asteroid. Brightness wiggles that are larger than the noise in your lightcurve and that reliably repeat from rotation to rotation are probably real, betraying genuine topographic features on the asteroid.

Faint background clutter can have a noticeable effect on your asteroid light-curves. For example, suppose that your asteroid is mag-14, and has a lightcurve amplitude of 0.25 magnitude peak to peak. As a purely arbitrary value, assume that the asteroid signal, at the exposure being used, is 100,000 ADU. A magnitude-16 field star would generate 15,846 ADU. If your asteroid happens to pass so close to that mag-16 star—barely one-sixth as bright as the asteroid—that both are within the measuring aperture, then you'll measure 115,846 ADU (combined asteroid plus star). That is, you'd think that the asteroid's brightness had grown to

$$\text{mag} = -2.5 \log(100{,}000 + 15{,}859) = 13.85$$

The presence of a 16th-magnitude star as clutter caused the apparent brightness of the asteroid to grow by 0.15 magnitude—a sizable fraction of the total lightcurve variation that you're trying to measure.

Now, you might recognize it if a star as bright as mag-16 was corrupting your photometry. You can work through the example using even fainter clutter sources. Suppose that an 18th-magnitude star wanders into your measuring aperture. You may not even recognize it as a stellar signal at all, but it would contribute 2,511 ADU. Adding these to your 14th-mag (= 100,000 ADU) asteroid, would make it appear as if the asteroid was mag-13.97 (i.e., 0.03 mag brighter than "truth"). It isn't unusual to see scatter of this amplitude (i.e., ±0.03 magnitude) in an asteroid lightcurve, as the asteroid passes over faint background clutter sources. You can see some evidence of that in the example lightcurve of Figure 4.13.

For most asteroid lightcurve projects devoted to determining the asteroid's rotation period and aspect ratio, this level of accuracy is quite acceptable. However, in cases where you're looking for 0.05-magnitude-class "wiggles", this calculation suggests several things you can do: (1) beware of background clutter, (2) if you're looking for small-amplitude wiggles or if you have an asteroid whose total lightcurve amplitude is small, censor your data to delete data points that may be affected by even very faint clutter sources, (3) use multiple nights, over several asteroid rotations, to confirm low-amplitude wiggles before you make too much of them, and in really critical cases, (3) take a long-exposure "deep field" of the FOV to help you find any faint clutter that may exist along the asteroid's path. On the other hand, if you've been careful with your imaging and data reduction, don't assume that data points are aberrant just because they fall off of the "normal" lightcurve. They might be indicating an important result! For example, binary asteroids have been discovered when such "aberrant" data points were found to have a periodicity indicating the orbital period of the pair.

4.5.4.6 Scintillation noise

For most variable star and asteroid lightcurve projects, your overriding concern regarding signal-to-noise ratio will be to get *enough* signal. Since Poisson noise is usually the dominant noise source, this means taking relatively long exposures (a minute or more).

However, there are cases where you have a bright target that provides a surplus of light. Then your challenge becomes one of staying within the linear range of your CCD. A 7th or 8th-magnitude star can saturate your imager in a surprisingly short exposure! This problem arises particularly in the project to detect the transits of extra-solar planets. These target stars tend to be bright (6th to 8th magnitude). The easiest way to limit the signal and stay within the linear range is to take a shorter exposure. Alas, at some point, the shortness of the exposure brings to the surface another source of noise: scintillation.

You are certainly familiar with the sight of stars twinkling, and most likely you know what causes it. There are small thermal fluctuations randomly distributed through the air in our atmosphere (a hundredth of a degree, or so). Little parcels of air that are slightly cooler or warmer than the overall average air temperature act as weak lenses, so that they either concentrate or disperse the light rays coming from the star. As these little parcels pass between your eye and the star, you see the star randomly brighten and dim—it twinkles (or, in scientific terminology, it scintillates). This model, simple as it is, explains several observations that you're probably aware of. First, planets rarely twinkle. That is because stars are seen as infinitesimal points of light, but planets present (comparatively) large disks. The planet's finite disk size has the effect of "averaging" over several parcels of air, thereby reducing the magnitude of the scintillation. Similarly, even when the stars are twinkling violently, if you observe with binoculars or a telescope, you don't see amplitude changes the star may be a blob, but its brightness seems quite constant in the eyepiece. In this case, the large aperture of the telescope is averaging over far more parcels of air than the little pupil of your eye can, and again the averaging effect dramatically reduces the perceived amplitude of the scintillation.

This is put onto a mathematical foundation by the following equation [11, 12]. The scintillation magnitude (RMS fluctuation relative to average signal) is reasonably predicted by:

$$\frac{\sigma}{S} = 0.09 \cdot \frac{X^{1.5}}{D^{2/3}\sqrt{2 \cdot t}} \cdot \exp(-h/h_0)$$

where X = airmass = $\sec \theta$ where θ is the zenith angle of the observation;
$\quad\quad D$ = telescope aperture (cm);
$\quad\quad\ t$ = exposure duration (sec);
$\quad\quad h$ = observatory elevation (km);
$\quad\quad h_0$ = atmospheric scale height (typically about 8 km);

and the constant 0.09 is based on a small number of measurements made at McDonald Observatory. It seems reasonable to guess that this constant may be somewhat larger at less optimal amateur observing sites.

This equation shows that as the exposure time is reduced, the relative magnitude of scintillation fluctuations increases. With my set-up ($D = 28$ cm, $h = 0.2$ km ASL) and an exposure time of 1 minute, even when looking pretty low in the sky ($\theta = 60°$), the predicted scintillation amplitude is $\sigma/S = 0.007$. That's small compared with many other noise sources, and hence is not a cause for worry. However, suppose that the target star is so bright that I have to reduce the exposure to 5 seconds. Then, σ/S can be larger than $\sigma/S = 0.02$. That is larger than the magnitude drop that is expected in an extra-solar transit, and hence it is a noise source to be concerned about in that project.

The moral of this story is that there is a limit to how short an exposure you want to use. If you find that keeping a bright target in the linear range of the sensor demands exposures shorter than 10 to 15 seconds, then you may be better off finding some other way of reducing the signal. Spectral filtering may be convenient. A bit of defocus has also been known to help.

4.5.4.7   Quantization noise (12-bit vs. 16-bit digitizers)

The readout that is used for photometric analysis is digitized, therefore it is always an integer number. Consider my ST-8XE imager with a gain of 2.3 electrons per ADU. If the pixel has gathered 2,300 photoelectrons, then the digital output will read 1,000 ADU. If the pixel collected one more electron, the signal would still read 1,000 ADU. A second additional electron wouldn't change the ADU count either. Not until a third photoelectron is added will the signal increment to 1,001 ADU. This effect is sometimes referred to as "quantization noise", because it is directly related to the fact that ADUs are read out as integer quantities. They don't come in fractional amounts. Two pixels that have received different signals (different by one or two photoelectrons) may read the same ADU signal (i.e., the readout isn't precisely the same as the signal [13]). That is the sense in which this is "noise".

The significance of this effect depends on two specification parameters of the CCD imager: the full-well capacity, and the number of bits in the digitization. Some older CCD imagers used electronics with 12-bit digitization. Suppose that such an imager had a full-well capacity of 200,000 electrons. Twelve-bit electronics can divide the 200,000 electrons into $2^n = 4,096$ distinct levels, equivalent to a gain of about $200,000/4,096 = 49$ electrons/ADU. A pixel involved with a faint star, collecting about 5,000 pixels, will be read out with a quantization uncertainty of $49/5,000 \approx 0.01$, roughly equivalent to a 0.01-magnitude uncertainty in the star's brightness.

Most modern CCD imagers use 16-bit electronics. A 16-bit A/D converter can divide the full-well capacity into $2^{16} = 65,536$ distinct levels. This is equivalent to a gain of $200,000/65,536 \approx 3$ electrons/ADU. That same faintly-illuminated pixel will be read out with a quantization uncertainty of $3/5,000 \approx 0.0006$, which is negligible compared with many other noise sources.

The moral of this story is: if you have a choice, avoid CCD imagers with high gain and the older 12-bit digitization. In general, if you use one of the modern commercial imagers that are available to amateurs, you won't need to worry about quantization noise. If you have an older imager with 12-bit electronics, use long-enough exposures to increase your signal, thereby making the signal large compared with the quantization noise.

4.5.5 Signal-to-noise ratio

One way to look at the electron count that you measure on a pixel is that it represents the sum of S, the "true" signal, plus a random number whose mean value is zero, and whose standard deviation is N, the noise. This view makes it clear that noise corrupts your photometric measurements, and that its effect depends on the ratio of signal to noise. If you have mostly signal, and very little noise, then your photometry will be quite accurate. If, on the other hand, the noise is comparable with the signal, then your photometry will be inaccurate—you can't be sure if what you're measuring is signal or noise.

The key metric is the signal-to-noise ratio, $\mathrm{SNR} = S/N$. Most photometric software makes an attempt to calculate the SNR of the star images that you have selected as the target and the comp stars, so you won't usually need to actually do the calculations. You can combine the concept that electron count = signal + noise, with the equation for translating ADUs into magnitudes, and calculate the uncertainty in magnitude that is caused by noise. The result is:

$$\sigma_{\mathrm{mag}} = -2.5 \log[1 + 1/\mathrm{SNR}]$$

A simple approximation that is valid when the SNR is reasonably large is:

$$\sigma_{\mathrm{mag}} \approx 1/\mathrm{SNR}.$$

That is a convenient rule of thumb: the photometric accuracy is about $1/\mathrm{SNR}$, so if you're attempting to get photometric accuracy of 0.02 magnitude, you need $\mathrm{SNR} \geq 50$. You have to interpret this rule of thumb carefully. For example, when you use differential photometry, you're actually combining the measurements made on two different stars—the target and the comp star. If each star has $\mathrm{SNR} = 50$, so that each star's magnitude is known to within 0.02 mag, then the delta-magnitude between the two (i.e., target minus comp) is known to $\sigma_{\text{diff}} = (\sigma_1^2 + \sigma_2^2)^{1/2} = 0.03$ mag. There may also be significant noise in the sky background counts that have been subtracted. Nevertheless, the rule of thumb is widely quoted, and is valuable, albeit perhaps a bit optimistic. If you are making the lightcurve of an asteroid whose amplitude is 0.1 magnitude, then you'll need to ensure that you can get an SNR high enough to be confident in your data, to about 0.02 mag, or else your lightcurve will be pretty ragged. So, try for at least SNR = 50, and preferably SNR > 100, in order to create a clean lightcurve. The same principle applies to variable star magnitude estimates: AAVSO requests an estimate of your accuracy, and 1/SNR is a reasonable estimate.

There are two general approaches to increasing your SNR for more accurate photometry: adjust your imaging exposure parameters to maximize the SNR of each image; or arrange to average several images, which has the effect of reducing the noise-induced variance.

Increasing the SNR in individual images depends on the nature of the dominant noise source. In general, the vast majority of variable star and asteroid lightcurve photometry images are driven by Poisson noise and by background clutter. The best ways to improve SNR in these situations are (a) eliminate images obviously corrupted by background clutter, and (b) get more signal. Take longer exposures, and (if allowed) image unfiltered instead of through a color filter.

4.6   PROJECT J: EXTRA-SOLAR PLANET TRANSITS

Over the past decade, astronomers have seen compelling evidence for the existence of planets orbiting stars other than the Sun. This evidence comes primarily from exquisitely accurate spectral studies. As the planet orbits the star, its gravity exerts a slight "tug" on the star, and the star's motion in response can be detected as a cyclic red- and blue-shift. Amazingly, with care and attention to the need for extremely accurate photometry, the amateur astronomer can not only confirm the observations, but also contribute to the study of these extra-solar planets [14].

The geometry of an extra-solar planet's orbit is illustrated in Figure 4.29.

The professional astronomer's Doppler measurements are excellent evidence for the existence of a planet, they define the orbital period, and they give good insight into the eccentricity of the orbit. However, Doppler shift does not allow the astronomer to calculate the mass of the planet. If the planet's orbital plane lies at an angle to the line of sight (LOS) to Earth, then the Doppler shift measurement tells us

$$\Delta\lambda = M \sin \varphi$$

Since we don't know the angle φ, this equation gives us only a lower-bound on the estimate of the planet's mass (M).

However, suppose that the LOS to Earth happens to lie exactly in the planet's orbital plane. In this unique circumstance, the planet will pass between Earth and the star, and if we're careful we may detect the brightness change when the planet blocks

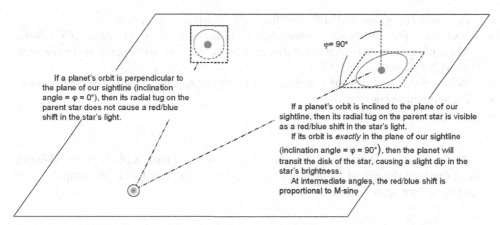

If a planet's orbit is perpendicular to the plane of our sightline (inclination angle = φ = 0°), then its radial tug on the parent star does not cause a red/blue shift in the star's light.

If a planet's orbit is inclined to the plane of our sightline, then its radial tug on the parent star is visible as a red/blue shift in the star's light.

If its orbit is *exactly* in the plane of our sightline (inclination angle = φ = 90°), then the planet will transit the disk of the star, causing a slight dip in the star's brightness.

At intermediate angles, the red/blue shift is proportional to M·sinφ

Figure 4.29. Geometry of extra-solar planet orbits. If a transit occurs, then we know that $\varphi \approx 90°$, and the red/blue Doppler shift data can be used to determine the planet's mass.

a bit of the starlight. The detection of such a transit is positive evidence that the planet's orbital plane is almost exactly $\varphi = 90°$. Knowing the orbital plane, astronomers can use their red-shift data to calculate the planet's mass. The details of the transit lightcurve can also be used to estimate the size of the planet. Then, knowing the size and the mass enables them to calculate the average density of the planet. So, detecting and measuring an extra-solar planet transit opens the door to learning a surprising amount about these distant worlds.

The trick, of course, is to detect the transit. As you can imagine from Figure 4.29, the vast majority of extra-solar planets will not create transits, because their orbits are inclined to our line of sight. Only those with an orbital inclination of 90 degrees will provide a transit. This project will require extraordinary care in your photometry, particularly in minimizing all sources of noise. Extra-solar planet transits create dips of 0.1% to 2% (0.001 to 0.02 magnitude) in the brightness of the target star. That is why I dragged you through the discussion of SNR in the previous section—to prepare you to achieve the level of meticulous care that is required to successfully detect extra-solar transits.

Because the probability that any given planet will transit its star is so small, this project is one that requires a high tolerance for negative results. Nevertheless, the value of the occasional positive result is so great, that it is worth your while to invest a few nights each time that there is a favorable transit opportunity for one of the roughly 200 candidate stars that have been identified.

4.6.1   Equipment required

The equipment required for this project is the same as that for variable star or asteroid photometry:

- Telescope.
- CCD imager.

- Personal computer and software for CCD control and operation.
- Software for differential photometry reduction of your data (CCDSoft, MaximDL, and AstroArt offer this capability, as does specialized software such as MPO Canopus and AIP).
- WWV receiver, GPS, or other source of accurate time to set your PC's clock.
- Planetarium program.
- (Optional) color filter.

Although it is not required, you may find it useful to have a photometric B-band filter, to reduce the signal from bright target stars, and thereby enable longer exposures (to reduce scintillation noise).

4.6.2   Conducting the observations

The purpose of this project is to monitor the star's brightness in order to detect—or conclusively rule out—the signature of an exo-planet transit. To do this, you follow essentially the same methods that you would use for conducting differential photometry to determine the lightcurve of a short-period variable star. Take a continuous (all-night-long) series of images of the target star, select an appropriate comp star in the image FOV, use your photometry software to determine $\Delta m = M_{tgt} - M_{comp}$ in each image, and then plot Δm vs. time. A successful result, showing the first half of a transit of the planet of HD 209458, is shown in Figure 4.30.

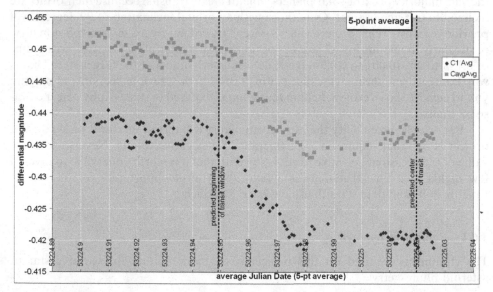

Figure 4.30. Typical exo-planet transit signature (HD 209458). Note that the amplitude is quite small: a bit less than 0.02 mag in this case.

There are two peculiar challenges for this project compared with a typical variable star project: (1) determining *which* star to monitor, and *when* to monitor it, and (2) achieving sufficient accuracy to have confidence in any apparent detection. Both of these challenges are more difficult than they may sound!

As you can probably imagine by studying Figure 4.29, the radial-velocity signal offers a good clue as to when to look for a transit. First, the transit will be centered at the time when the radial-velocity signal is zero. Second, by combining knowledge of the star's color-index (spectral type) and brightness, astronomers can estimate the star's distance, and physical size. Third, the orbital period of the planet is also provided by the radial-velocity signal. With a few simplifying assumptions about the shape of the planet's orbit, all of this information can be combined to bracket the time-interval in which a transit is likely to occur (if, in fact, the planet's orbit is at an inclination of 90 degrees to our line of sight).

The good news is that you will not need to do the necessary calculations. Predictions of transit windows for all of the extra-solar planets known from radial-velocity measurements have been calculated [15] by Greg Laughlin (University of California, Santa Cruz) and posted on the website of TransitSearch.org (an enterprise initiated by Tim Castellano, of NASA Ames Research Center). The bad news is that, even though you can easily download the predicted window, you still have to confirm that the star is, in fact, observable from your location during the indicated window. For example: is it above the horizon for at least several hours? During the hours of darkness? When there isn't interference from moonlight? You will use your planetarium program to address these questions. For most cases, you will be able to monitor only a portion of the "transit window" on any single night. Subsequent nights may provide the opportunity to monitor other portions of the transit window. For many candidate objects, you'll need to monitor several transit windows in order to check the entire width of the window.

An example of the database available at TransitSearch.org is shown in Figure 4.31.

The upper panel is an excerpt of data from the TransitSearch.org website. The full file available at their website lists all of the stars known to have planets. For each star, the columns provide:

- The star's identity (usually its HD-catalog number, unless it has a more common name such as 51 Peg).
- The planet's identity (most are "b", but a few stars are known to have more than one planet, in which case you'll see listings for planets "b", "c", etc.).
- The planet's orbital period, in days.
- The *a-priori* probability that the planet transits the star. (Note that most of the values in the full table are less than 10%—the geometrical requirements for transit are pretty restrictive, and most planets will not transit their stars, as viewed from Earth. The only way to know for sure is to do this project, for each and every star. Those few that are discovered are extraordinarily valuable!)
- The approximate RA and Declination of the star.

Star	Planet	Period (days)	P (%)	R.A.	DEC.	Depth (%)	Next Center (UT)	Window	Ephemeris	Results
TrES-1	b	3.03	11	19:04	+36:38	1.64	5/20/2005 12:33	in	Ephemeris	Results
HD209458	b	3.525	12.1	22:03	+18:53	0.89	5/23/2005 10:56	in	Ephemeris	ults
51_Peg	b	4.231	10.9	22:57	+20:46	0.79	5/23/2005 2:57	out	Ephemeris	Res

```
Predicted Transit Epochs: HD209458_ b

Predicted Duration of a Central Transit:   199.028045 Minutes

Begin Transit Window           PREDICTED CENTRAL TRANSIT     End Transit Window
                                      All Times UT
                               HJD          Year  M  D  H  M
  2453732.42 2005 12 27 22  4   2453732.49 2005 12 27 23 46   2453732.56 2005 12 28  1 29
  2453735.94 2005 12 31 10 40   2453736.02 2005 12 31 12 22   2453736.09 2005 12 31 14  5
  2453739.47 2006  1  3 23 15   2453739.54 2006  1  4  0 58   2453739.61 2006  1  4  2 40
  2453742.99 2006  1  7 11 51   2453743.07 2006  1  7 13 33   2453743.14 2006  1  7 15 16
```

Figure 4.31. Example "Candidate List" and "Ephemeris" from TransitSearch.org. (Used with the kind permission of TransitSearch.org)

- The predicted depth of the transit. "Depth" is listed as a percentage change in light, but for the slight brightness drops predicted an acceptable approximation is $\Delta m = D$. That is, a 1% depth is equal to a brightness change of $\Delta m = 0.01$ magnitude.
- The date and time of the center of the next transit window.
- An indication of whether the star is "in" or "out" of the transit window at the time you examine the website.
- Hyperlink to a detailed ephemeris for the selected star/planet.
- And a hyperlink to a discussion of results so far on that particular star; with recommendations of whether it should continue to be monitored.

The lower panel in Figure 4.31 shows an example of a detailed ephemeris, in this case for the star HD 209458. This detailed ephemeris provides:

- The identity of the target star, and the predicted duration of a total, central transit.
- The time the predicted transit window opens. This is given both as JD, and as UT YYYY-MM-DD-HH-MM.
- The predicted time of central transit.
- The time the predicted transit window closes.

It is essentially impossible that a transit will occur outside of the predicted transit window (the window encompasses the 3σ probability interval for a transit). So, your observing strategy should be to begin observing an hour or so before the window opens, and continue observing until an hour or so after the window closes. If the window is more than a few hours long (as most are), then your strategy should be to monitor the star during all nighttime hours within the window when the star is above about 30 degrees elevation.

Now, let's assume that you have identified an attractive target and upcoming transit window. The second challenge is to ensure that you can provide sufficiently accurate differential photometry. That challenge is not for the faint of heart. The example shown in Figure 4.30 shows that the transit signature of HD 209458—one of the *greatest* depths predicted—is not quite 0.02 magnitude, peak to peak. That is a very small signal! It is, in fact, smaller than the measurement uncertainty that you'd be quite happy to achieve in typical variable star and asteroid lightcurve photometry. Achieving this level of quality in your photometry requires attention to most of the sources of "noise" that were discussed in the previous section. Of particular importance are Poisson noise, gain/linearity, scintillation noise, and guiding errors.

All of the extra-solar planet candidate stars are quite bright. That provides the potential for excellent Poisson SNR, which is absolutely required in order to achieve the quality required. Since you're looking for a signal that may be 0.01 mag or smaller, it is reasonable to try for photometric noise of $\sigma \approx 0.003$ or less. That, in turn, requires an SNR ≥ 450 on both the target and comp stars (assuming that they are equal brightness). It isn't difficult to achieve such a high Poisson SNR for these bright target stars, but it may be challenging to find a suitably bright comp star that can be placed in the same FOV. Relatively short focal-length optical systems, providing relatively wide FOV (or use of focal-reducers on longer focal-length systems) will help in this regard.

Since you're imaging a bright star, and aiming for as high as possible an SNR, you need to be very careful that you don't exceed the linear range of your imaging system. The linearity requirement will probably set the limit on how long an exposure you can use, and hence the limit on the Poisson SNR that you can achieve.

If you are using unfiltered or V-band photometry, with typical amateur CCDs and telescopes you are likely to find that the "linearity" requirement limits you to exposures of a few seconds. That presents a serious "scintillation" noise concern—both target star and comp-star brightness will appear to fluctuate from image to image, and there is no guarantee that they will fluctuate in synchrony. With exposures of less than about 10 seconds, this scintillation noise can easily be the dominant noise source in the photometry, significantly larger than the Poisson noise. If you find yourself in this situation, the best way to reduce scintillation noise is to use a filter that "throws away" a fair amount of light, thereby enabling you to use a longer exposure. For HD 209458 (at V-mag = 7.7), I found that using a photometric B-band filter was a good solution: the combination of filtering and the relatively low sensitivity that most CCDs have in B-band enabled me to use 30-second exposures. That long an exposure virtually eliminates the scintillation noise while still achieving a very high Poisson SNR in the linear range of my imager. In a pinch, other amateur astronomers have found it useful to slightly defocus the image, thereby reducing the signal on the brightest pixel and permitting a longer exposure.

At the level of accuracy that this project requires, the effect of "tracking error noise" (see Section 4.5.4.4) can be significant if you attempt unguided imaging. You will almost certainly need to autoguide during your exo-planet transit imaging run, to ensure that the target and comp-star images stay on essentially the same pixels throughout the night.

Similarly, you should not move your filter wheel at all during the night's sequence of images, including flat-field images. The very slight registration error that exists in all commercial filter wheels can lead to slight changes in the dust donuts and vignetting conditions, at the 1% photometric level. If you've moved your filter wheel between the science images and the flat-field images, then this level of residual error is likely to exist (incomplete flat-fielding).

You can see that achieving the desired high-quality photometry entails a tricky tradeoff between your target star and comp-star brightness, separation between the two stars, and your equipment (FOV, linear range, telescope aperture, and available filters). The best way to deal with this tradeoff is to conduct some experiments a few nights before the start of the transit window. Make a series of exposures of different duration, through different filters, and plot the peak pixel ADU vs. exposure, in all available filters. Find your best combination that will: provide high Poisson SNR, stay well within your sensor's linear range, and have long enough exposure to keep scintillation noise to a minimum. Then, using your two or three "best" combinations, take a series of a few dozen images, and examine the differential photometry to estimate your achieved accuracy. If your consistency across a couple dozen images is ± 0.01 mag, or $\sigma \approx 0.005$ mag that is quite satisfactory. If your consistency is much worse than ± 0.03 mag, or $\sigma \approx 0.015$ mag, then you probably want to see if you can improve things by adjusting imaging parameters, or perhaps using a different telescope/image combination. By the way, for this check of the internal consistency of your differential photometry, plot your actual differential photometry of the (constant-brightness) target star vs. time. Do not rely on the "predicted" standard deviation that your photometry software calculates based in the star image SNR.

4.6.3   Reducing and analyzing your data

Use carefully selected imaging parameters to gather your photometry of the target star during the transit window. Be sure to take good flats and darks for this condition. Reduce your images and perform photometric analysis in the usual way. For plotting the data, you may find it useful to export your photometry data (time, delta-mag) into a spreadsheet program. The spreadsheet allows you to adjust the plot parameters, to make any transit signature more visible, and also enables you to do some statistical filtering ("averaging") to improve the SNR.

If there is no transit signature, then the plot of delta-mag vs. time will be a straight line (i.e., constant brightness throughout the observing run). Determining the mean and standard deviation of the data gives you an indication of the quality of your data (e.g., ideally, you'll have a standard deviation of 0.01 mag or less). Determining the best-fit line through the data (which most spreadsheets can do quite automatically) will tell you how close to "constant" your data is (perfectly constant, on average, will give slope $= 0$).

If there appears to be a differential-photometry signature, you should examine it carefully to decide if it is a real transit signature, or if it is some sort of accident. You may see occasional, isolated data points that lie more than 3σ above or below the average. These probably represent some sort of error, such as a cosmic ray hit that is unique to that image. You may also see long sequences of data points that seem to be "off the trend line" for some accidental reason. Atmospheric conditions are the most common, such as a faint contrail or wisp of cloud passing through the FOV. To first order, these don't affect differential photometry, but at the level of quality that we require here, second-order effects can be significant. If the target and comp stars are of noticeably different color, then a wisp of cloud can create differential (second-order) extinction of a few hundredths of a magnitude. Recognizing such groups of discordant data points is a bit of an art. Plotting the raw instrumental magnitude of the comp star vs. time will give you important clues as to the quality of the night, and will also identify times where wisps of cloud may have interfered. Plotting comp-star magnitude vs. airmass will point out any "oddities" that happened as the field got low in the sky. (There are a variety of effects that can mess with your photometry if you have to go below about 30 degrees elevation angle.) Examine both your data and your notes from that night, to be sure that you don't accidentally report a transit that later has to be retracted when other observers can't replicate the observation!

If your transit signature is uncertain, it may help to do a little mathematical filtering or averaging of your data, particularly to reduce scintillation noise that often plagues transit data. There is nothing particularly wrong with averaging every three consecutive data points, or performing a top-hat filter in which each data point X_i is replaced by $\mathcal{X}_i = (X_{i-2} + X_{i-1} + X_i + X_{i+1} + X_{i+2})/5$, as ways to get a better picture of your results. Sometimes such filtering will make the signature of a transit easier to see in a noisy data set. However, sometimes the filtering will create the illusion of a smooth transit signature, when in fact there isn't one. If you do decide to examine your data with some smoothing filter, there are two points to remember. First, averaging should be done on *intensity* data, not *magnitude* data. Averaging intensities is a linear process, but averaging logarithms (and magnitudes are, after all, logarithmic) isn't. So if you're going to do some averaging or smoothing, you should first convert your data from magnitudes into intensity, do the averaging, and then convert back to magnitudes. Second, in some quarters it is considered bad form to submit "smoothed" photometry unless the raw data points are also displayed.

4.6.4 Test cases

Because of the degree of meticulous care and accuracy that this project requires, it is almost mandatory that you demonstrate your ability to detect a known transit before you attempt to monitor candidate stars for potential transits. The two best test cases are HD 209458 (in Pegasus), and TrES-1 (in Lyra).

HD 209458 has become the *de-facto* standard "test case" for the majority of TransitSearch.org participants, since it has relatively frequent transits (its period is 3.52 days), its transits are reasonably deep (nearly 2%), and it has a convenient comp star (HD 209346, V-mag = 8.3) only 12 arc-min to the west.

You should devote at least one "non-transit" night, and one or two "transit" nights to HD 209458, to confirm the adequacy of your photometry. On the "non-transit" night, your differential photometry plot should be a night-long series of data points that lie within ±0.005 magnitude of a constant-brightness line (i.e., slope = 0).

The "transit" night photometry should show a clear signature of the transit. If it does, you will have confirmed your ability to perform differential photometry at the 1% level, and you're ready to start monitoring other candidate stars for extra-solar planet transits.

4.6.5   Reporting your results

The central organization for coordinating extra-solar transit search photometry results is TransitSearch.org. They often coordinate observing campaigns of particular stars with the AAVSO. If your observations were part of an AAVSO campaign, then your report should be directed to them. If your observations were part of an independent project based on TransitSearch.org data, then you should report your data and results directly to TransitSearch.org.

With your first report to TransitSearch.org, you should also include as reference material your results from previous monitoring of a test case (probably HD 209458). A "non-transit" night's flat-line, and a successful "transit" signature from this star will be used to confirm the quality of your photometry. This is particularly important because most of the reports on candidate stars will be "negative"—that is, no transit observed during a particular observing window. It is important to know that this is a valid negative report, not the result of inadequate data quality!

Your report should include:

- Description of your equipment (telescope and imager).
- Observing location.
- Brief description of your observation procedure, including exposure duration, cadence, and spectral filter used.
- Time interval during which observations were undertaken (preferably expressed in UT).
- Brief description of data reduction method, the photometry software used, and highlighting any part of the night(s) where data was censored due to bad conditions.
- Data plot of delta-mag vs. time. The time base can be UT, heliocentric JD, or other convenient time base, but be sure it is clear exactly what time is plotted.
- Data file (ASCII or spreadsheet format) of the raw photometric data for "positive" reports (this is not normally required for "negative" reports).

4.7   PHOTOMETRY ON THE STANDARD B-V-R SYSTEM

The preceding sections described how you can acquire very useful science data using differential photometry for variable star studies and asteroid lightcurves. With differential photometry, you've determined the magnitude *difference* between your target and the comp star(s), but you haven't determined the *actual* magnitude of the comp stars. In the case of asteroid lightcurves, you didn't even attempt to determine the asteroid's true brightness: the brightness change relative to the comp stars contains all of the information that is needed to determine the asteroid's rotation period and shape. In the case of variable stars, you did link the target's brightness to its "true" V-magnitude because the AAVSO star chart provided you with the "true" magnitude of the comp stars.

For the next project—asteroid phase curves—you'll need to determine the actual "true" V-magnitude of your comp stars and use them to link the asteroid's brightness to the standard photometric system. Unfortunately, none of the star databases used in the popular planetarium programs can be relied upon for photometry. The stellar positions in these catalogs are excellent, but their photometry is marginal. Errors of a tenth of a magnitude between the catalog-reported brightness vs. the "true" brightness of stars are typical. Therefore, you must make some special measurements and data analysis to link your asteroid brightness to the standard magnitude system. This section describes the necessary background on how this will be done.

4.7.1 The standard B-V-R photometric system

In the introductory section of this chapter, I mentioned in passing that the star Vega is the fundamental reference star of most photometric systems—it is not just "a magnitude-zero star", it is *the* definition of magnitude-zero brightness.* The brightness of every other star in the sky—the ones we know as "3rd magnitude" or "8.5 magnitude"—was determined by measuring their brightness relative to Vega (either directly, or indirectly), and applying the fundamental magnitude equation:

$$\Delta m = (m_1 - m_2) = -2.5 \log(I_1/I_2)$$

If you like algebra, consider what happens when I_1 is the brightness of our target object, and $I_2 = I_{\text{Vega}}$ is the brightness of Vega. Since Vega is defined as magnitude zero, you can set $m_2 = m_{\text{Vega}} = 0$. Substituting all this into the fundamental equation, you get:

$$m_1 = -2.5 \log(I_1/I_{\text{Vega}})$$

With this equation you can determine the magnitude of your target on the conventional magnitude scale—not just its magnitude difference from an arbitrary "comp star".

Of course, normally it isn't practical to actually use Vega as your "comp star"— it's so bright that it will probably saturate your CCD image, it isn't always conveniently placed in the sky, and even if it is visible, it is likely to be far away from your target, which means that you see it through a very different atmospheric path than that through which you see your target object. So, even though Vega is the fundamental photometric reference, you'll probably never actually take an image of it. The situation is similar to the one faced by surveyors: they know that the Greenwich

---

*Vega can be considered as the "Ur-star" for most descriptive purposes. The zero point of V-magnitude on the UBVRI system works out such that Vega is declared to be $V = +0.03$, which is still close enough to "zero" for the purposes of this discussion. Vega's color index is (B-V) = (V-R) = 0.0 in the UBVRI systems of Johnson and Morgan, and of Landolt.

meridian is the fundamental zero point of longitude, but they hardly ever place their tape measures at Greenwich to start a survey. Instead, they start at some more convenient marker whose position has been accurately determined. In order to create a network of conveniently-placed photometric references, several generations of astronomers have conducted meticulous surveys and measurements to define "standard stars", anchoring their brightness to the zero-magnitude reference, so that you can conduct your photometry relative to a network of these standards [16].

The gold standard of photometric reference stars is the set known as the "Landolt standard stars", acknowledging the meticulous work that Dr. Arlo Landolt has done to provide a self-consistent photometric reference frame [17]. Their brightnesses are nicely matched to CCD imaging projects (they range from about V-magnitude 11.5 to 16). They are arrayed along the celestial equator, so that they are easily visible from any observatory. Despite the pre-eminence of Vega, when someone refers to a "photometric standard star" in the visible spectrum, he is almost certainly referring to a Landolt standard.*

Recall that our discussion of variable star CCD photometry assumed that you were using a V-band filter, and that the AAVSO chart had provided you with accurate V-magnitudes and color indices for the comp stars. You may have wondered how those V-magnitudes and color indices were determined. This section will describe how it was done. They were carefully measured, and compared with the Landolt (or similar) standard stars. In that way, the magnitude and color index of each "comp star" was anchored to the standard magnitude system. For that reason, they are sometimes referred to as "secondary standards". Many are also known as "Henden field" stars, in recognition of the extensive work done by Dr. Arne Henden for the USNO and AAVSO, in conducting the photometry of comp stars.

The challenge of putting your photometry onto the standard photometric system, to determine the actual magnitude of a target that doesn't have predefined secondary standards, is essentially the same challenge as that of creating secondary standards. It requires a special diligence in both your observing procedure and your data reduction, but once you've gained the necessary skill, you can contribute to some very useful projects that few amateurs (or professionals) are tackling today. The special complications are: transforming your measurements to the standard system, and dealing with atmospheric extinction. We'll cover atmospheric extinction first.

4.7.2 Atmospheric extinction

For differential photometry when your comparison stars are in the same field of view as your target, you don't need to calculate the effect of atmospheric extinction.

*The complete listing of Landolt standard stars, including their photometry, positions, and finder charts, are contained in [16]. A table of the Landolt stars is available on the internet at *http://www.noao.edu/wiyn/queue/images/tableA.html* If you use TheSky as your planetarium program, you can download a database file containing the Landolt stars from *http://www.bisque.com/tom/data/landolt.htm* If you use MPO Photored for photometric reductions, it is shipped with a database containing the Landolt stars plus many Henden fields.

Regardless of the effect, it was doing the same thing to the target as it was to the comp stars, so it had no net effect on your calculated target delta-mag. If you intend to determine the standard photometry of a target for which secondary standards aren't available (e.g., asteroid "phase curve" determination), then you do need to deal with the effect of atmospheric extinction.

However, your target asteroid probably isn't in the same field of view as a Landolt standard star. In order to compare the asteroid's brightness to the Landolt standard's brightness, you need to point your telescope first at one, and then at the other. Odds are, since they are viewed through different atmospheric paths, the extinction effect will not be identical for the two. You need to figure out how to determine the extinction in each direction, and then adjust for it.

From your experience at visual observing, you know that the atmosphere can make significant changes to a star's brightness. The atmosphere absorbs and scatters starlight, so that we receive less light in our ground-level observatory than was present at the top of the atmosphere. The longer the atmospheric path, the greater the loss of light. From any particular location, looking straight up (toward the zenith) presents the shortest possible atmospheric path, and hence the lowest atmospheric extinction. As shown in Figure 4.32, the lower the elevation of the line of sight, the longer the atmospheric path, and hence the greater the effect of atmospheric extinction. The length of the atmospheric path is described by the "airmass", X. For most situations, the airmass is calculated by

$$X = \sec(\theta)$$

where θ is the zenith angle of the line of sight ($\theta = 0$ degrees at the zenith, and $\theta = 90$ degrees at the horizon). The airmass is $X = 1.0$ when you are pointed straight up, at the zenith. Along that line of sight, your observations pass through the full thickness of the atmosphere above you. If you point your line of sight lower, then the length of your observation path through the atmosphere gets longer, so X gets larger. At 45 degrees from the zenith (halfway between the zenith and the horizon) $X = 1.4$. At a zenith angle of 60 degrees (i.e., 30 degrees above the horizon) $X = 2.0$: you're looking through twice as long an atmospheric path as you were when pointed at the zenith.

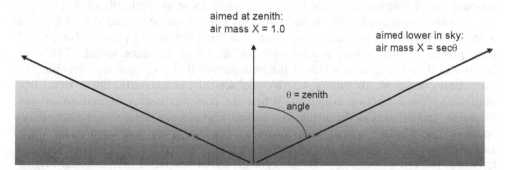

Figure 4.32. Atmospheric extinction is a function of "airmass". Airmass depends only on zenith distance, *if* the sky is uniformly clear and stable all night.

The equation given for airmass is valid down to zenith angles of about 60 to 70 degrees. At greater zenith angles (i.e., viewing closer to the horizon), there are additional complicating effects that you will not want to deal with, so stay above 30 degrees elevation ($\theta \leq 60$; i.e., $X < 2.0$) for all of your CCD photometric projects.

At any single wavelength, the effect of atmospheric extinction is:

$$m = m_0 + k \cdot X$$

where $m =$ the magnitude you observe (your "instrumental magnitude"), and m_0 is the magnitude that you *would have observed* if you could have placed your telescope outside the atmosphere (your "exoatmospheric instrumental magnitude"). The interpretation of this equation is that the star gets fainter as the airmass increases. The constant k is called the "extinction coefficient". It is measured in magnitudes per airmass. For typical observing sites under good conditions, k ranges from 0.1 to 0.3 mag/airmass.

Remember back to your high school algebra class. That equation describing atmospheric extinction is the equation of a straight line. If you plot instrumental magnitude (m) vs. airmass (X), you'll get a straight line whose slope is the extinction coefficient (k). You can actually see this equation in action in your data from an asteroid lightcurve project, where you took a night-long series of images of the asteroid and its comp stars. As your field of view rose, culminated, and then set, you have monitored your comp stars at a wide range of airmass. Take your data from a clear, stable night and calculate the zenith angle and airmass for each image. Then, select one comp star and plot its instrumental magnitude in two ways: (1) instrumental magnitude vs. time, and (2) instrumental magnitude vs. airmass. What you'll see will be similar to Figure 4.33.

As expected, the star gets brighter as it approaches the zenith. Note that this change in instrumental magnitude amounts to about 0.15 magnitude. That is comparable with the brightness variation of a typical asteroid lightcurve (which is why you used differential photometry for your lightcurve project, instead of just plotting instrumental magnitude vs. time). It is much larger than the level of accuracy that you will be aiming for in the "asteroid phase curve" project. That is why you must account for atmospheric extinction when you are doing absolute photometry.

The extinction coefficient changes with sky conditions, so you need to determine it for each night that you take data in an absolute photometry project. Happily, if your project entails making a long series of images of the same target, then your comp-star data has hidden within it the information that you need to determine the extinction coefficient. In Figure 4.33(b), you see the characteristic linear trend line implied by the extinction equation. When the star was near culmination, it was brightest, and it faded as the airmass increased. The slope of the line is the extinction coefficient, k. Modern spreadsheets make it a trivial exercise to determine the slope and intercept of the best-fit trend line for such a set of data. If you extend this line leftward, to zero airmass ($X = 0$), you get the exoatmospheric instrumental brightness of the star—that is, the magnitude that you would have measured if there were no atmosphere. Equivalently, having determined the extinction coefficient, you can

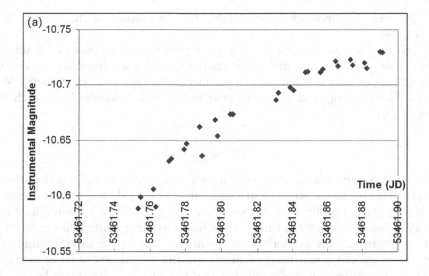

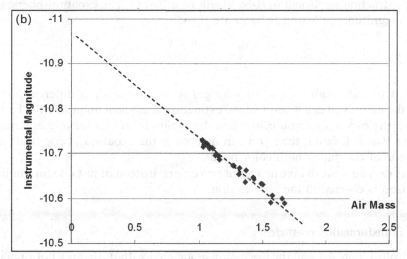

Figure 4.33. Effect of atmospheric extinction. (a) Instrumental magnitude vs. time: the star appears faint near the horizon, and brightens as it approaches the zenith. (b) Instrumental magnitude vs. airmass. The same data as above shows a linear trend line. The slope of the line is the extinction (k) in magnitudes/airmass. Extending the line to airmass = 0 gives the exo-atmospheric instrumental magnitude of the star.

re-arrange the magnitude-vs.-airmass equation to calculate the exoatmospheric magnitude:

$$m_0 = m - k \cdot X$$

That equation is strictly valid at a single wavelength. Your photometric V and R-band filters are narrow enough that you can act as though they can be treated as if they gave you "monochromatic" light, for the purpose of the projects in this book [18, 19].

Recall that in aperture photometry your photometric software adds up all of the ADUs in the star image and subtracts the contribution from sky background. Suppose that you take two images of your target, through your photometric V and R-filters. The resulting measurements, your instrumental magnitudes in each filter, are defined by:

$$v = -2.5^* \log(\text{ADU}_v)$$

$$r = -2.5^* \log(\text{ADU}_r)$$

Atmospheric absorption and scattering are more severe in the blue than they are in the red. This is why the setting Sun looks red: the blue light has been scattered and absorbed by the atmosphere, so it never reaches our eyes. The V-band falls roughly mid-way between blue and red bands. Therefore, you have a situation similar to that shown in Figure 4.6: the extinction coefficient depends on the spectral band of the sensor (v, or r), and you'll need to determine the v-band and r-band extinctions separately. The extinction coefficient that you measure in V-band will be called k_v, and the extinction coefficient in R-band will be called k_r. The exoatmospheric instrumental magnitudes in V and R-band are then:

$$v_0 = v - k_v \cdot X$$

$$r_0 = r - k_r \cdot X$$

What this all means is that by taking a series of images at different airmasses during the night, you can determine the extinction coefficient and therefore be able to calculate the exoatmospheric instrumental magnitude. If you make these images in both the V and R-band, then you can determine the exoatmospheric instrumental magnitude of the star in both color bands.

Why do you want to take images in two colors, instead of just one (or unfiltered)? The reason is covered in the next section.

4.7.3   Transformation coefficients

Recall from Figure 4.4 and the discussion associated with it, that it's not sufficient to simply measure the target, then measure the standard star, and apply the magnitude equation. For precision photometry, we need to be making these measurements using the standard spectral band pass (also known as the standard filter definition). If my spectral sensitivity curve is different from the standard, then I'll measure different (erroneous) photometry. Professional astronomers have expended great efforts to define the standard sensitivity curves—the "UBVRI" color system.

What if your system's spectral sensitivity curve isn't quite identical to the B, V, and R-band definitions? Begin by looking back at Figure 4.4, and consider the effect of having a different spectral sensitivity. In particular, suppose that your instrumental V-band is slightly to the red, compared with the standard V-band—that is, you are a bit more sensitive to red light, and a bit less sensitive to blue light, than the standard

V-band definition allows. If the star is "white" (same brightness in B, V, and R-bands), then the slight discrepancy won't matter at all—you'll measure the same brightness as the "standard V-band" indicates. If the star is "red", you will see it as being brighter than the "standard V-band" system does, because you are more sensitive to that red light. Similarly, if the star is "blue", then your system (with its reduced blue sensitivity compared with the standard V-band definition) will measure the star as being a bit fainter than the standard V-band system indicates. In general, your measurements are in error, and the error depends on the color of the star being observed.

Unfortunately, we can be quite confident that my system has a different spectral sensitivity curve than yours, and that it is different from the standard sensitivity curve—we're using different cameras and telescopes, different filters, and we have different atmospheric conditions. Therefore, each photometrist needs a way to transform his observations to the standard system, so that his data can be combined with that of other astronomers around the world. You can do that by determining the "transformation coefficients" of your system.

The conventional nomenclature is that raw "instrumental" magnitudes are shown in lower case (v, r), and "standard" magnitudes are shown in upper case (V, R). By definition, the standard magnitudes V and R are exoatmospheric magnitudes—that is, the amount of light within the standard pass band that entered the atmosphere before being reduced by atmospheric extinction. The "exoatmospheric instrumental" magnitudes are denoted by the subscript "0" (i.e., v_0 and r_0).

The previous section described how to correct your raw instrumental magnitudes for atmospheric extinction, to get the exoatmospheric instrumental magnitudes. These represent the amount of light within *your* (non-standard) spectral band pass that entered the atmosphere. You will transform your exoatmospheric instrumental magnitudes to standard magnitudes by using the equations

$$(V - R) = T_{vr} \cdot (v_0 - r_0) + Z_{vr}$$

and

$$(V - v_0) = T_v \cdot (V - R) + Z_v$$

In these, T_{vr} and T_v are the "transformation coefficients", and Z_{vr} and Z_v are the "nightly zero points".

If you're mathematically inclined, you'll recognize that what's being done here is a Taylor-series expansion of the color index and V-mag in terms of the color indices, with only the first-order (linear) terms being used. Experience has shown that this linear approximation is quite sufficient to achieve the accuracy required for almost all projects, and that there is no benefit in attempting to account for higher-order terms.

You determine the transformation coefficients T_{vr} and T_v for your system by imaging a field that contains several standard stars (i.e., a Landolt standard field). Take images of this field through both your v and r-band filters. The Landolt or Henden field gives you a batch of stars whose color index $(V - R)$ and V-magnitude (V) are known on the standard system. Your images give you the *instrumental* color $(v - r)$ and *instrumental* magnitude (v) of those same stars. Enter that data set into a spreadsheet program (such as Excel), as a table of V, R, v, and r for each star. In the same spreadsheet, calculate the standard color index $(V - R)$ and instrumental color index $(v - r)$ for each star. Then, create graphs of the "standard minus instrumental" color vs. standard color, and "standard minus instrumental magnitude" vs. standard color. An example of these graphs is illustrated in Figure 4.34.

star	Std Magnitude V	Std Magnitude R	instrumental mag v	instrumental mag r	calculations (V-R)	calculations [(V-R)-(v-r)]	calculations (V-v)
1633 d	13.760	13.350	-6.0940	-6.7580	0.410	-0.2540	19.854
1633 e	13.780	13.250	-6.0680	-6.8535	0.530	-0.2555	19.848
1633 f	14.240	13.720	-5.5900	-6.3790	0.520	-0.2690	19.830
1633+099	14.397	14.490	-5.5265	-5.7485	-0.093	-0.3150	19.924
1633+099B	12.970	12.380	-6.8425	-7.7225	0.590	-0.2900	19.813
1633+099C	13.230	12.610	-6.5870	-7.4925	0.620	-0.2855	19.817
1633+099D	13.690	13.370	-6.1605	-6.7500	0.320	-0.2695	19.851

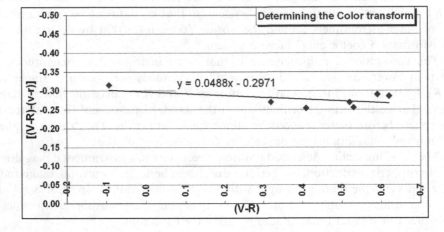

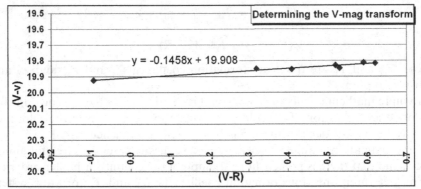

Figure 4.34. To determine your system's transforms, image a field of standard stars in two colors (V and R-band); then plot standard minus measured color and magnitude vs. standard (catalog) color index. In this example, color transform is $T_{vr} = 1/(1 - 0.0488) = 1.05$, and the magnitude transform $T_v = -0.1458$. Knowing your transforms, and the nightly zero points, you can convert measured color index $(v_0 - r_0)$ and magnitude (v_0) of any star into its magnitude (V) and color $(V - R)$ on the standard system.

The first graph is a plot of the following equation:

$$[(V - R) - (v - r)] = m \cdot (V - R) + b$$

All of the data points should fall reasonably close to a straight line. Your spreadsheet will have a command that determines the best-fit line through your data. The slope of that line is "m", and the y-intercept is "b". With the slope in hand, you can calculate your color-transform coefficient:

$$T_{vr} = 1/(1 - m)$$

The second graph that you make is a plot of:

$$(V - v_0) = T_v \cdot (V - R) + Z_v$$

The best-fit line through your $(V - v_0)$ data points will have a slope that equals T_v. These two constants, T_{vr} and T_v, are the keys that transform your instrumental colors and magnitudes into standard colors and magnitudes, so that you can determine an object's color and brightness on the standard VR system.

You will want to carefully determine your system's transforms about once per year. The reason for the annual check is that some things do change slowly: as optical coatings degrade or mirrors get dirty, their transmission characteristics may change, and it has been observed that some types of filters may gradually change. Of course, you also need to re-determine the transforms any time that you change an element of the system (e.g., get a new telescope, or exchange a filter). As a practical matter, I've found it useful to image a standard-star field on most nights, and keep track of the resulting transforms—if they are noticeably different from the norm, then I know that something odd happened on that night, or on that set of images.

By the way, you may have noticed that in Figure 4.34 I plotted the raw instrumental magnitude (v and r) instead of exoatmospheric magnitude (v_0, r_0). When you're measuring your transforms, if you use one of the Landolt fields that contains several standard stars, and get all images (in both bands) done in a short period of time near culmination (so that the airmass doesn't change noticeably over the imaging period), then you don't need to correct for atmospheric extinction. This is a bit counter-intuitive, but if you work through the equations you'll see that the extinction effect is collected into the term that describes the y-intercept. It doesn't affect the slope of the line, which is where the transform is found [20].

4.7.4 Putting photometry onto the standard system

The business of transformation coefficients may seem a bit "picky". After all, they never came up in the discussion of visual observing of variable stars (in Chapter 3), and none of the CCD photometry projects that have been described so far demand the use of transformation coefficients. Part of the difference is that CCD observations have the potential for such high accuracy compared with visual observations that color effects are both detectable and correctable. Typical visual observations might be accurate to ±0.2 magnitudes, and for practical purposes we can't detect the different spectral response between different people's eyes. Your CCD differential photometry can achieve accuracy of ±0.05 magnitude or better. At this level of accuracy, transformations are both worthwhile and valuable for certain projects.

Transforms aren't mandatory for most variable-star photometry, because the AAVSO or BAA charts give you comp stars that have well-defined V-magnitudes, and you use a photometric V-band filter that (presumably) tailors the spectral response of your CCD to be pretty close to the standard passband. They aren't required for asteroid lightcurve determination, because in that project you don't attempt to determine the "true" brightness of the asteroid—only its delta-brightness from (uncharacterized) comp stars is determined.

There are, however, situations where you'll want to determine the "true" magnitude of a target that doesn't have predefined comp stars. In such a case, you need a way to transform your instrumental magnitudes into the standard system, and in that case you'll use your transformation coefficients. The next project—asteroid phase curves—requires that you determine the V-magnitude of an asteroid on several nights. To do that, you'll need to know your transforms, and you'll also need to follow some complex data-collection and data-analysis steps.

The procedure for putting your photometry onto the standard system has three steps: (1) determining the transformation coefficients for your system, (2) taking each night's data, and (3) reducing your data to the standard system. The step called "taking each night's data" has three components: each night, you will gather data that will be used to determine the atmospheric extinction and the nightly zero point, and gather photometry on your target object. So the cycle of activity will look like this:

● About once per year, determine your transformation coefficients.
● Each night, take data that will allow you to:

    ○ determine the atmospheric extinction coefficient in v and r-bands;
    ○ determine the nightly "zero point" in both v and r-bands;
    ○ collect your target photometry, in v and r-bands.

● Perform data reduction to put your target and comp-star magnitudes onto the standard V, R-magnitude system by:

    ○ determining the V-mag and color index of each comp star (for each night);
    ○ determining the V-mag and color index of the target. We will use all these steps as part of the next project.

4.8   PROJECT K: ASTEROID PHASE CURVES

You know from experience that the brightness of moonlight is a strong function of the Moon's phase. Our Moon is very bright when it is full, but casts relatively little illumination when it is a thin crescent. A similar sort of effect happens with asteroids. Consider the geometry illustrated in Figure 4.35. When an asteroid is at opposition, we are seeing it fully illuminated (analogous to a full moon). Ignoring for a moment the rotational lightcurve variations, we expect that the asteroid will be fainter when it is only partially illuminated (analogous to a gibbous moon). Of course, the asteroid's brightness will also fade as it moves away from Earth. If its orbit takes it farther from the Sun, it will receive less sunlight, so again its brightness will fade. The effect of Sun- and Earth-distance on the asteroid's brightness over the course of an apparition is a straightforward geometric effect. The effect of solar phase angle (i.e., the fraction of the asteroid's surface that is illuminated as seen from Earth) is more complicated, and contains some interesting information about the asteroid. In the professional litera- ture, you will find plots of an asteroid's brightness vs. phase angle for some asteroids. Most of them show that as the asteroid approaches phase angle $= 0$ (fully illumi- nated), the brightness surges noticeably. Relatively few asteroids have good "phase curves" describing this effect because of the relative difficulty of making the necessary measurements and the very few people (amateur or professional) who undertake the project.

Second-order extinction

You may wonder why you are using V and R-band, but not B-band. After all, most historic professional studies of star colors used (B-V) as the color index. The reason is that atmospheric extinction effects in the B-band are a bit more com- plicated than described in the main text. The scattering of light in the atmosphere is much more pronounced at short wavelength (blue light) than it is at longer wavelengths (red light). The sky is blue because it preferentially scatters blue light; and the warm yellow Sun becomes orange–red at sunset because its blue light has been reduced by scattering, so mostly red light is left to reach your eyes. That effect also occurs with starlight. Atmospheric absorption and scattering do not just reduce the star's light intensity ("first-order extinction"), they also change the star's color. A star's blue light is reduced more significantly than is its red light. Further, the variation of extinction with wavelength is more pronounced at short wavelength than it is in the V or R-band. This color-dependent extinction effect is called "second-order extinction".

In order to account for this effect, the extinction equation for the B-band becomes:

$$b = b_0 + k' \cdot X + k'' \cdot (b - v) \cdot X$$

where k' is the "first-order extinction" coefficient (magnitudes/airmass) and k'' is the "second-order extinction" coefficient.

The second-order extinction coefficient is measured in "magnitudes/airmass/ magnitude of color". In the spirit of getting you going as rapidly as possible on photometry on the standard system, I've elected to recommend using V and R-band imaging.

Both theory and experience have shown that second-order extinction is only significant in the B-band. The second-order extinction coefficients for V and R-bands are almost always small enough that they can be ignored.

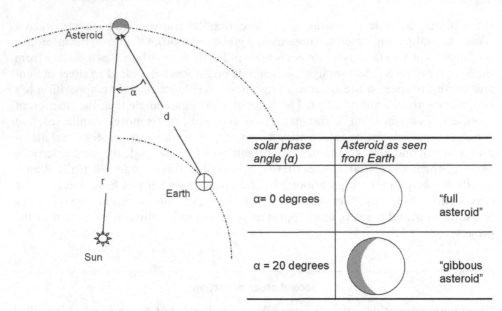

solar phase angle (α)	Asteroid as seen from Earth	
α= 0 degrees		"full asteroid"
α = 20 degrees		"gibbous asteroid"

Figure 4.35. As the Earth and asteroid orbit the Sun, we see the asteroid at different solar phase angles.

This project starts with observations to accurately determine the lightcurve and its V-magnitude on the standard magnitude system. You should strive for an accuracy of about 0.03–0.05 magnitude, which requires excellent SNR and careful attention to your photometry methods. In order to unravel the effect of the lightcurve variation, you need to create a densely-sampled, accurate lightcurve for the asteroid.

If you follow the asteroid over the course of an apparition, you will see it at a range of solar phase angles (from waxing gibbous, to "full" near opposition, and on to waning gibbous later in the season). Careful photometry over the course of an apparition can determine the way that the asteroid's brightness changes with solar phase angle. The solar phase angle and the asteroid's distance from Sun and Earth are all controlled by the asteroid's orbit. In order to separate out the phase-angle effect, you need to unravel the effect of the changing Sun- and Earth-distance. This is done by determining the asteroid's V-magnitude brightness, and then converting the brightness to "reduced" magnitudes by:

$$V_R = V - 5 \cdot \log(r \cdot d)$$

where V = the measured V-magnitude;
         d = the distance of the asteroid from the Sun (in AU);
         r = the distance of the asteroid from Earth (in AU).

The reduced magnitude takes out the effect of the constantly changing distance from the Sun and Earth, so that the only remaining reasons for brightness change are the rotational lightcurve, and the changing phase angle.*

An asteroid phase curve that was created from your own data is pretty neat to see—an accomplishment that puts you among a very small group of accomplished asteroid photometrists. Beyond the inherent "neatness" of seeing the opposition surge of your asteroid, the shape of the asteroid's brightness-vs.-phase curve turns out to have several important uses.

First, the brightness at zero phase angle defines the asteroid's "absolute magnitude", H. The absolute magnitude of an asteroid is defined as the brightness that the asteroid would have when fully illuminated (phase angle $\alpha = 0$), at a distance of 1 AU from the Sun, and also 1 AU from the observer. As Dr. Alan Harris [21] has pointed out, this would give the observer very hot feet. Happily, it is used as a theoretical definition, not a prescription for actual observations! The absolute magnitude, H, is found by plotting the measurements of V_R vs. α, and extrapolating the curve to the intercept point where $\alpha = 0$.

The curve of V_R vs. α has a specific shape [22], defined by the "slope parameter", G. The slope parameter offers some clues about the surface texture of the asteroid. Within some broad limits, the slope parameter also provides information about the asteroid's albedo. If you know the albedo and the absolute magnitude, you can estimate the physical size of the asteroid (i.e., its projected area).

This is a challenging and complex project. It requires all of the usual asteroid lightcurve discipline, plus two-color imaging and knowledge of your system's transformation coefficients, plus use of your planetarium program to calculate the asteroid's Earth and Sun distances and solar phase angle for each night of observation. You'll devote quite a few nights to an individual asteroid. In return, you get to make a uniquely valuable contribution to asteroid science, one that few amateurs have undertaken.

---

*It may not be immediately obvious why the factor "5" appears in the equation for reduced magnitude. It comes about as follows. The intensity of a light source falls off with the square of its distance. An asteroid is faint both because it is far from the Sun (one "R-squared" effect) and because it is far from us (a second "R-squared" effect).

Suppose that we observe that the asteroid appears as a light source with intensity I, when it is at distance r (AU) from the Earth and d (AU) from the Sun. If we could magically move it so that it was exactly 1 AU from the Sun and also 1 AU from the observer, then its intensity would become:

$$I_R = I \cdot (r^2 d^2)$$

Recall that the magnitude of a light source was defined as $M = -2.5 \log(I)$, so:

$$M_R = -2.5 \log(I_R) = -2.5 \log(I \cdot r^2 \cdot d^2)$$
$$= -2.5 \log(I) - 2.5 \log[(r \cdot d)^2]$$
$$= M - 5 \log(r \cdot d)$$

Phase curve equations

The standard equation for asteroid magnitude vs. phase angle, promulgated by the International Astronomical Union, is:

$$V_R(\alpha) = H - 2.5 \log[(1 - G)\Phi_1 + G\Phi_2]$$

where H is the absolute magnitude, and G is the "slope parameter". (In the absence of a good value for the slope parameter, the "default" value $G = 0.15$ is often used to predict asteroid brightness, or to estimate the absolute magnitude from a single brightness measurement.)

The functions Φ_1 and Φ_2 are given by:

$$\Phi_1 = W \cdot S_1 + (1 - W) \cdot L_1$$

and

$$\Phi_2 = W \cdot S_2 + (1 - W) \cdot L_2$$

where:

$$W = \exp[-90.56 \cdot \tan^2(\alpha/2)]$$

$$S_1 = 1 - \frac{0.986 \cdot \sin(\alpha)}{[0.119 + 1.341 \cdot \sin(\alpha) - 0.754 \cdot \sin^2(\alpha)]}$$

$$L_1 = \exp\{-3.332 \cdot [\tan(\alpha/2)]^{0.631}\}$$

$$S_2 = 1 - \frac{0.238 \cdot \sin(\alpha)}{[0.119 + 1.341 \cdot \sin(\alpha) - 0.754 \cdot \sin^2(\alpha)]}$$

$$L_2 = \exp\{-1.862 \cdot [\tan(\alpha/2)]^{1.218}\}$$

In all of the above, $\alpha =$ solar phase angle, in radians.

The program MPO Canopus contains a tool that does a least-squares fit to these equations, to determine H and G from our data.

4.8.1   Equipment needed

In order to determine an asteroid's phase curve, you need everything that you used for asteroid lightcurve differential photometry, plus photometric V and R-band filters:

- Telescope.
- CCD imager, with photometric V and R-band filters.
- Personal computer.
- Software for CCD control and operation.
- Planetarium program.
- Software for photometry reduction of your data.
- WWV receiver, GPS, or other source of accurate time to set your PC's clock.

- GPS or other means of accurately determining your observing location (latitude, longitude, and elevation).
- Spreadsheet program (e.g., Microsoft Excel) for analyzing and plotting the results.

You also need a long run containing clear, stable ("photometric") dark nights, in order to get good calibration of the photometry—color index and magnitude—of your target and comp stars. For many amateur observing sites, that's a real problem! You can work around it by making lightcurve data in two colors on all nights (even those that aren't perfectly clear and stable), and then devoting one or two genuinely "photometric" nights to calibrating your comp stars, so that they can be used as "secondary standards".

The data reduction for the project places a special demand on your planetarium program: you must determine the asteroid's distance from Earth, distance from the Sun, and its solar phase angle. SkyMap Pro (in its "asteroid/ephemeris" tool) will give you all three parameters. TheSky will give you the Earth and Sun distances, but you will have to go elsewhere to determine the solar phase angle. One excellent source is the "Minor Planet Ephemeris Service" at the Minor Planet Center (on the web at *http://cfa-www.harvard.edu/iau/MPEph/MPEph.html*).

The photometric V and R-band filters are a modest expense, assuming that you've already invested in all of the other equipment (about $150 each for 1.25-inch filters, or a few hundred dollars each for 2-inch filters). Unfortunately, the photometric R-band filter's pass band is different from the "astro-imaging" R-filter that you use for R-G-B color imaging.

Relatively few asteroids reach really small phase angles in any given year— perhaps a dozen or so bright enough for accurate amateur photometry will reach minimum phase angles less than 1 degree in any given year. You can find predictions of candidate "low phase angle" targets in the *Minor Planet Bulletin* early each year. This will enable you to plan your phase-curve project timetable and fit it into your other project schedules.

4.8.2 Making the observations

Beginning as early in the apparition as practical (i.e., starting when you can gather at least a few hours of photometry on the target asteroid), gather photometry in the same way that you would for determining the asteroid's lightcurve, with four special requirements:

- Make your observations in a photometric filter band (i.e., V or R, instead of using unfiltered images).
- Twice each night, take a V-R-R-V set of images; make one set when the asteroid is near the horizon, and the next set when it is near culmination.
- At least once each night, take a V-R-R-V set of images of the Landolt standard field that is closest to your target asteroid. This should be done when the asteroid is near culmination.

● (Optional) once each night (also near culmination), take a V-R-R-V set of images of the field that contains the comp stars that you used on the first "good" night of observations of this asteroid. (You can use these measurements of the night #1 comp stars as a check on your photometric accuracy.)

You repeat this set of observations throughout the asteroid's apparition, in order to get data over a wide range of phase angles. The very-low-phase angle conditions (near opposition) are the most critical nights, because the most dramatic opposition surge occurs in the range $\alpha = 0$ to about $\alpha = 2$ degrees. Strive to image the asteroid all night on every night from phase angle -3 deg to $+3$ deg. Schedule your other nights to fill in the phase curve, getting data every few degrees of phase angle, to as large a phase angle as possible—at least $\alpha = 10$ degrees, and preferably to $\alpha = 20$ degrees.

This data-gathering effort will span one to three months, depending on whether you cover both the negative and positive phase angles, or just one0half of the phase curve.

4.8.3   Reducing and analyzing your data

Image reduction will be done as you would for any other photometry project: dark-subtraction, flat-fielding and (if you're using scaled darks) bias subtraction.

Each night's images are analyzed in the same way that you would for an asteroid lightcurve project: select comp stars, follow the asteroid throughout the night and monitor its brightness relative to the comp stars (see Section 4.4 for the details). It is best to reduce each night's data promptly, so that you can determine the asteroid's rotation period as early in the project as possible. If it's a slow rotator, you'll probably need more nights than you would for a typical asteroid, just to determine the lightcurve period.

For each night, plot the instrumental magnitude vs. airmass for one or more comp stars, to determine the atmospheric extinction for that night. You'll do this separately for the V and R-bands, since the V-band extinction (k_v) will be different from the R-band extinction (k_r). (That is why you took the V-R-R-V images twice, once near the horizon and once near culmination. You need to have the R-band images at two different airmasses, so that you can plot the results to determine the R-band extinction.)

Each night's data also includes V and R-band images of one or more Landolt standard stars. Calculate the airmass of that observation, and determine the exo-atmospheric instrumental magnitudes of the standard star:

$$v_{0,std} = v_{std} - k_v \cdot X$$

and

$$r_{0,std} = r_{std} - k_r \cdot X$$

With those calculations done, you can determine the nightly zero point in each color. This is the term that relates instrumental magnitudes to true magnitudes. Since the true V and R-magnitudes of your standard star is known from the Landolt catalog (call them V_{std} and R_{std}), and you know your transforms (because you took my earlier advice and carefully determined them before starting this project!), you can calculate the zero points from the following equations:

$$(V_{std} - R_{std}) - T_{vr} \cdot (v_{0,std} - r_{0,std}) = Z_{vr}$$

and

$$V_{std} - v_{0,std} - T_v \cdot (V_{std} - R_{std}) = Z_v$$

Now you know everything that you need to know to translate any object's instrumental magnitude to the standard photometric system: you know the atmospheric extinction (in both bands), your transforms, and the nightly zero points. So, you can determine the magnitude of your target on the standard system by applying the same set of equations to each data point in your lightcurve:

$$(V_{ast} - R_{ast}) = T_{vr} \cdot (v_{0,ast} - r_{0,ast}) + Z_{vr}$$

and

$$V_{ast} = v_{0,ast} + T_v \cdot (V_{ast} - R_{ast}) + Z_v$$

Whew! By doing this, your lightcurve is no longer expressed in terms of "delta-brightness from the comp stars", but rather it is in terms of standard V-magnitude.

Each night gives you a new piece of the lightcurve, and (once you have enough data to accurately determine the period and shape of the lightcurve), you can examine the "midpoint" or "lightcurve average" brightness each night. (In this way, you are in effect eliminating the periodic variation due to the asteroid's rotation, and instead concentrating on its average brightness). Over a few weeks, you'll see that the asteroid's average brightness changes—getting brighter as it approaches opposition, and fading after it passes opposition.

Create a simple spreadsheet table, whose columns are: Date (JD), V_{avg}, solar phase angle (α), Earth distance (d, in AU), and Sun distance (r, in AU), and enter the data from each night into it. For each data point, calculate the reduced magnitude

$$V_R = V_{avg} - 5 \cdot \log(r \cdot d)$$

and plot V_R vs. solar phase angle. If all has gone well, you'll get a curve similar to that in Figure 4.36: a nicely-done asteroid phase curve, showing the opposition surge in brightness, and capable of being fitted to the standard phase-curve equation to determine the H and G parameters [23].

Why did I recommend that each night you also make V-R-R-V images of your original (first-night) comp stars? Doing so gives you a set of reference points that you can use to double-check your magnitude calculations—you can cross-reference all data to these first-night comp stars, in addition to the Landolt standards, in effect using the first-night comp stars as secondary standards. Doing so will give you warning if something "odd" happened on a particular night. If the cross-check doesn't check, then you can either investigate what happened, or simply delete that night's data from the phase curve. There are several things that can go wrong. From most amateur observing sites, the most common problem is unstable atmospheric conditions: we aren't often blessed with "photometric" conditions all night. If the atmospheric extinction changes over the course of the night, then your V-magnitude

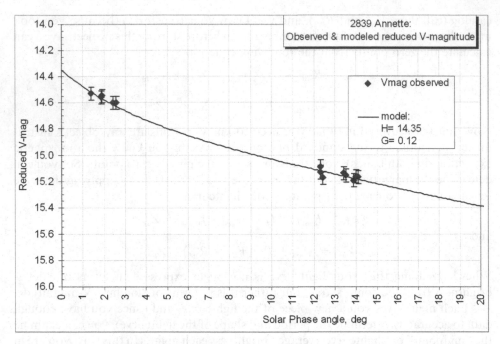

Figure 4.36. Example of an asteroid phase curve: mean "reduced" V-magnitude vs. solar phase angle.

calculations will be thrown off. Examine your plot of instrumental magnitude vs. airmass: if it isn't a straight line, or if the extinction is different before and after culmination, then conditions changed during the night!

4.8.4   Reporting your results

Phase curves should be reported in the *Minor Planet Bulletin*, in the same short-article format that is used to report the asteroid lightcurve that you have determined in the course of this project.

4.9   SAS AND CBA

Two organizations that are particularly devoted to amateur photometry and to facilitating amateur–professional partnerships in photometric projects are the Society for Astronomical Sciences (SAS), and the Center for Backyard Astrophysics (CBA). If you are interested in trying your hand at photometric projects, you should join one or both of these organizations.

The principle SAS activity is the annual "Symposium on Telescope Science". It is an intense two or three-day meeting where both amateur and professional astronomers describe some of their projects and results. While not exclusively devoted to photometry (you'll also hear papers on everything from fireball cameras to radar observations and spectroscopy), the annual Symposium is a fabulous way to be simultaneously educated, motivated, and excited by the research activities that your fellow amateur astronomers are involved in. Check out the SAS website at: *http:// www.socastrosci.org* The Proceedings from recent symposia, and the periodic newsletter, are available for free download.

The Center for Backyard Astrophysics is a worldwide network of amateur and professional astronomers who are devoted to the study of cataclysmic variable stars using CCD photometry. Despite its light-hearted name, the group routinely gathers research-grade data that is used by professional astronomers to generate papers in peer-reviewed journals. In acknowledgment of the caliber of the observations, the amateur observers are often credited as co-authors in these research papers. Check out the website at *http://cba.phys.columbia.edu/*

4.10 RESOURCES

This chapter will get you started, and enable you to successfully complete some important photometry projects. As your skill and experience increases, you'll want to learn more about photometry than what I've given you here. There is a wealth of material available in books, articles, and websites for your further education. Two of my favorite books are:

Henden, A.A. and Kaitchuck, R.H., *Astronomical Photometry: A Text and Handbook for Advanced Amateur and Professional Astronomer*, Willmann-Bell (1990).
Warner, B., *A Practical Guide to Lightcurve Photometry and Analysis*, Springer (2006).

There is also an active Yahoo® group dedicated to asteroid topics (photometry, astrometry, and related discussions)—the Minor Planet Mailing List. This group's membership is a mix of amateur and professional astronomers, so the discussions can be very educational. The Minor Planet Mailing List is located at *http://groups. yahoo.com/group/mpml/*

4.11 REFERENCES

[1] Bessell, M.S., "Standard Photometric Systems", *Annual Review of Astronomy and Astrophysics*, vol. 43, p. 293 (2005).
[2] Skiff, B., *Asteroid Photometry and the Catalogues in general use*, e-mail message to the Minor Planet Mailing List (September 18, 1998).
[3] Bessell, M.S., "The Hipparcos and Tycho Photometric System Passbands", *Publications of the Astronomical Society of the Pacific*, vol. 112, p. 961 (July, 2000).

[4] Miles, R., "Asteroid Phase Curves: New Opportunities for Amateur Observers", *Proceedings for the 24th Annual Conference of the Society for Astronomical Sciences* (May 24–25, 2005).

[5] Kaasalainen, M. and Torppa, J., "Optimization Methods for Asteroid Lightcurve Inversion. I. Shape Determination", *Icarus*, vol. 153, p. 24 (September, 2001).

[6] Harris, A.W., "Fourier Analysis of Asteroid Lightcurves: Some Preliminary Results", *Abstracts of the Lunar and Planetary Science Conference*, vol. 18, p. 385 (1987).

[7] Pravic, P., Harris, A., and Michalowski, T., "Asteroid Rotations" in Bottke, W., Cellino, A., Paollicchi, P., and Binzel, R. (eds.), *Asteroids III*, University of Arizona Press (2002), p. 113.

[8] Benner, L.A.M., "Contributions by Amateur Astronomers to Support Radar Imaging of Near-Earth Asteroids", *Proceedings for the 23rd Annual Conference of the Society for Astronomical Sciences* (May 24–25, 2004).

[9] Durech, J. and Kaasalainen, M., "Asteroid Shape Models Refined by Stellar Occultation Silhouettes", *Bulletin of the American Astronomical Society*, vol. 36, p. 143 (December, 2004).

[10] Beaulieu, J.-P, "Discovery of a cool planet of 5.5 Earth masses through gravitational microlensing", *Nature*, vol. 439, p. 437 (January 26, 2006).

[11] Young, A.T. "Photometric Error Analysis VI: Confirming Reiger's Theory of Scintillation", *The Astronomical Journal*, vol. 72, no. 6, p. 747 (August, 1967).

[12] Dravins, D., Lindegren, L., and Mezey, E., "Atmospheric Intensity Scintillation of Stars. III. Effects for Different Telescope Apertures", *Publications of the Astronomical Society of the Pacific*, vol. 110, p. 610 (May, 1998).

[13] Howell, J., "Introduction to Time-Series Photometry Using Charge-Coupled Devices", *Journal of the AAVSO*, vol. 20, p. 134 (1991).

[14] Seagroves, S., Harker, J., Laughlin, G., Lacy, J., and Castellano, T., "Detection of Intermediate-Period Transiting Planets with a Network of Small Telescopes: transitsearch.org", *Publications of the Astronomical Society of the Pacific*, vol. 115, p. 1355 (December, 2003).

[15] Tim Castellano, personal communication (June, 2006).

[16] Johnson, H.L. and Morgan, W.W., "Fundamental Stellar Photometry for Standards of Spectral Type on the Revised System of the Yerkes Spectral Atlas", *The Astrophysical Journal*, vol. 117, no. 3 (May, 1953).

[17] Landolt, A.U., "UBVRI Photometric Standard Stars in the Magnitude Range $11.5 < V < 16.0$ around the Celestial Equator", *The Astronomical Journal*, vol. 104, no. 1, p. 340 (July, 1992).

[18] Henden, A.A. and Kaitchuck, R.H., *Astronomical Photometry*, Willmann-Bell, 1990.

[19] Buchheim, R.K., "The magnitude and constancy of second-order extinction at a low-altitude observing site", *Proceedings for the 24th Annual Conference of the Society for Astronomical Sciences* (May 24–25, 2005).

[20] Warner, Brian, *A Practical Guide to Lightcurve Photometry and Analysis*, Bdw Publishing, Colorado Springs CO (2003).

[21] Harris, A.W, in a note on the Minor Planet Mailing List (yahoo.groups.MPML)

[22] Bowell, E., Hapke, B., Domingue, D., Lumme, K., Peltoniemi, J., and Alan, W., "Application of Photometric Models to Asteroids", in Binzel, R., Gehrels, T., and Matthews, M. (eds.), *Asteroids II*, University of Arizona Press (1989), p. 524.

[23] Lupishko, D.F. and Belskaia, I.N., "Surface, shape and rotation of the M-type asteroid 16 Psyche from UBV photometry in 1978 and 1979", in *Asteroids, comets, meteors; Proceedings of the Meeting, Uppsala, Sweden, June 20-22, 1983*, Uppsala, Sweden, Astronomiska Observatoriet, 1983, pp. 55–61.

5

CCD astrometry

When the Lord placed the stars in the sky, he did not glue them to their places. Pretty much everything in the universe is in motion, including the Earth from which we observe the heavens. Astrometry is the science of measuring the positions and motions of celestial objects. Astrometry has a long and wonderful history. In diverse times and places, it has provided the evidence that drove mankind's ever more sophisticated understanding of the universe, and our place in it.

For millennia, stargazers have noted the positions, and monitored the motions, of the celestial objects—rudimentary but in some cases surprisingly accurate astrometry. As the astrometric data became more accurate, the theories to explain the motions became ever more complex and cumbersome. In 1543, Copernicus published his hypothesis that the Earth wasn't the immobile center of the universe, and showed how the familiar motions of celestial objects could be explained by placing the Sun at the center of the solar system. The Sun rose and set because the Earth rotated once around per day; the stars changed with the seasons because the Earth orbited around the Sun once per year; and the planets' complex paths across the sky could be simply explained by their motion around the Sun, as viewed from the moving platform of the Earth. Aside from the elegant simplicity of the heliocentric model, compared with the competing "epicycle" theories, there wasn't much in the way of hard evidence pointing to one theory vs. the other. Then Galileo observed the motion of Jupiter's bright satellites around that planet—definitely they weren't revolving around the Earth—and he concluded that this supported the notion that there were many centers of motion. Since the Earth was not necessarily at the center of the universe, it seemed more reasonable to place the Sun at the center. There was legitimate controversy on whether this Sun-centered universe represented reality or was just a convenient mathematical trick. If the Earth really does move around the Sun, why don't we see the stars waggling slightly back and forth due to the parallax effect? Galileo spent considerable effort trying to observe such parallax, in order to vindicate his opinion that the Earth did move. He failed in this astrometric project, with unfortunate consequences.

Nevertheless, over the next couple of centuries it became commonly accepted that the Earth did in fact revolve around the Sun. Newton's law of gravity did a fine job of explaining how and why all of the planets, and Jupiter's moons, move the way they do. In 1729 J. Bradley recognized that a peculiar annual cyclic motion of stars (first observed around 1680) could be explained by the velocity of the Earth's motion around the Sun. The effect is now known as the "aberration of starlight", and it is much larger than true parallax. Bradley's explanation encompassed two important ideas: the Earth does move, and the speed of light is finite. Happily, in 1838 F.W. Bessel did a very meticulous astrometric study and succeeded in measuring the parallax of a nearby star (61 Cygnii). That astrometric result—of quite delicate precision, considering that the star's parallax is just 0.3 arc-sec—pretty well closed out any lingering doubts about the reality of the Earth's motion.

Unfortunately, continued progress in astrometric accuracy and precision led to the recognition that something was wrong with Mercury's motion—it didn't quite follow the path predicted by Newton's laws. In 1915 Einstein came to the rescue with the general theory of relativity. His equations correctly predicted the observed details of the motion of Mercury. Alas, his theory also made some really bizarre predictions. For example, for a long, long time it had been well-known by everybody that light traveled in a straight line. Everybody except Einstein, that is, because his theory predicted that light rays would be bent when they passed through a gravitational field. Astrometry was used to put this idea to the test, and the predicted displacement was verified by careful measurements of the positions of stars seen near the Sun's limb during the total solar eclipse of November, 1919.

So, astrometry has been a critical discipline throughout the history of astronomy, and its importance is undiminished today. We send spacecraft into the cosmos, and accurate astrometry is needed to determine both the location of the destination and the current position of the spacecraft. A host of asteroids and comets are sailing around the Solar System, and we need accurate astrometry of each of them so that we'll know where to find them in the future. In the case of the near-Earth asteroids, astrometry tells us whether or not they present the threat of an impact on Earth. Stars move, and identification of high-proper-motion stars is of interest as a clue to which stars are close to our Sun. Binary stars orbit their center of mass; accurate measurements of their orbital paths can tell us what their masses are—a critical piece of information to understand the properties and evolution of stars.

With modern amateur telescopes, the wide availability of CCD imagers, and easily-used software, the amateur astronomer can make astrometric measurements that are accurate enough to contribute to all of these areas [1]. In addition, some skill in astrometry is a necessary adjunct to the "discovery" projects described in Chapter 6. It would be really embarrassing to discover something new but not be able to tell anyone where it is! So, it is worth adding some astrometric projects to your astronomical agenda. You can contribute to the advancement of science, and probably have some real fun along the way.

5.1   BASIC PRINCIPLES OF ASTROMETRY

The basic (somewhat oversimplified) concept of determining an object's RA and Dec from your CCD image is pretty straightforward.

Suppose that you have a CCD image of a field that contains a target object such as an asteroid. Further suppose that your image also includes a few stars whose positions are very accurately known from one of the modern high-accuracy astrometric catalogs. It is reasonable to suspect that a little geometry and algebra should be sufficient to use the "astrometric reference stars" as benchmarks and, from their known positions, determine the position of the asteroid. The concept is illustrated in Figure 5.1. The idea is simple: given the known image positions and RA, Dec coordinates of an array of "astrometric" reference stars in the image, you can calculate the position of the target. I emphasize that this figure is illustrating only the *concept*. The actual calculations to deal with real-world complexity are a lot more involved, but happily you won't have to do them. Software packages that perform the necessary calculations are readily available. If you're a CCD astro-imager, chances are that you already own one of them, so you can get started doing some astrometry without any additional purchases.

This concept, and the illustrative example of Figure 5.1, should raise several questions in your mind:

- How do you identify the "astrometric" stars in the image?
- Where do those "astrometric catalogs" come from?

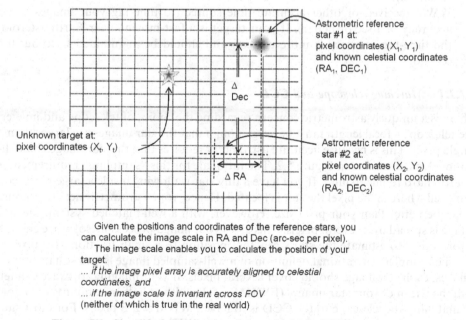

Given the positions and coordinates of the reference stars, you can calculate the image scale in RA and Dec (arc-sec per pixel).
The image scale enables you to calculate the position of your target.
... *if the image pixel array is accurately aligned to celestial coordinates, and*
... *if the image scale is invariant across FOV*
(neither of which is true in the real world)

Figure 5.1. Simplified description of the principle of CCD astrometry.

- How accurate can I expect my astrometry to be?
- Are there special imaging techniques (exposure time, filter, CCD binning, etc.) to use, or to avoid?
- Are there special requirements for "matching" the CCD to the telescope?
- Are particular CCDs and/or telescopes better than others?

Those questions are where the details are found that make astrometry so challenging, interesting, useful, and—dare I say—fun.

5.1.1   Equipment needed

Astrometry is a CCD imaging project, for which your standard CCD set-up will be used. The unique requirements are software for astrometric data reduction, and accurate position and time data for your observing site. The equipment you need is:

- Telescope (equatorially mounted, polar aligned, and tracking).
- CCD imager.
- PC, with CCD-control software and planetarium program.
- Astrometry reduction software.
- Catalog of astrometric reference stars (i.e., a database of well-measured star positions near the target object).
- GPS receiver, or other method of accurately determining the position of your observing site (to an accuracy of about 100 feet in latitude, longitude, and elevation).
- WWV receiver, or other accurate time source (to time-tag your images to an accuracy of 1 sec or better). If your target is a fast-moving near-Earth asteroid, the time-tag accuracy is more critical, and should be maintained to about 0.1 second.

5.1.1.1   Matching telescope and CCD

There is a uniquely astrometric concern regarding the choice of telescope and imager: the telescope's focal length must be long enough that the star images are larger than a single pixel. This is because accurate astrometry requires that the star images be well-sampled, as illustrated in Figure 5.2. The rationale for this "sampling" requirement is not too hard to understand. If you have a tiny star wallowing inside a large pixel, you can't tell where in the pixel the star is located. Hence, your astrometric accuracy could never be better than your pixel size. However, with a well-matched system, the star image is spread over several pixels. This gives your image-processing software enough information to estimate the star's centroid position to a small fraction of a pixel.

The simplest operational definition of a well-sampled image is "no square stars". That is, each star image should cover several pixels. If your imaging software can tell you the width of your star images (FWHM = full width at half-maximum), your goal of matching the telescope to the CCD is to have FWHM $\approx$ 2–3 pixels. For most sites and most amateur equipment, the star image's FWHM is set by atmospheric seeing

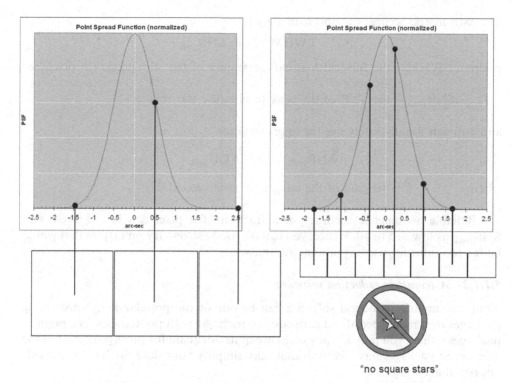

Figure 5.2. Image sampling. A poorly-sampled image (left) yields astrometry that is not much better than ±0.5 pixel. An adequately-sampled image (right) contains enough information to calculate the star's centroid to a small fraction of a pixel.

conditions rather than by the telescope. "Seeing" of 2–4 arc-sec is typical, so that you'll want to select your focal length to give you about 1–2 arc-sec per pixel. If you are blessed with 1-arc-sec seeing, then you'll want to use smaller pixels—or longer focal length—to take full advantage of your excellent site.

In general, you will want to avoid "binning" of pixels in your astrometric images—get the best sampling (i.e., smallest pixels) that you can. Similarly, you may want to avoid using a focal reducer.

There is, of course, a penalty for using small pixels and/or long focal length. If you spread the star's light over several pixels, then naturally only a portion of the signal lands on any single pixel. If you read the discussion of signal-to-noise ratio in Section 4.5, you will not be surprised to learn that astrometric accuracy of a well-sampled image is dependent on the signal-to-noise ratio. However, the relevant metric for astrometry is the SNR of the brightest pixel in the star image, called SNR_{peak} (to distinguish it from the total integrated signal-to-noise ratio used in the discussion of photometry). The higher the SNR_{peak}, the better your astrometric accuracy will be (all other things being equal). You may, however, be surprised by the relatively modest SNR_{peak} requirements for astrometry, when compared with photometry.

The astrometric accuracy is approximately [2]

$$\sigma_{ast} = FWHM/(2.36 \cdot SNR_{peak})$$

where   FWHM = the full-width-at-half-maximum of the star image (in pixels, or
                    arc-sec);
         SNR_{peak} = the SNR of the peak pixel in the star image;

and you can usually safely use the approximation

$$SNR_{peak} \approx \sqrt{g \cdot ADU_{peak}}$$

where g = the gain of the imager (in electrons/ADU).

Reliable astrometric accuracy of a fraction of a pixel can be achieved with SNR_{peak} as low as 5 to 10. Hence, you can do good astrometry on targets that would be far too faint for useful photometric study.

5.1.1.2   Astrometric reduction software

Your astrometric reduction software can be one of the popular image-processing packages or a more specialized software program. As with photometry, the popular packages (which you may already own) are quite adequate for most projects; but the specialized packages have features that may simplify your data analysis or provide improved accuracy.

All star charting or planetarium programs contain an internal database of stars. The database entry for each star has its position (RA and Dec), proper motion (maybe), magnitude (usually), and color (sometimes). Most popular programs use the *Hubble Space Telescope Guide Star Catalog* (GSC) as their faint-star database. This catalog provides accurate positions for all of its stars, and can be used for most amateur astrometric projects. In general, any planetarium or CCD image-processing program that is capable of performing astrometric reductions will automatically select the astrometric stars to use, without bothering you with the details. Examples include CCDSoft + TheSky, MaximDL, and AstroArt.

Some programs can use other (and more accurate) stellar catalogs, such as the USNO SA-2 or the UCAC2 catalogs. If your first few astrometric projects infect you with the "astrometry bug", then you will most likely want to upgrade to an astrometric star catalog, and the software that can take advantage of it (examples include MPO Canopus and IRAF). Use the software that you have for your first couple of astrometry projects. Then, if you decide that you enjoy those projects, make the modest investment in specialized software.

The key functions of the astrometric software are to (a) determine the position $[x, y]$ of each star in the image to a fraction of a pixel, (b) find the function that relates pixel position to RA–Dec across the image (the so-called "plate constants"), and finally (c) use the plate constants to determine the RA–Dec position of your target. It is also a good thing if the software provides you with a table of the "residuals" between actual vs. calculated reference star positions. The residuals give you an indication of how good your solution was, and an estimate of the accuracy of the astrometry of your target.

If you would like to know a little bit about what your astrometric software is doing "behind the scenes", refer to Appendix B. There, you'll find some background information on the celestial coordinate system and a description of the modern astrometric reference catalogs.

5.1.2　Making the observations

You gather your images for astrometry in the same way that you would for any other CCD imaging project. Select an exposure duration that provides a good signal on both the target and the reference stars, but do not allow the target or the astrometric reference stars to be saturated in the image. The better your focus and guiding, the better your astrometry will be, so use your usual care in focusing, and autoguide if possible. In general, it is wise to take three or four identical images in succession, so that you aren't at risk of a cosmic-ray hit on your target or any of the myriad other things that can go wrong in a single image. With two or three good images, you can check the internal consistency of your results to give you more confidence in them. Depending on your target, you may need to take several series of images a few hours (or a few days or months) apart, to determine the motion of your target.

There are no special requirements regarding spectral band—using unfiltered imaging is fine. If possible, make your astrometric observations when the target is more than 30 degrees above the horizon. If you must observe your target at lower elevations, using a spectral filter (e.g., photometric V or R-band, or imager's G or R-filter) will help to minimize some of the confusing atmospheric effects that become significant when imaging at large airmass.

Although your astrometric software can probably read most CCD image file types, it is best to save your images in FITS format. If your imager control software permits, it is useful to have it write the telescope pointing direction (RA, Dec), image time, and exposure to the FITS header. Take your science images in "raw" FITS format, without any auto-dark or auto-flat applied. This recommendation is less important for astrometry than it is for photometry, but it is still good practice. If anything should go wrong during image processing, it's nice to be able to return to the raw, unadulterated science image. Of course, before moving anything in your optical train, take good flat-field images to use for reduction of your science images.

5.1.3　Reducing and analyzing your results

The routine disciplines of dark- and flat-frame reduction should be applied to your astrometric images. Do not use any nonlinear algorithms: no "sharpen", "image enhancement", or similar processes are permitted. They will corrupt your astrometry.

See your user's manual for the details of opening, reducing, and displaying results with your astrometric software. In general, there is very little for you to do manually—the software will take care of just about all of the calculations. You may have to tell it the details of your imaging system (focal length, pixel size, etc.), and your location. If you didn't write the RA–Dec to the FITS header of the images you will have to tell your software the approximate pointing direction in order to help it match the image to the star catalog.

The astrometric analysis software will match your image to the star catalog and compute your plate constants. Most of the software packages will display the "residuals" from this matching operation. Examine the residuals carefully. If one or two stars show residuals that are significantly larger than the norm, then try deleting the offending stars from the calculation. Most programs allow you to do this "censoring"—see your user's manual. Why might a star show unusually large residuals? That star may have been saturated on your image, or it may have had a very low SNR, there may be a defect in your image at that point (a cosmic-ray hit, or a dead pixel), or the star may have a large proper motion (if your catalog doesn't include proper motion information).

With the image properly and accurately matched to your astrometric reference catalog, your software will instruct you to pinpoint the object in the image that is your "target". It will then calculate the centroid of the target to a fraction of a pixel, and use the plate constants to determine its RA–Dec location.

Then, you must format your results to match the reporting requirements for the project you are doing. Each project (and each central coordinating organization) has its own unique format requirements, which will be discussed in the following sections.

5.2   PROJECT L: ASTEROID ASTROMETRY

Astrometry of asteroids provides an excellent "learning project" to develop and demonstrate your skill at astrometry. Occasional position measurements of asteroids are needed to maintain precise orbits for these objects. The longer the time base over which astrometric data is taken, the more accurately the orbit is defined. Small changes in orbital elements over the years may betray evidence of close approaches between asteroids (rare, but not out of the question [3]), or of non-gravitational forces acting on these bodies. Astrometric measurements of newly-identified or poorly-observed near-Earth asteroids are often "time-critical" because the asteroid is only bright enough for accurate astrometry during a short interval when it is close to the Earth. The need to promptly determine precise orbital information for these NEAs makes the amateur's contribution of special importance to this activity.

The hard truth is that professional astronomers are not desperately awaiting your routine astrometric data on well-measured main-belt asteroids, but they can and will make good use of your data. A possibly harder truth is that if you are tantalized by the asteroid discovery project (described in Chapter 6), then the ability to make accurate astrometry is a skill that you absolutely must learn and demonstrate before your "discovery" reports can be taken seriously. (You don't want to be embarrassed by reporting the re-discovery of a well-known asteroid!). Amateur astronomers provide a great service to the professional community by conducting follow-up astrometry on suspected or newly-discovered near-Earth asteroids [4].

This project has four steps: Finding the asteroid, imaging it, determining its astrometry, and reporting your data. If you decide that asteroid astrometry is something that you'll do more than once or twice, you will also want to request a Minor Planet Center "observatory code".

5.2.1   Finding the asteroid

In order to prepare for your observing session, you should first "find" the asteroid in your planetarium program. The central source for asteroid orbital elements is the Minor Planet Center (physically located at the Smithsonian Astrophysical Observatory, and—perhaps more usefully for most of us—cyber-located at *http://cfa-www.harvard.edu/iau/mpc.html*). You can download the orbital elements of all 100,000+ asteroids in a file called "MPCORB" (about 20 MB). The MPC also offers the MPCORB data formatted for entry into most of the popular planetarium programs, such as TheSky and SkyMap Pro, etc.

You will want to have reasonably up-to-date orbital elements in your program, but the position uncertainty for most asteroids is much smaller than the typical amateur CCD FOV. For most purposes, updating your orbital element file once a year is more than adequate. However, if you're following recently-discovered near-Earth asteroids, then you'll need to download very current elements. For the very closest asteroid passes, you'll want to download daily updates of the orbital elements and ephemerides to take advantage of the most recent observations. For this sort of special case, you can download the current elements of a single asteroid and paste it into your planetarium program's database. This is much more efficient than downloading the entire MPCORB in order to add one more object to your file.

With the asteroid's elements in your planetarium program, you can accurately point your telescope to the correct location in the sky. Then, in order to locate the asteroid in your images, you take a pair of images about 15 minutes apart so that you can identify the asteroid by "blinking" the images (as described in Section 6.2.4). If the asteroid is a well-measured one (as most asteroids brighter than about magnitude 18 are), and if your planetarium program is a deep/complete one, then you may be able to simply identify the asteroid from its position relative to the background stars. Even if this is the case, it is usually a wise idea to do the "blink comparator" routine, to be sure of your identification (and to see if there are any unidentified "fast-movers" in the field).

5.2.2   Asteroid imaging routine

The Minor Planet Center has developed very specific requirements for your imaging routine. In order for your reports to be acceptable, you need to follow their rules. These are: (1) make two or three astrometric images of your target each night, separated by an hour or two; and (2) make astrometric images of the asteroid's position on two nearly-consecutive nights.

Strictly speaking, each "measurement" requires only a single image. However, I'm conservative and painfully experienced in many of the embarrassing things that can go wrong in any single image. So, my recommendation is to take three images per "measurement" (i.e., 3 images per measurement × 2 measurements per night × 2 consecutive nights = 12 images per asteroid). The redundancy gives you a way of comparing the internal consistency of your observations and reductions.

The "two nearly-consecutive nights" requirement is imposed by the Minor Planet Center in order to support their automated analysis of the consistency of your data. It is very difficult to judge the accuracy of single-night observations (spanning at most 6 hours or so). A two-night run provides a long enough baseline that the consistency of the results (i.e., the reasonableness of the orbital arc) can be determined. The nights need not be exactly consecutive, although that is the usual practice. If you've meas-ured some asteroids on night #1, but it's cloudy on night #2, then attempt to measure the same asteroids on night #3 so that you have data that can be linked when you submit it ...

I mentioned above that you want to use exposures that are long enough to get a good signal-to-noise ratio, but not so long as to saturate any of your astrometric reference stars. For most amateur set-ups, and most main-belt asteroids, exposures of one to a few minutes are appropriate. The typical main-belt asteroid moves across the sky with an angular rate of about 10 to 70 arc-sec per hour; hence it moves only a few arc-seconds in a three-minute exposure. This is small enough that the "trail" of the asteroid won't confuse your astrometry software's centroid algorithm. For practical purposes, the asteroid will still be very "star-like" in your image.

The typical angular rate of 70 arc-sec/hr for a main-belt asteroid also helps you understand the need for pretty accurate time-sync in your images. You are striving for astrometric accuracy of a fraction of an arc-second. The asteroid may move an arc-second in less than a minute. So, if you have a timing error of 60 seconds, that by itself will use up your entire error budget. As a practical matter, you should strive for timing accuracy that will make this error source negligible. At the start of each night, sync your PC's clock to WWV to within about ±0.5 second for main-belt asteroid observations. This isn't too hard to do. However, see Appendix A for a discussion of some potential pitfalls in regard to "PC time".

5.2.3 Special considerations for near-Earth asteroids

In the case of near-Earth asteroids, you may be faced with a more difficult tradeoff regarding exposure duration. These objects tend to be pretty small and, therefore, faint. They are bright enough to see and measure only when they're fairly close to Earth; but then their angular rates can be very, very high. I remember in May, 2001 watching (visually!) asteroid 66391 (1999 KW4) in my 6-inch Newtonian, at modest power with about a 0.5-degree field of view. The asteroid was about magnitude 10.5—easy to see—as it sailed past Earth at the quite-close distance of about 0.03 AU. Its angular rate was nearly 1 arc-min per minute. I watched it enter, sail across, and then exit my field of view in the course of about 20 minutes! So, it was easy to see this rock if you could find it, but it was tricky to find because it didn't stay in one place for long. Attempting CCD astrometry on such a fast-mover will drive you to pretty short exposures to keep the asteroid's trailing down to tolerable levels.

Using a short exposure may result in a low signal-to-noise ratio, obviously not good for your astrometric accuracy. Unfortunately, you can't increase the SNR_{peak} by taking a longer exposure, because the asteroid keeps moving on to the next pixel. So, the best you can do is strike a balance between exposure duration and trailing, and take many images so that most error sources can be "averaged out" in the data analysis.

The high angular rate of NEAs also puts additional emphasis on your time-sync accuracy. Think about that asteroid that was moving at 1 arc-min per minute. That's equivalent to 1 arc-sec/second. In order for your timing accuracy to be a negligible contributor to your astrometric error, your timing accuracy should be better than 0.1 second. It is possible to do that, but it isn't easy. So, don't try for a fast-moving NEA until you've succeeded in measuring some more-normal asteroids, and don't ignore the really significant challenge of achieving excellent time accuracy when you do go after an NEA.

5.2.4 Determining the astrometry

With good images, good time-synchronization, good astrometric software, and a good database of astrometric reference stars, the process of determining the target's position in each image is pretty much a "turn the crank" operation of your astrometric software. There isn't much that you can do to interrupt the calculations, and no good reason to attempt to do so.

After your results are computed, there are a few things that you'll want to check: What was the signal-to-noise ratio of your target image? If it is too low (less than $SNR_{peak} \approx 10$, or so), then look carefully at the consistency between your individual measurements before accepting them. Did your astrometric software report the "observed minus calculated" discrepancy, and is this discrepancy consistent from measurement to measurement? The "O – C" discrepancy is the distance between your measured position and that predicted by the orbital elements. Do not worry about the magnitude of O – C (unless it's really large—several arc-seconds on a low-number asteroid, for example). What's important is: Does O – C vary significantly from measurement to measurement? If it is approximately the same from measurement to measurement (within an arc-sec or so), then all measurements are consistent. If one measurement is an "outlier", then look closely at the image involved, and the results derived from that image, to see if you can discover why it is discordant (e.g., a cosmic-ray hit near the object image, or a faint asteroid landing on a dead pixel, or an interfering background star).

You may also want to plot RA–Dec vs. time, to confirm that all of your measured points from a two or three-night observing session lie on a smooth curve. For main-belt asteroids, the curve will be virtually indistinguishable from a straight line. For a near-Earth asteroid, there may be a slight curvature to the apparent path. The purpose of this plotting exercise isn't to determine the "true" path, it's to confirm that there aren't any discordant data points coming out of your measurements. The calculators at the Minor Planet Center will take care of determining the true

arc and updating the asteroid's orbital elements based on the new data that you
provide.

5.2.5   Reporting your data

The Minor Planet Center accepts astrometric reports by e-mail, in a standardized
format. Their required format is shown in Figure 5.3. All of the popular programs
that perform astrometry have built-in routines that will write your results into this
MPC format, so check your user's manual.

5.2.6   Applying for a Minor Planet Center "observatory code"

Fixed observatory locations that conduct astrometric observations of asteroids and
comets are assigned an "observatory code" by the Minor Planet Center. If you decide
that asteroid astrometry is your cup of tea, then you'll want to apply for your own
observatory code. The procedure is pretty simple. At your fixed-base observing site,
carefully determine the accurate geographic coordinates (latitude, longitude, elevation), to as good an accuracy as possible (within 100 feet or so, in each direction).
Second, make astrometric measurements on several asteroids over a two or three-
night observing run. Third, decide what name you want for your observatory. (You
can find the MPC's guidelines on observatory names, and the list of existing observatories, at the MPC website).

Submit your observations by e-mail, along with your site's accurate location
(latitude, longitude, and elevation), instrument description, observer contact information, and your desired observatory name. The MPC will check your observations

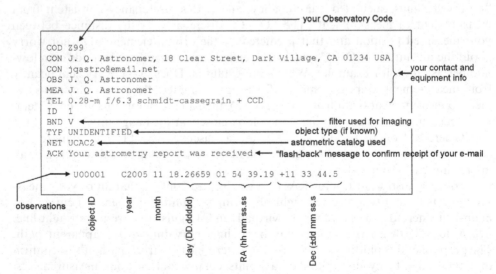

Figure 5.3. Minor Planet Center asteroid astrometry report e-mail format.

for accuracy and consistency. If they are of acceptable quality, then you will receive your observatory code by return e-mail (usually within a couple of weeks).

Once you have an observatory code, you use it for future submissions from the same location. If your observing site is an astronomy club observatory, you may still want to apply for your own observatory code, since your observing routine, reduction process, and instrumentation are likely to be different from those of other club members who use the same physical site. These differences justify treating each observing program as a separate "observatory code".

5.3 PROJECT M: SEPARATION AND POSITION ANGLE OF BINARY STARS

Binary stars provide important laboratories for determination of a variety of fundamental stellar properties. The widely-spaced "visual binaries" are particularly important because we can hope to observe their orbital motion, thereby measuring the star's paths around their common center of mass. That information can be used to determine the masses of the two stars. It is one of the very few unambiguous, direct ways of determining stellar mass. It thus provides an important anchor to sophisticated models of stellar spectroscopy and stellar evolution.

You may have seen diagrams showing well-determined orbital motion of binary stars. Figure 5.4 is a typical example. Considering the long history of binary-star observation, and the fundamental importance of determining their orbits, you may be surprised at how few binary star orbits are considered "well determined". Astronomy is often seen as a science filled with big numbers, including great mountains of data. I've see diagrams that show the red shift of nearly 1 million galaxies [5]. Over 100,000 binary stars are listed in the latest *Washington Double-Star Catalog*, but the current *Sixth Orbit Catalog of Binary Stars* contains fewer than 2,000 "well-determined" binary star orbits.

Even more surprising (to me, at least) is the list of "neglected" binaries. These are star systems with separation >3 arc-sec and brighter than 11th magnitude (hence most are within the range of a careful amateur CCD astronomer) that have not been measured for more than 20 years. The "neglected binaries" list amounted to over 6,400 star systems, as of this writing [6]. There is room for significant amateur astrometry contributions to this important field!

Your well-taken and processed CCD image of the field containing a double-star can be analyzed by your astrometry software to determine the separation and position angle of a binary pair. You use your processed image and software to determine the RA–Dec position of the primary star of the pair, and then repeat the process to determine the RA Dec position of the secondary star. These positions allow you (or the software) to determine the separation (ρ) and position angle (φ) of the secondary, relative to the primary.

Since many popular CCD image-processing programs will output the coordinates of the two stars, but will not calculate the separation and position angle, here are

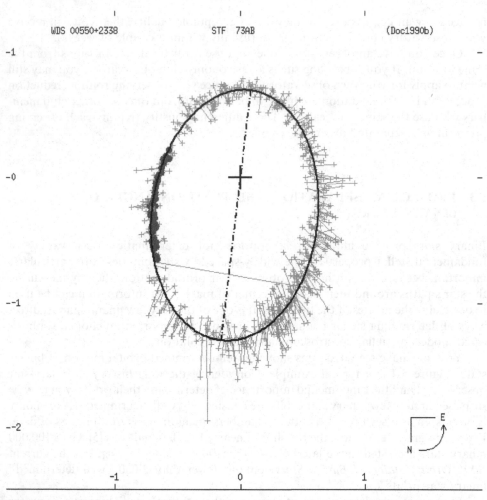

Figure 5.4. Example of a "well-determined" binary star orbit. (Used with the kind permission of the United States Naval Observatory)

the necessary equations [7]:

$$RA_P = \text{RA of primary star (in radians)}$$

$$D_P = \text{Declination of primary star (in radians)}$$

$$RA_S = \text{RA of secondary star (in radians)}$$

$$D_S = \text{Declination of secondary star (in radians)}$$

The separation is given by

$$\rho = \sqrt{(RA_P - RA_S)^2 \cdot \cos(D_P) + (D_P - D_S)^2}$$

and the position angle is calculated by

$$\varphi = \tan^{-1}\left[\frac{D_P - D_S}{(RA_P - RA_S) \cdot \cos(D_P)}\right]$$

Some astrometry software will give the separation and position angle directly, so that you don't have to do the calculations. Check the user's manual of the software that you use.

5.3.1   How close can you go?

There are a couple of special considerations that you'll run into when doing CCD astrometry of double stars: selecting the exposure, and determining your closest-measurable separation.

The problem of selecting the exposure to use is that you'll have to balance two competing criteria. On the one hand, you want to get as many astrometric reference stars as possible in the image, with as good an SNR as possible (to get the most accurate astrometric solution of your "plate constants"). That obviously encourages you to use reasonably long exposure times. On the other hand, you must keep the star images comfortably within the linear range of your CCD. Most of the stars in the *Washington Double Star Catalog* are pretty bright, by CCD imaging standards—mag 12 or so—and many are as bright as mag 8. That will force you to limit your exposure duration, to keep them comfortably in the linear range of your imager. The astrometric reference stars, on the other hand, are likely to be quite a bit fainter. So, you'll want to make a few test images of each pair on your observing list, to determine the best exposure to use. It is also wise to use the photographer's habit of "bracketing" exposures—take a few at the "best" exposure, plus a few at a shorter exposure and a few at a longer exposure. That way, you have some options when you're doing your data reduction.

There is a limit to how closely-spaced your double-star target can be: too close, and you won't get a meaningful measurement. Despite the fact that your astrometric software can determine the position of a single, isolated star to a fraction of a pixel, your double star's components must be separated by several times the FWHM in order for a meaningful separation measurement to be made. Figure 5.5 provides an idealized view of the situation. If the two stars are of equal brightness and are well-separated, then you'll have no trouble placing your measurement aperture over first one, and then the other. Your astrometry software will then give you the RA and Dec of each star individually, and you can calculate the separation and position angle using the equations above. (Or, if your software is set up for double-star measurements, it may give you the separation and position angle directly; you merely need to tell it which star is the "primary").

However, if the stars are too close together, then it isn't practical to make a measuring aperture so small that it doesn't encompass some light from the other star. As a result, when your astrometric software calculates the centroid of the light intensity inside the measuring aperture, the centroid is displaced toward the other star, and the calculation will indicate a separation that is smaller than the "true"

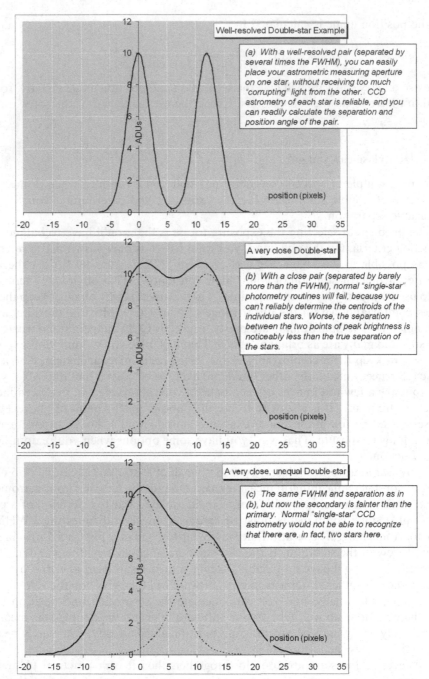

Figure 5.5. Idealized examples of the challenge of double-star CCD astrometry. In each graph, the intensity profiles of the individual stars are shown as dotted lines, and the combined intensity distribution of the pair is shown as a solid line.

separation between the two stars. The boundary between "wide enough" pairs and those that are "too close" depends on the brightness difference between the two stars. If they are of equal brightness, then you may be able to measure pairs that are separated by as little as two to three times the FWHM of your stellar images. If the stars are significantly different in brightness, then they will need to be more widely separated in order for you to reliably determine their separation.

5.3.2   Testing yourself

When you start out at double-star astrometry, you'll want to do a couple of experiments to learn about the limits of your equipment in this project. First, find a convenient field with plenty of widely-spaced stars (a sparse open cluster might do nicely), and take a series of short-exposure images (10 seconds or so). Use your image-processing software to do the usual dark-frame and flat-field reduction, and then determine your typical and "best" image FWHM. This may be noticeably smaller than the FWHM that you typically achieve in long-exposure deep-sky astro-images, since with these short exposures you probably won't have any guiding errors, and you can select the images taken in the moments of best seeing. The resulting FWHM gives you a starting point for selecting double stars that are within the capability of your equipment and your site. In general, your observing list should concentrate on pairs that are spaced by more than two to three times your FWHM.

The second experiment is to measure a few pairs that have well-known characteristics from recent measurements. A good source is the list of "calibration pairs" available at the WDS website. Select a few pairs ranging from "wide" (10 times your FWHM) to "close" separation (a few arc-seconds), whose components aren't too different in brightness (i.e., less than one or two magnitudes difference between the two components). Take several images of each of these pairs, and use your astrometric software to determine the separation and position angle of each pair. Then do two checks of the quality of your results. First, examine the results that you got for the widest pair. Since you took several images, you have several independent estimates of that pair. Do these estimates agree with one another, to within a fraction of an arc-second in separation, and within 5 degrees or less in position angle? They should. Then, compare your results with the ephemeris (if this is a "calibration" pair), or with recent measurements (if this is a star with a well-determined orbit). Again, your results should match the published characteristics to within an arc-second or less in separation, and a few degrees or less in position angle. In general, the wider the pair, the better the consistency, and the better the match to "known" parameters should be. A typical "wide-pair" result is shown in Figure 5.6.

Continue this examination of both the internal consistency and the match to known parameters for each pair in turn. You will probably recognize a point where your data becomes a poor match to the known parameters. That's the lower limit to separations that you can reliably measure. An example of this sort of study is illustrated in Figure 5.7. For my set-up, separations down to about 4 to 5 arc-sec are reliably measurable, if the stars aren't too different in brightness. Greater brightness

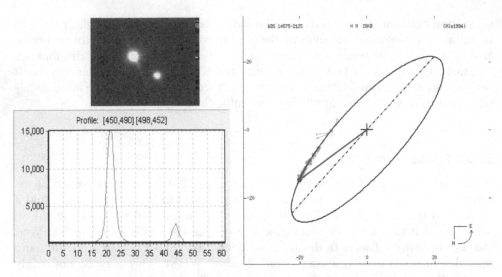

Figure 5.6. Pairs are excellent targets for double-star CCD astrometry if they are separated by several times the FWHM, and their magnitudes aren't too different. Here, the stars differ by about 2 magnitudes, but are separated by seven times the FWHM. (a) Detail of a 3-sec exposure CCD image of the wide-pair HD 131977 (WDS 14575–2125). It is simple to place measuring apertures around each star! (b) Intensity profile through the image. The star image's FWHM ≈ 3 pixels and the separation is ≈ 22 pixels. (c) The measured separation and position angle (from MPO Canopus) are consistent with recent published measurements. (Used with the kind permission of the United States Naval Observatory)

difference requires a larger separation for reliable measurement of the separation distance.

5.3.3   Selecting your target pairs

The best source of candidate pairs that are in need of fresh data is the *Washington Double-Star Catalog*, available on the web at *http://ad.usno.navy.mil/wds/* (Yes, that's a US Navy website; the catalog is maintained for the IAU by the US Naval Observatory.) Items that may be of use to the amateur researcher are highlighted in Figure 5.8. Spend an hour or so scanning through the available data, and then select some well-measured pairs that are within your capability in terms of magnitude, magnitude difference, and separation. Use these for your first projects. Then begin working your way through the "neglected pairs"—there are far more such pairs needing measurement than there are astronomers watching them!

5.3.4   Reporting your results

The publication that is dedicated to disseminating double-star measurements in the USA is *The Journal of Double Star Observations*, published by the University

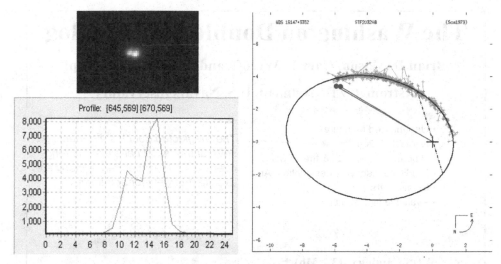

Figure 5.7. The close-pair HD 121325 (WDS 13550−0804) is just past the limit of my system's capability. (a) The star images are barely separated in this 3-sec CCD image, making it difficult to place a measuring aperture on one star that isn't "corrupted" by light from the other star. (b) The intensity profile through the stars shows a situation very similar to Figure 5.5(c): the stars are about 0.5 magnitude different in brightness, and are separated by only 1.75 FWHM. (c) Because the stars are so close, the resulting astrometry under-estimates their separation, compared with recent published measurements. (Two estimates, using slightly different conditions of measuring aperture are shown. If the result is critically dependent on the size/shape of the measuring aperture, then you have another cause for doubting the reliability of your measurement.) (Used with the kind permission of the United States Naval Observatory)

of South Alabama. Typical articles are descriptions and tabulations of quarterly or annual results from individual double-star observers. It also contains useful articles describing observing and measurement methods. The *JDSO* welcomes both professional and amateur contributions. You can download recent issues (no charge) from *http://www.jdso.org* Submissions to *JDSO* are reviewed by the experts at the US Naval Observatory for possible addition to the *WDS Catalog*.

Double-star measurements are a unique opportunity for a very modestly-equipped, but careful, amateur astronomer to make a very valuable astrometric contribution to stellar science. Do try a few double-star measurements, if only as a "full moon" diversion from your deep-sky imaging!

5.3.5 Other techniques

In this project I've described the simplest form of CCD double-star astrometry, because it uses equipment and software that many CCD astro-imagers already own. There are other techniques that can overcome some of the limitations of this "simplest" method. One is to use more advanced image-analysis software, such as IRAF, that can disentangle closely spaced pairs. Another is the time-honored filar

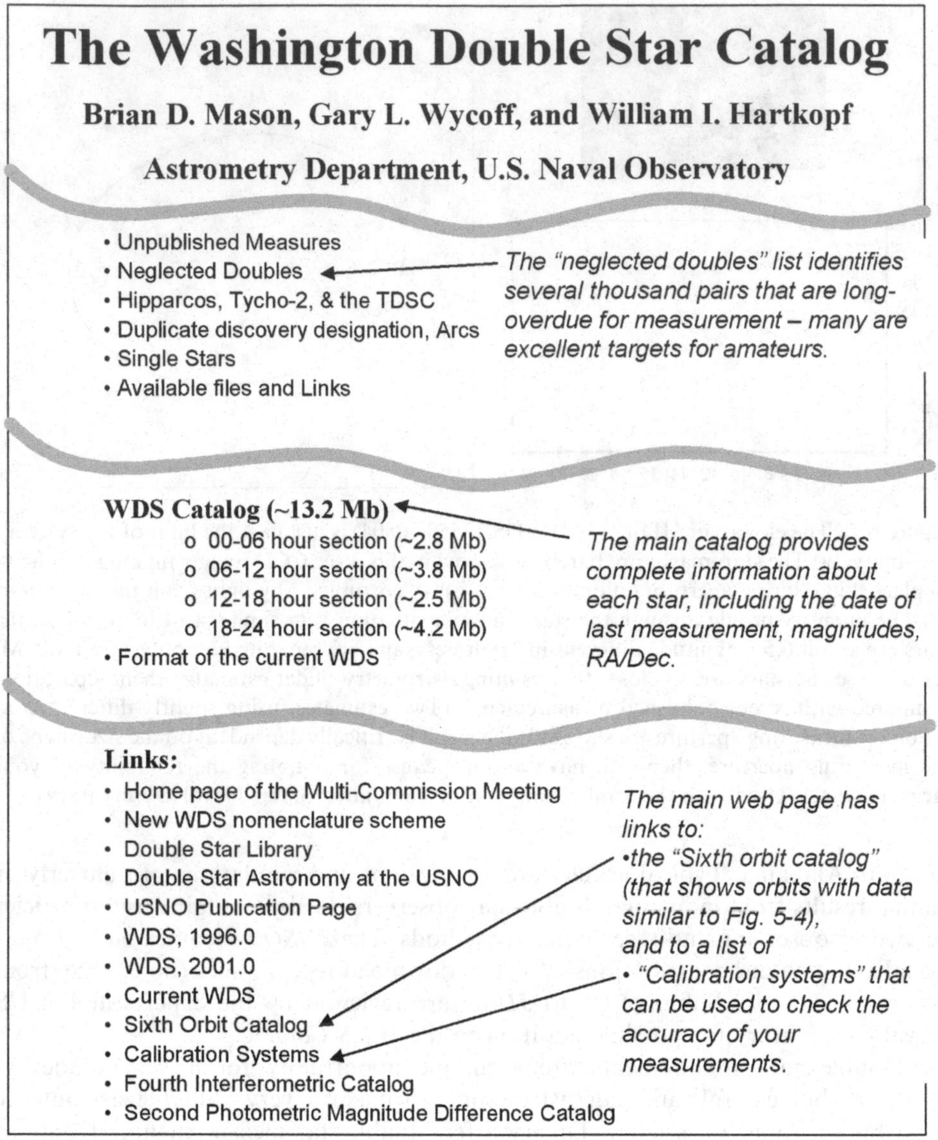

The Washington Double Star Catalog

Brian D. Mason, Gary L. Wycoff, and William I. Hartkopf

Astrometry Department, U.S. Naval Observatory

• Unpublished Measures
• Neglected Doubles ◄——— *The "neglected doubles" list identifies*
• Hipparcos, Tycho-2, & the TDSC   *several thousand pairs that are long-*
• Duplicate discovery designation, Arcs *overdue for measurement – many are*
• Single Stars   *excellent targets for amateurs.*
• Available files and Links

WDS Catalog (~13.2 Mb) ◄
   o 00-06 hour section (~2.8 Mb)   *The main catalog provides*
   o 06-12 hour section (~3.8 Mb)   *complete information about*
   o 12-18 hour section (~2.5 Mb)   *each star, including the date of*
   o 18-24 hour section (~4.2 Mb)   *last measurement, magnitudes,*
• Format of the current WDS   *RA/Dec.*

Links:
• Home page of the Multi-Commission Meeting   *The main web page has*
• New WDS nomenclature scheme   *links to:*
• Double Star Library   *•the "Sixth orbit catalog"*
• Double Star Astronomy at the USNO   *(that shows orbits with data*
• USNO Publication Page   *similar to Fig. 5-4),*
• WDS, 1996.0   *and to a list of*
• WDS, 2001.0   *• "Calibration systems" that*
• Current WDS   *can be used to check the*
• Sixth Orbit Catalog ◄   *accuracy of your*
• Calibration Systems ◄   *measurements.*
• Fourth Interferometric Catalog
• Second Photometric Magnitude Difference Catalog

Figure 5.8. The website of the *Washington Double Star Catalog* contains a wealth of information for double-star observers. It is located at *http://ad.usno.navy.mil/wds/* (Used with the kind permission of the United States Naval Observatory)

micrometer and visual measurements of separation and position angle. The filar (or bi-filar) micrometer isn't a common accessory in the amateur astronomer's arsenal, but if you catch the double-star bug you may want to invest in one, and add visual measurements to your list of projects.

5.4 RESOURCES

There are several valuable texts covering many details of double-star observing and methods of making the measurements. Two of the most respected are:

Argyle, R. (ed.), *Observing and Measuring Visual Double Stars*, Springer-Verlag (2004).
Jones, Kenneth Glyn (ed.), *The Webb Society Deep-Sky Observer's Handbook, Volume 1, Double Stars*, Enslow Publishers (1986).

5.5 REFERENCES

[1] Stafford, T., "Measuring the Distance to Teegarden's Star", *Sky & Telescope*, vol. 111, no. 2, p. 73 (February, 2006).
[2] Raab, H., "Detecting and Measuring Faint Point Sources with a CCD", in *Proceedings of the Meeting on Asteroid and Comets in Europe, 2003, May 1–4, 2003, Costitx, Majorca, Spain*.
[3] Galad, A., "Asteroid Candidates for Mass Determination", *Astronomy & Astrophysics*, vol. 370, p. 311 (2001).
[4] Stokes, G.H., Evans, J.B. and Larson, S.M., "Near-Earth Asteroid Search Programs", in Bottke, W., Cellino, A., Paolicchi, P., and Binzel, R. (eds.), *Asteroids III*, University of Arizona Press (2002), p. 45.
[5] Springel, V., Frenk, C.S. and White, S.D.M., "The Large-Scale Structure of the Universe", *Nature*, vol. 440, p. 1137 (April 27, 2006).
[6] Mason, B.D., Wycoff, G.L., Hartkopf, W.I., Douglass, G.G., and Worley, C.E., "The 2001 US Naval Observatory Double Star CD-ROM. 1. The Washington Double Star Catalog", *Astronomical Journal*, vol. 122, p. 3466 (December, 2001).
[7] Warner, B.D., "Observing Visual Double Stars with a CCD Camera at the Palmer Divide Observatory", in Warner, B. D., Foote, J., Mais, D., and Kenyon, D.A. (eds.), *Proceedings for the 25th Annual Conference of the Society for Astronomical Sciences* (2006).

6

Discovery projects

There are many interesting things in the universe that haven't been discovered yet, or which are so poorly known that each new observation offers the chance to discover more about them. Some of these discovery opportunities are well within the range of amateur equipment and amateur efforts to bring them to light. For some types of discoveries, amateur astronomers can effectively compete with the professionals because we have a lot more freedom to use our telescope time on projects that entail a lot of "just looking around". Imagine trying to present that idea to the scheduling committee of a major observatory—"I'm not sure if I'll find anything, but I'd like to schedule a month of nights to just look around in the sky ..."—not likely!

In a broad generalization, the discovery projects that amateurs have been successful with fall into two categories: "intentional searches" and "accidental or serendipitous discoveries". In terms of reducing and analyzing the observations, there is no difference between these two categories. Both types require a fair amount of diligence, and both types require that the observer treat his/her observations as "raw data" that needs to be well-documented, analyzed in a careful, consistent, quantitative way, subjected to independent confirmation, and reported appropriately. The difference lies in what you do at the telescope. If you're intentionally searching for supernova or asteroids, for example, you'll establish a search pattern and image sequence that maximizes your probability of success. On the other hand, you may be intending to make a beautiful portrait of some nebula, in which case you will frame your image and establish the imaging sequence primarily for that goal. But as a scientist, you won't stop there. In addition to preparing your image for artistic display, you should also take the trouble to examine it for "unusual" features (a moving object? an object whose brightness changes? an object that wasn't there last night, or last week?).

Amateurs can discover comets, asteroids, novae and supernovae, and variable stars. Amateur astronomers are conducting searches for asteroids in our own Solar System, nova in the Milky Way, and supernovae in distant galaxies. Some of these searchers have designed and built special telescopes tailored for the project, organized teams of observers and analysts, and logged an astonishing record of success.

The approach to all of these "discovery" projects is simple, in concept: you observe (and/or make an image of) a particular FOV, wait a while, and examine the field again. You then compare the two observations, looking for differences or changes.

Simple! But in order to increase your chances of success, you need to select a strategy that is matched to your equipment, and to your objectives:

- "take an image ..." What field of view should you use? What exposure should you use? Or, equivalently, what limiting magnitude are you striving for?
- "... of a particular FOV ..." Where should you look? Are there preferred parts of the sky, or preferred times of the year?
- "... wait a while ..." How long between images? What effect does the wait have on your probability of success, and the value of your discovery?
- "... take another image ..." One? Or a series? Should you return to the original FOV only once, or should the "take, wait, take" series be run several times, or as a continuous sequence for weeks or months or years?

The development of an appropriate search strategy depends very much on your intended quarry. It must also be adjusted to match the equipment that you have, or can reasonably acquire. Here are some projects, guidelines to get you started, and stories of true-life adventures.

6.1 PROJECT N: LUNAR METEOR IMPACTS

It is now well-accepted that virtually all of the craters on the Moon are the result of meteor impacts. It is also obvious to any dedicated star-gazer that the Earth is continually sweeping up meteors. Most are no larger than sand-grains, but occasionally a big, bright bolide reminds us that there are sizable chunks still sailing through near-Earth space.

So where are the observations of meteors hitting the Moon, and blasting out new craters? One possible observation dates from medieval times. On June 18, 1178, some English monks reported in the chronicle of Gervase of Canterbury: *"During a bright new Moon, the horns were tilted towards the east. Suddenly, the upper horn split in two. From the midpoint of the division, a flaming torch sprang up, spewing out fire and hot coals and sparks"* [1]. The veracity of this is hard to judge from so great a distance in time. It may have been a huge meteor impact on the Moon, but there are other plausible explanations, including over-active imaginations. In 1953 Dr. Leon Stuart made a photograph of the Moon that showed a small, bright spot near the lunar terminator [2]. Close examination argues that it wasn't an atmospheric meteor or a film flaw; there weren't satellites in Earth orbit back then, so it wasn't one of those; and the image gives evidence of being associated with the lunar surface. This flash clearly wasn't imaginary, but there is no consensus on just what it was.

Until the mid-1990s there was only one widely-accepted observation of an impact on the Moon, and that was not a meteor, but the Soviet rocket Lunik II. In that 1959 event, observers reported visually detecting the cloud caused by the impact [3]. Of course, they had the advantage of knowing exactly where and when to look, but nevertheless their success gave a tantalizing hint that lunar impacts should in principle be detectable. Later, in the 1990s the Japanese spacecraft Hiten was crashed into the Moon, and several teams reported observations of that impact [4]. A meteor impact on the Moon could be expected to be more visible, since it should liberate a lot more energy, considering the relatively slow impact velocity of the spacecraft compared with the enormous speed of a typical meteor.

Again, where are the observations? Aside from a smattering of single-observer "lunar transient phenomena", nothing really compelling turned up until the Leonid meteor storm of 1999. Dr. David Dunham (a man with many hats, including leadership of the IOTA and expert in spacecraft navigation at Johns Hopkins University) pointed out that the situation of that meteor shower was especially favorable for a project to search for lunar impacts. The Leonid storm promised a very high density of meteors. The velocity vector of the meteors was such that they would impact on the dark portion of the Moon, and the dark portion of the gibbous moon would be nicely placed for observing for several hours in the evening. He conducted continuous video observation of the lunar surface. The net result of several hours of observation and video recording was detection of 11 very brief point-like flashes on the surface of the Moon. The characteristics of the flashes matched the expected optical signature of a modest-size meteor impact. Seven of these flashes were simultaneously observed by more than one station, which pretty well ruled out such explanations as satellite glints, aircraft, meteor flashes in the Earth's atmosphere, and "noise" in the video cameras.

Even better, the equipment that Dr. Dunham and his partners used was not particularly sophisticated: it was the same sort of small telescope and video camera set-up that many observers use for monitoring lunar and asteroid occultations.

The Leonid meteor shower was selected for the initial test because the predicted density of the meteor stream maximized the odds of some sort of impact occurring. However, more-regular monitoring of the lunar dark-side surface may very well be a useful addition to meteor-shower planning. On November 7, 2005, during the (low-rate) Taurid meteor shower, scientists at NASA-Marshall Space Flight Center (MSFC) observed a very similar flash, with similar equipment [5]. Their assessment was that they most likely recorded the impact of a meteor about 8 inches in diameter.

A sustained effort to observe additional lunar impacts is beginning at NASA-MSFC [6], and coordinated observations by amateur astronomers will be of great value for confirming mutual observations, increasing the amount of time during which the lunar surface is under scrutiny, and helping to accurately locate the impacts. The organized observation of the lunar surface, with all observers using consistent techniques and all results (positive and negative) entered into a single database, will be a valuable asset to astronomers who study meteor impacts. This project can contribute data that will help answer such questions as:

- "How common are visible impacts?" (The frequency of impacts might correlate with meteor showers, or it might not.)
- "What is the distribution of brightness of the impact flashes?" (Their brightness probably tells us something about the size and/or energy of the meteor.)
- And "What is the distribution of the duration of the flashes?" (Most observed flashes have been very brief, but a few reports suggest that quite long flashes—up to several tenths of a second—may occasionally occur.)

This is a team project. These days, there are so many things flying around, from aircraft to low-Earth-orbit satellites and debris, to geosynchronous satellites, and (occasionally) spacecraft in lunar orbit, that you'll want confirmation of any suspected meteor impact events to separate them from the surprisingly frequent situations where a satellite passes between you and the Moon. Because of the low probability of observing a genuine lunar impact, and the many possible types of "false alarms", observations by independent observers must match quite closely in time and location on the Moon in order to provide valid confirmation. Two observers separated by 30 miles who detect similar flashes at the same time, can pretty well rule out most types of false alarms and spurious flashes (such as video noise, cosmic rays, satellite glints, or "point" meteors).

In order to assist observers by correlating observations, the Association of Lunar and Planetary Observers (ALPO) has taken on the role of being the central coordinating organization for impact observations, through their Lunar Meteorite Impact Search project There are two ways to participate in this project. One is to organize a team of observers in your local club. That way you'll be able to ensure that at least two observers are monitoring the Moon at selected times. Alternatively, you can be a "lone observer", and rely on the ALPO to confirm your observations by matching them with those of other observers. The drawback of the "lone observer" approach is that it may turn out that you were, in fact, the only person monitoring the Moon at the time of the event. In that case, your observation will forever be a "suspected" rather than a "confirmed" impact.

6.1.1   Equipment needed

Theoretically, the necessary observations can be made visually. In fact, at least one of the original 1999 Leonid flashes was seen by a visual observer, confirming the video record. As a practical matter, however, if the visual observation were the only detection, I doubt that anyone would have placed high confidence in its reality. There are too many ways that an observer could, through fatigue or other accidental effect, make a "false" observation. This sort of extended, continuous, low-probability-of-event monitoring cries out for video recording, which is immune to fatigue, allows for re-play during analysis of the observations, and provides a permanent record of the event.

The required equipment for this project is almost identical to the set-up that you would use for video monitoring of lunar and asteroid occultations as described in Chapter 2:

- Telescope.
- Video-recording system.
- WWV receiver (or other accurate time reference).
- GPS receiver (or other method of accurately determining your location).
- Planetarium program.

Not required, but nice to have, is a PC with a video-capture board and video-processing software to use during image analysis. And, if you already have a time-code-insertion generator that writes precise time onto the video image, by all means take advantage of it!

The telescope to use for this project must be driven (preferably at "lunar" rate), and an equatorial mount is preferred, so that the image won't rotate during the observing session. A "Go-To" mount is a definite advantage for the step of recording "comparison stars". Any type of telescope is acceptable, but since you'll be monitoring the "dark" portion of the Moon, you'll want to beware of the risk of glare from the brightly-lit lunar crescent. A well-baffled refractor is probably less susceptible to glare than a reflector or Schmidt–Cassegrain type, but use of a refractor is certainly not a firm requirement. There are plenty of confirmed or suspected impact flashes in the ALPO database that were detected using Newtonian and SCT telescopes.

The video-recording system will consist of a video camera, an adapter that allows you to mate the camera to your telescope's focuser, a video recorder (preferably with a sound track), and a video monitor. A camcorder provides the camera and recorder in a single package, but it may be too heavy to mount onto your telescope, and its monitor may be too small for accurate focusing and tracking. A TV–VCR combo is a convenient package for both recording and viewing the imagery, but of course you need a stand-alone video camera to feed it. Low-cost, high-sensitivity video cameras are readily available. These have been well-used for lunar and asteroid occultation observations and are perfectly appropriate to use for lunar-impact searching.

For lunar-impact recording you'll want to monitor the entire "dark" portion of the Moon, since the impact may happen anywhere at any time. You'll also want to be able to see some lunar features in the video, so that the location of any impacts can be determined. That means that you'll want to keep the sunlit portions of the lunar surface out of the field of view, because the bright glare will tend to saturate the video camera, washing out the details on the lunar surface. The camera you select should give you the option to disable the Automatic Gain Control (AGC). That way, you won't have to worry about the AGC being driven into wild fluctuations if a portion of the sunlit crescent enters the image.

A field of view of about 0.5 degrees is a good choice, allowing you to image the entire earthshine-illuminated portion of the Moon, while keeping the sunlight-illuminated crescent out of the image. Assuming a $\frac{1}{3}$ inch chip in the video camera (e.g., Supercircuits PC-23C), this is achieved with a 4-inch F/10 telescope. (By the way, this is almost exactly the type of set-up that was used for the original Leonid 1999 discoveries.) If you use a larger-aperture, longer-focal-length telescope, then a focal reducer can be used to widen the field of view while preserving sensitivity.

Either an analog (tape) or a digital video recorder can provide good results. Some users report that the quality of the data is better with a digital recorder; and there is some advantage to putting the data into digital format for analysis and archiving.

6.1.2   Planning your observations

There are three situations that provide appropriate nights for conducting this project. These are:

- Near a meteor-shower peak (i.e., within a couple of nights of the maximum zenith hourly rate).
- Evenings when the waxing crescent Moon is visible for more than an hour or so (i.e., between new moon and first quarter).
- And pre-dawn hours when the waning crescent Moon is visible for more than an hour or so (i.e., between third quarter and new moon).

It seems logical to select meteor-shower nights for monitoring the lunar surface for impact flashes, and that is indeed when most monitoring has occurred. However, there are good reasons for not restricting yourself to "meteor-shower" nights. The seismic stations that were installed on the lunar surface by Apollo astronauts recorded about 1,700 meteor impacts during the $5\frac{1}{2}$ years that the network operated [7]. Overall, the average rate of detected impacts was a little less than 1 impact per day. Curiously, this seismic record of meteor impacts—which was only sensitive to large events, from meteorites weighing more than about 100 grams—does not show particularly pronounced peaks in activity during meteor showers. The more consistent trend is that the number of seismic signals tended to be greater in the period April through July than it was at other times of the year [8].

Therefore, in addition to making a special effort during meteor-shower nights, it is a worthwhile investment of your observing time to monitor the Moon's earthshine-illuminated portion every month, both on the evenings between new moon and first quarter, and on the early mornings between third quarter and new moon. In general, you'll want to monitor when the Moon is higher than about 20 degrees elevation. The rationale for this rule is that one of the pieces of data you will try to gather is the approximate brightness of the flash. Below about 20 degrees elevation, atmospheric extinction is quite significant at most sites, and may throw your brightness estimates off. (Refer back to Section 4.7.2 for a discussion of atmospheric extinction.) If you're blessed with a particularly crystal-clear night, good seeing, and/or an excellent site, then you can stretch this rule and go as low as 15 or even 10 degrees above the horizon.

Not every meteor shower presents a good opportunity to monitor for lunar meteor impacts. A particular set of geometric criteria must be met in order for shower meteors to be visible. First, the flashes will only be visible against the "dark" portion of the lunar disk. This means that the Moon must be in a position such that the dark side faces toward the radiant. Second, the portion of the dark side that is accessible to meteorite impacts must also be visible from Earth.

The calculations that determine if a particular meteor-shower night is attractive for lunar-impact monitoring are pretty complex. Happily, you won't have to do them. The ALPO lunar-impact website includes plots that show you the impact circumstances for the annual meteor showers. Therefore, your first stop in planning an observing campaign should be the ALPO lunar-impact website at *http:// www.lpl.arizona.edu/~rhill/alpo/lunarstuff/lunimpacts.html* Figure 6.1 is an example of the type of plot that you'll find there. This particular example is for the Geminid meteor shower of 2006. This plot contains several pieces of useful information. First, the calculations were done for the predicted time of maximum ZHR on Earth (in the case of this example, that is December 14, 2006 at 05:34 UT). Second, the peak impact rate on the Moon will occur when the Moon (not the Earth) passes through the densest part of the meteor stream. This will normally be a different time than the time of peak ZHR on Earth. The caption gives the time difference. A negative number means that peak rate on the Moon occurs *before* peak ZHR on Earth; and a positive number means that peak impact rate on the Moon is predicted to occur *after* the peak

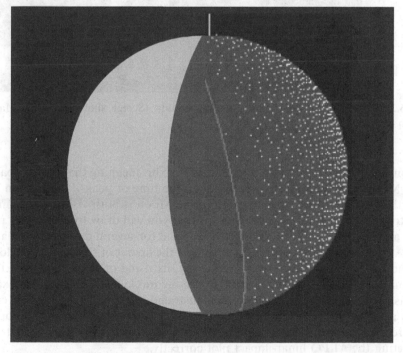

GEM Dec 14, 2006 05:34 UT -0.5 hrs Moon rises 4.8 hrs before sunrise
zhr = 120   42% impacts on unlit near side w/polar graze = 11 deg

Figure 6.1. Example of a "lunar-impact circumstances" plot from the ALPO website. (Used with the kind permission of the Association of Lunar and Planetary Observers)

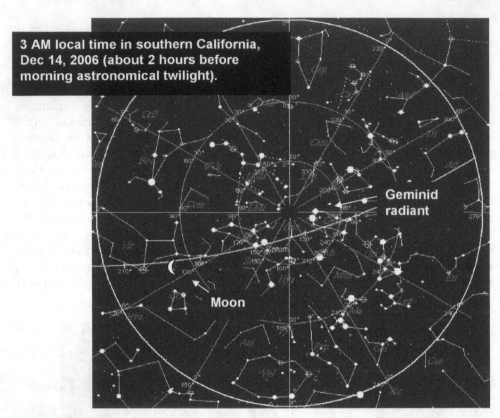

3 AM local time in southern California, Dec 14, 2006 (about 2 hours before morning astronomical twilight).

Geminid radiant

Moon

Figure 6.2. Situation in the sky during 2006 Geminids. (Screen shot from Chris Marriott's SkyMap Pro)

ZHR on Earth. In this example, the value is -0.5 hr, meaning that peak-impact rate on the Moon occurs a half-hour *earlier* than the time of peak-ZHR rate on Earth. Third, the graphic shows which portion of the Moon is both visible to us, and also facing toward the radiant. The conclusions that you will draw from this plot regarding the area of the Moon being impacted are valid for several days around that time.

Use your planetarium program to examine the layout of the sky at your location and time zone: the Moon's position in the sky, its rising or setting time, the time of astronomical twilight, and the hours that are favorable for observing. An example, again using the 2006 Geminid shower, is illustrated in Figure 6.2. Desirable circumstances are that the sky be dark (aside from the Moon), and that the Moon be at 20 degrees or greater elevation angle. Your all-sky plot will confirm that you're interpreting the ALPO lunar-impact plot correctly.

Your planetarium program will also help you identify a few fields where you can record "comparison stars". These will be used to estimate the brightness of any impact flashes that you do observe. These comparison stars should not be too far from the Moon, and should be positioned so that you can record them at about the same elevation angle that you recorded the Moon. They should range from about 4th to 9th magnitude. Depending on the situation, one field might suffice, or you may have to locate and record several fields to get a good range of comparison star brightness. Since it is reasonable to guess that lunar-impact flashes are relatively bright in the infrared, try to avoid using blue stars as your comparison stars. The accuracy goal in estimating the brightness of a lunar-impact flash is modest—about half a magnitude—so these "comp stars" do not need to be photometric standard stars. Any convenient fields will do. Note that in order for these comparison-star brightness measurements to be meaningful, it is essential that your camera's AGC be "off" during both the lunar-impact and comp-star monitoring!

Using these two plots, you can see that observations for this example must be scheduled for the pre-dawn hours, since this is a waning moon that rises a few hours before sunrise. The constellation of Gemini (the radiant) is just west of the zenith, and the Moon is rising above the southeastern horizon, so the meteors will be generally coming at the Moon from the northwest, hitting its dark side. This particular example looks like a grand opportunity to monitor for lunar meteor impacts: you have a couple of hours when the Moon is well above the horizon, the dark side is facing into the meteor stream, and roughly half of the meteors that do hit the Moon will hit on the side that is visible to us.

Finally, beware that there are any number of things that can go wrong in a telescope–video effort. You will definitely want to set up and use your system on a non-meteor-shower night, to be sure that you know where everything goes, and become familiar with its operation. If possible, record the lunar surface a month prior to the scheduled meteor-impact observations (when lunar conditions are nearly the same as will be encountered on impact night). You will also want to do some experiments to learn how to adjust your camera's level, gain, and contrast for best visibility of the dark lunar surface. These experiments will also show you just how far the bright portions of the Moon can be allowed to intrude into the image without adversely affecting the results. The key objective of these experiments is to confirm that the recorded video can be played back and examined satisfactorily.

6.1.3 Conducting the observations

Do not underestimate the delicacy of what you are attempting to accomplish in this project. The typical impact flash will span only three or four video fields (i.e., barely a 15th of a second), it may appear anywhere on the lunar surface, and at its peak it isn't likely to be brighter than magnitude 4 or 5. Such an event will only be visible against the "dark" (earthshine-illuminated) portion of the lunar surface. In order to max-imize your probability of recording an impact flash, the dark portion of the Moon should fill most of your imager's frame. Your video imager's brightness/contrast should be set so that lunar features are visible, to enable you to determine the location of any suspected flashes. The sunlit portion of the Moon must be kept out of the frame, to avoid the glare that can easily wash out your view of the earthshine-illuminated portions of the lunar surface.

The requirement to keep the brightly-lit portions of the Moon out of the field of view (or at most, just barely in one edge of the FOV) means that you'll need to be careful with tracking, and monitor the position of the Moon's image in the video field of view regularly throughout the recording session. Many commercial mounts offer a "lunar tracking rate" in addition to the more-standard sidereal rate. This will help to keep the Moon in the FOV with minimum operator intervention. Check your user's manual and control instructions.

Because it is critically important that apparent impacts be correlated between two or more observers, accurate time ticks must be included on the video record. As discussed in Chapter 2, there are two convenient ways to do this. One is to play a WWV receiver's signal into the video's audio track. The other is to directly inject a time-code onto the video. The "audio track" approach is much less expensive—free if you already have a WWV receiver. A video time-code generator is likely to cost a few hundred dollars, but it makes data reduction to identify the exact time of any event much easier.

The observations are gathered in two steps: monitoring the lunar surface, and recording fields with "comparison stars" that will be used for estimating the magnitude of any flashes that are detected. If you're monitoring the waxing Moon (i.e., evening observations), then you'll take the comp-star fields after you've completed your lunar monitoring. If you're observing in the pre-dawn hours (waning Moon), then you should record the comparison-star fields before you begin monitoring the Moon (so that your comp-star record isn't affected by morning twilight). In either case, the comp stars should be at similar range of elevation angles as the moon was when it was being monitored, so that atmospheric extinction effects are similar.

With your video camera and telescope set up, the procedure for gathering data is pretty simple. Start your camera and video recorder, adjust the volume on your WWV receiver so that its tones are recorded on the audio track of your videotape, and make a few introductory comments on the audio track (date, time, location, sky conditions, and the equipment you are using). Point your telescope so that the Moon (including the sunlit portion) is captured on the video record, then slowly pan until the sunlit portion is moved to just outside the field of view. Confirm that you can recognize a few features on the earthshine-illuminated lunar surface, and that the dark limb is visible. Then, keep the recorder running until the Moon is no longer favorably placed. If you're observing in the evening (between new moon and first quarter), you'll stop monitoring the Moon when it drops below about 20 degrees elevation. If you're observing in the pre-dawn hours (such as the 2006 Geminid example), you'll stop when twilight interferes with observing the dark lunar surface. Before you leave the Moon, do a slow pan to bring the sunlit portion back into the image (as a reference to the location of the terminator).

Without changing anything in your imaging set-up nor the camera's AGC (which should still be "off"), slew to the fields of comparison stars that you have selected. It's a good idea to use the audio channel as well as your notebook to record the identity of each field of comp stars. Keep the WWV radio singing into the audio track during this step, also. Take 15–30 seconds of imagery of each field of comparison stars.

While your video camera and recorder are monitoring the Moon, you should also be visually monitoring the lunar surface, either with a separate telescope, or by watching the video record as it is recorded. The idea here is that if you have the opportunity to visually monitor the Moon, then you may be able to see an impact flash, and note the time. You can then pay particular attention to that portion of the video record, and you have a weak form of confirmation if your visual observation is borne out by the video record. Alternatively, if you have a single telescope, then attentively watch the video monitor as the data is being recorded. You're going to have to watch the tape several times during your data analysis, and "real time" is a good time to make your first pass. If you suspect a flash, note the time and then you can carefully review that portion of the tape during your data analysis.

6.1.4   Analyzing, reducing, and reporting your results

Let's suppose that you and a partner have recorded several hours of lunar video from different locations. How do you reduce it? Slowly, meticulously, and (on the first pass) independently. You are looking for a ~mag-5 "flash" that covers only a few pixels, and which lasts only a few video frames. Played back at "normal" speed, this flash may be very hard to detect. I have watched videos of the 1999 Leonid flashes, and even after being told in advance where in the image the flash would occur, and at what time, I missed them on the first showing. A casual viewer is not likely to notice these subtle but important flashes! Plan on going through your videos several times at "normal" speed. This can be done visually, with patience and attention. That is the approach that most observers use. If you have the option of playing the record back at a slow speed, your ability to detect the flash should improve.

Some observers are experimenting with frame grabbers and digital image-processing methods to pre-screen the video, (hopefully) simplify the detection process, and (again hopefully) improve the odds of detecting flashes that are embedded in the video. If these developments are successful, you'll see reports on the ALPO website. If you are experienced with this technology, you should definitely consider bringing it to bear on your video data.

Each candidate flash must be examined carefully to see if it betrays the distinguishing marks of a false alarm vs. the characteristics of a genuine "impact flash". The most common false alarms are cosmic-ray point flashes on the video sensor. These tend to be single-frame events, and may be either single-pixel or oddly-shaped extended artifacts in the image. A good candidate impact flash will last two or more video fields (i.e., it's not a single-video-field event), will show a star-like point spread function (i.e., it's not a single-pixel event, and it's not the sort of odd-shaped thing that you commonly record from cosmic-ray hits). For each candidate event, record the time, duration, and location on the Moon. Use the images of your "comparison stars" to estimate the brightness of the flash (in its brightest frame), and record its brightness along with any other distinguishing features.

After each observer has examined his or her tape, they should compare the times of any detected flashes, and their location on the Moon. Flashes that appear on one record but not the other should be examined carefully on both video records. It may be that one observer simply missed it during the initial examination of the video. A flash that is found at the same time, and appears in the same location on the lunar surface, on both records, is a candidate lunar impact. The criteria for confirmation are quite stringent. The times of the two detections must match within ±2 seconds, and their locations on the Moon must match within 2 degrees in latitude and longitude in order to be considered a "confirmed" detection.

Both "lone observers" and local teams should submit the complete record of their observations to the ALPO, so that observations from many sites can be combined, compared, and permanently recorded. The ALPO Lunar Meteorite Impact Search project is coordinated by Mr. Brian Cudnik. The project's website is located at: *http://www.lpl.arizona.edu/~rhill/alpo/lunarstuff/lunimpacts.html*

Your report to the ALPO coordinator should consist of two parts: a data sheet and a lunar map. The format for the data sheet is shown in Figure 6.3. It is self-explanatory. You must report who you are, where you observe from, your observing method ("video recording"), the times during which you monitored the Moon's surface, sky conditions (seeing, transparency, and clouds). For the video-recording method described here, you do not need to worry about "reaction time". Then, list all the events you observed. For each event, include the UT date and time, duration and magnitude of the event, and any other comments or unique features of the observation. (If you're not familiar with Universal Time, see Appendix A.)

The second part of your report is a map of the lunar surface, showing the location of the terminator, and of your observed flashes. An example of this is shown in Figure 6.4. Note that it is critical that you accurately cross-reference each "flash location" on your map with the corresponding "event #" on your data table.

If you have organized a local team, each observer's data should be separately reported to the ALPO, so that their database is kept accurate. For flashes that were confirmed by another member of your "local team", note the confirmation in the "comments" column of your report.

Because part of the value of this project consists of gathering data that will give a better understanding of the statistics of lunar impacts and the associated flashes, it is important that you record in your notebook, and report to the ALPO, *all* of your observing sessions, even if some of them do not yield any observed or suspected flashes. This way, the approximate rate of flashes can be determined. If everyone were to report only their "successful" observations, then this important rate information would be lost.

As in other projects, be sure to archive the tapes (or digital copies) that show impact flashes—they may be requested by astronomers doing further analysis of the events.

6.1.5 ALPO

The Association of Lunar and Planetary Observers has a much broader scope than this lunar-meteor-impact project. The ALPO is the central organization for a wide array of amateur observing projects, covering just about everything in the Solar System. ALPO members conduct both visual and CCD monitoring of activity on

```
┌─────────────────────────────────────────────────────────────────────┐
│                  A.L.P.O. Lunar Meteoritic Impact Search              │
│              Observation Report Form #1 (Revised 21 May 2004)         │
│                                                                       │
│   Observer Name & Address: _____   │
│   _____   │
│   _____   │
│   Location of Observation:_____   │
│   Long:_____(E or W) Lat:_____(N or S) Elev. (if known).:_____(ft or m) │
│                                                                       │
│   UT Date:_____ UT Start:_____ UT End:_____ │
│   UT Date:_____ UT Start:_____ UT End:_____ │
│   UT Date:_____ UT Start:_____ UT End:_____ │
│   UT Date:_____ UT Start:_____ UT End:_____ │
│                                                                       │
│   Instrument(s): _____   │
│                                                                       │
│   Observing method(s) used (video, visual with tape recorder & WWV, other): │
│   _____   │
│   _____   │
│   _____   │
│                                                                       │
│   Seeing (1-10):_____ Transparency (1-6):_____                │
│                                                                       │
│   Clouds (none, haze, very few, scattered, variable, broken, overcast): │
│   _____   │
│                                                                       │
│   Was reaction figured into the time of event? (y or n) _____         │
└─────────────────────────────────────────────────────────────────────┘
```

A.L.P.O. Lunar Meteoritic Impact Search
Observation Report Form #1 (Revised 21 May 2004)

Observer Name & Address: _____

Location of Observation:_____

Long:_____(E or W) Lat:_____(N or S) Elev. (if known).:_____(ft or m)

UT Date:_____ UT Start:_____ UT End:_____
UT Date:_____ UT Start:_____ UT End:_____
UT Date:_____ UT Start:_____ UT End:_____
UT Date:_____ UT Start:_____ UT End:_____

Instrument(s): _____

Observing method(s) used (video, visual with tape recorder & WWV, other):

Seeing (1-10):_____ Transparency (1-6):_____

Clouds (none, haze, very few, scattered, variable, broken, overcast):

Was reaction figured into the time of event? (y or n) _____

List event(s) observed and / or suspected. Use the Observation Report Form #2 to pinpoint where on the moon the event(s) was (were) seen. Make sure to cross-reference these events (#1, #2, etc.) Include the following: Date & Time (UT, to the nearest second), Confidence (%), Comments (estimated magnitude, apparent color, duration of event, different method used, etc.). Continue on back if necessary.

event #	UT date	UT time	duration (sec)	magnitude	comments
1					
2					
3					

Figure 6.3. ALPO Lunar Impact reporting form. (Used with the kind permission of the Association of Lunar and Planetary Observers)

the planets (especially Mars, Jupiter, and Saturn), so that professional astronomers (and in some cases, spacecraft) can be alerted to unusual events that are worthy of closer study. Every amateur researcher who is involved in Solar System observations and projects should be a member of the ALPO. Membership information, and a description of the many activities that the ALPO members are involved in, is available at: *http://www.lpl.arizona.edu/alpo/*

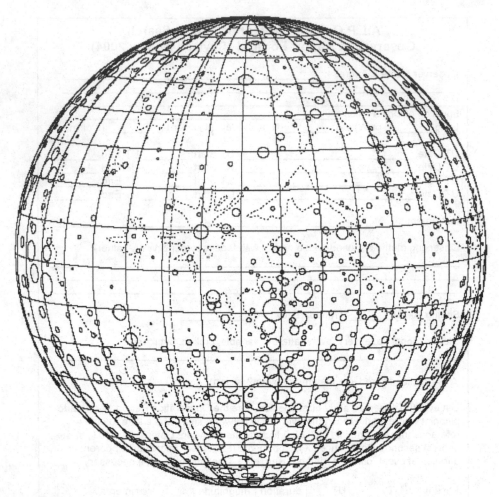

Figure 6.4. ALPO lunar map for identifying the location of observed impact flashes. (Used with the kind permission of the Association of Lunar and Planetary Observers)

6.2   PROJECT O: ASTEROID HUNTING AND DISCOVERY

There was a time, back at the dawn of the amateur CCD era (the mid-1990s), when it was surprisingly common for amateur astronomers to discover new asteroids. Dedicated asteroid hunters with modest-size telescopes could be reasonably confident of "bagging" one or two new discoveries per year, and of earning the right to name their discoveries. This project has gotten quite a bit harder since then, for several reasons. First, as more and more asteroids were found, only the fainter ones remained awaiting discovery. In the late 1990s the typical amateur discovery was about mag 16–17. Today, the typical "new asteroid" is about magnitude 19 at discovery (essentially all of the brighter ones having already been found). Second, the concern over potentially-hazardous near-Earth objects has prompted the construction and operation of several professional surveys. These feature large-aperture telescopes, sensitive, wide field of view CCDs, and all-night every-night automated operation. They are formidable competitors to the amateur star-gazer!

Nevertheless, there is still potential for amateur discoveries. A corny proverb says that "when the elephants get up to dance, the mice scurry to the corners of the room." In this context, we amateurs are the mice, and the professional surveys are the elephants. Don't squander your energy by going head-to-head against LONEOS or the Sloan Digital Sky Survey! A central element of asteroid-discovery success today is in planning your observing strategy—deciding *where* and *when* to search, and *how* to search. This guidance comes from the experience of Andrew Lowe [9], who has honed it to a fine art, and applied it to achieve over 100 credited asteroid discoveries in the course of 2004–2005. As of March, 2006, 80 of his discoveries are multi-opposition objects, and 15 have been numbered. So, if you have a burning desire to discover (and name) your own asteroid, take heart, and refine your strategy.

The theory of asteroid detection is to establish a search pattern of six or nine images, with slight overlap between the images, such as that illustrated in Figure 6.5.

Your choice of a six-FOV or a nine-FOV (or larger) search pattern will depend to some extent on your chosen image exposure, and the accuracy of your mount. The idea is to have a search pattern that you can complete in about 60 to 90 minutes.

Take your images (or series of sub-exposures) for each FOV #1 through #9, in sequence. Call this "sequence A". When this "sequence A" is complete, about an

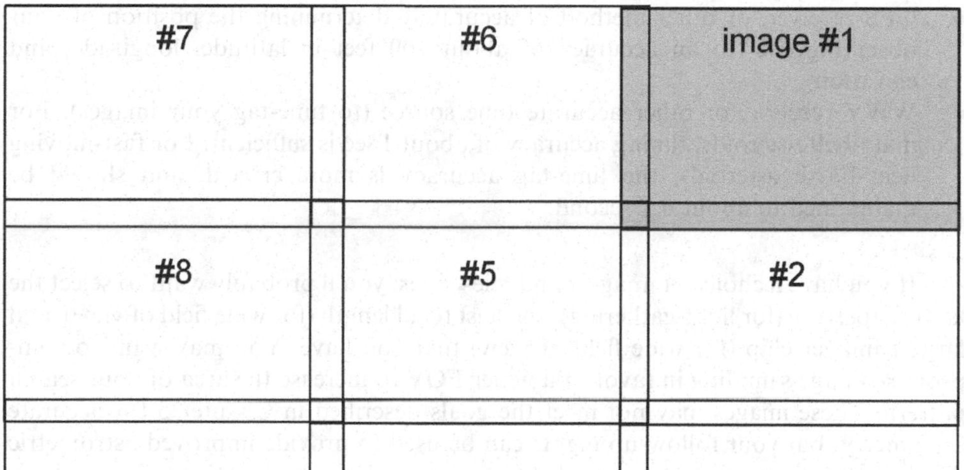

#7	#6	image #1
#8	#5	#2
#9	#4	#3

Figure 6.5. Typical CCD search pattern.

hour after you started it, you'll begin a second series of images of these same FOVs, #1 through #9. Call this "sequence B". Finally, take a third set of images of FOVs #1 through #9 ("sequence C"). You now have three images of each FOV, spaced about an hour apart. If you have sufficient energy, and a sufficiently long night, you might establish a second (or even a third) search pattern on which to collect images.

The exact orientation of the images (relative to RA/Dec directions) isn't particularly important. However, there is some (modest) benefit to aligning the long axis of your FOVs, and of your search pattern, in the direction parallel to the ecliptic (which is approximately the direction in which the asteroids will move).

6.2.1   Equipment needed

This is a CCD-imaging project, for which your standard CCD set-up will be used. It requires the same assets that you would use for asteroid astrometry; in fact, your ability to perform accurate astrometric reduction of your discovery images is an essential skill for this project. Equipment needed is:

- Telescope (equatorially mounted, polar aligned, and tracking).
- CCD imager.
- PC, with CCD-control software and planetarium program.
- Astrometric reduction software.
- Catalog of astrometric reference stars (database of well-measured star positions near the target object).
- GPS receiver, or other method of accurately determining the position of your observing site (to an accuracy of about 100 feet in latitude, longitude, and elevation).
- WWV receiver, or other accurate time source (to time-tag your images). For main-belt asteroids, timing accuracy of about 1 sec is sufficient. For fast-moving near-Earth asteroids, the time-tag accuracy is more critical, and should be maintained to about 0.1 second.

If you have a choice of imagers and telescopes, you'll probably want to select the largest aperture (for light-gathering), shortest focal length (for wide field of view), and largest imager chip (for wide field of view) that you have. You may want to compromise image sampling in favor of a wider FOV to increase the area of your search pattern. These images may not meet the goals described in Chapter 5 for accurate astrometry, but your follow-up nights can be used to provide improved astrometric data on any discoveries.

6.2.2   Planning your search pattern

Planning your search for asteroids involves "where", "how", and "when". Regarding *where*: there are several well-equipped and well-funded professional surveys busily searching the sky for moving objects every clear night. Therefore, you want to avoid the portions of the sky that they have examined in the previous few nights. This is the essential step in planning your search pattern: don't expend time and energy re-plowing ground that the big boys have recently dug up!

The best resource in this regard is the "Sky Coverage" data that is readily accessible from the Minor Planet Center's website (*http://cfa-www.harvard.edu/iau/mpc.html*). There's an amazing wealth of information available on that site. If you scroll down near the bottom of the page (as shown in Figure 6.6), you'll find the hyperlink to "Sky Coverage".

That hyperlink sends you to the form shown in Figure 6.7, at which you can request the areas covered by the major surveys in recent days. An example of the resulting plot is also shown in Figure 6.7. This shows the areas that are potentially fruitful for your searching, since they haven't been mined by the professional surveys recently. Armed with this information, you can select a search region that hasn't been picked over, that is well-placed for your observing site, and is (hopefully) not too close to the Moon's glare.

How and *when* to search are both driven by the fact that you're searching for objects that are mag 19 or possibly fainter. This will force you to use fairly long exposures, probably stacking them, in order to reliably detect objects this faint when you're blinking your images. The demand for sensitivity places a premium on large aperture (larger is better, of course). The asteroids are moving, which limits the effective exposure that you can use in striving for increased SNR. When the exposure has gotten long enough that the asteroid has moved through a couple of "seeing disks", or crossed more than a few pixels (whichever is greater), then increasing the exposure is not effectively adding to its visibility. For most amateur sites, the seeing disk is a few arc-seconds. The pixel scale is determined by your imager and telescope:

$$\Delta\theta \approx 206 \cdot S/f \text{ (arc-sec)}$$

where S = pixel size, in μm;
          f = telescope focal length, in mm.

The factor 206 converts the result to arc-seconds. For most amateur set-ups, $\Delta\theta$ is of the order of a few arc-seconds. Depending on the exact orbit, and position relative to Earth, main-belt asteroids typically present rates of 10 to 70 arc-sec/hr = 0.17 to 1 arc-sec/minute. This suggests that exposures in the range 5 to 15 minutes will be the useful limit. Longer exposures will lead to excessive trailing in the asteroid image. If your search pattern is centered near the area where a main-belt asteroid would be entering or leaving retrograde motion, your quarry will be more nearly motionless and hence you can use longer exposures without trailing. Your mileage may vary, so you'll want to run a few experiments against known asteroids of around magnitude 19, to confirm that your imaging protocol is capable of reliably detecting such objects.

When should you search? Since you're striving to reach mag 19 or deeper in your images, weather and moonlight will certainly factor into your scheduling. This project also requires that you observe the same search pattern on two consecutive

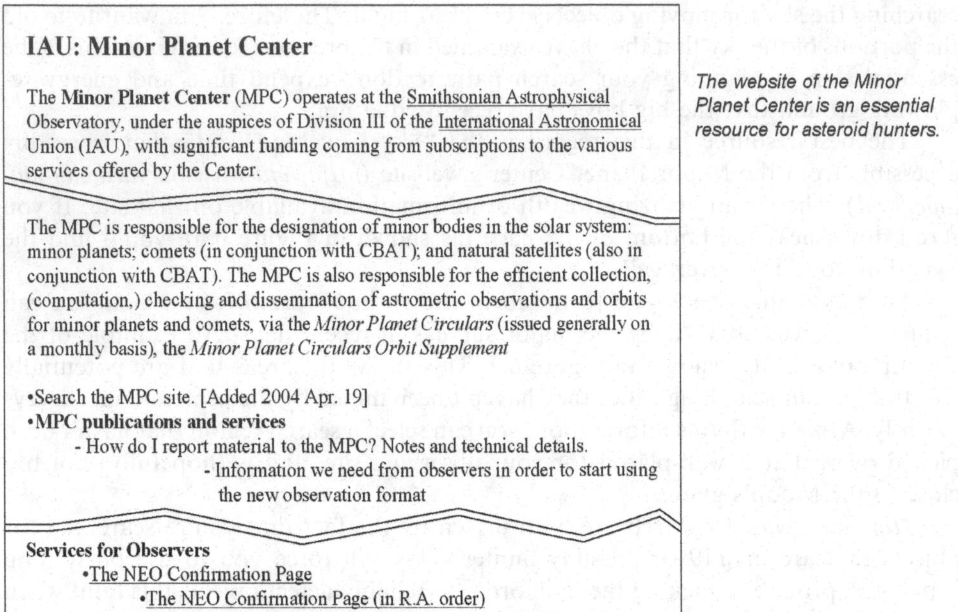

IAU: Minor Planet Center

The **Minor Planet Center** (MPC) operates at the <u>Smithsonian Astrophysical</u> <u>Observatory</u>, under the auspices of Division III of the <u>International Astronomical</u> <u>Union</u> (IAU), with significant funding coming from subscriptions to the various services offered by the Center.

The MPC is responsible for the designation of minor bodies in the solar system: minor planets; comets (in conjunction with CBAT); and natural satellites (also in conjunction with CBAT). The MPC is also responsible for the efficient collection, (computation,) checking and dissemination of astrometric observations and orbits for minor planets and comets, via the *Minor Planet Circulars* (issued generally on a monthly basis), the *Minor Planet Circulars Orbit Supplement*

•Search the MPC site. [Added 2004 Apr. 19]
•**MPC publications and services**
    - How do I report material to the MPC? Notes and technical details.
          - Information we need from observers in order to start using the new observation format

Services for Observers
    •<u>The NEO Confirmation Page</u>
        •<u>The NEO Confirmation Page</u> (in R.A. order)
        • NEA candidate last added/updated (or page modified) on Mar. 17.83 UT.
        •New comet needs astrometric followup. (Mar. 17.83 UT).
        •There are 11 NEA candidates in need of confirmation at the present time (Mar. 17.83 UT).
    •<u>The NEO Page</u>.
        •The <u>NEA Observations Planning Aid</u>.
        •<u>Date of Last Observation of NEOs</u> not seen in a while, with easy access to ephemerides.
        •Lists of bright and faint recovery opportunities for old one-opposition objects.

    •<u>Guide to Minor Body Astrometry</u> Notes for new or potential astrometric observers [Updated 2006 June 1]
    •<u>Dates of Last Observation of Distant Objects</u>
        •An MPES-aware listing of the dates of last observation of Centaurs, TNOs and SDOs.
    •A page allowing you to prepare customized versions of the observable-object lists. [Added 2003 Jan. 29]
    •<u>Sky coverage</u>
        •See which parts of the sky have been searched recently and add your own coverage information.

The website of the Minor Planet Center is an essential resource for asteroid hunters.

The "Guide to Minor Body Astrometry" is an excellent description of everything you want to know about identifying, locating, and reporting your asteroid discoveries.

The "Sky Coverage" link sends you to a form where you can request a plot that shows you where the major surveys have been searching recently.

Figure 6.6. The website of the Minor Planet Center provides a variety of essential information for asteroid hunters. (Used with the kind permission of the Minor Planet Center)

(or "nearly consecutive") nights. Therefore, you'll want to keep that requirement in mind when you examine the weather forecast and your non-astronomical schedule commitments. Of course, the more nights you devote to searching, the more likely it is that eventually *something* new will cross your field of view!

Data entry form (at left) allows you to specify the time interval of interest, and which surveys to consider (normally, you'll select the default "all reporting observers").

Sky coverage plot (below) shows where the major surveys have searched recently.

Sky Coverage Plots

Use this form to generate plots of the sky coverage obtained by some or all of the surveys searching for NEOs. Plots may be returned in a number of different formats (GIF, B&W PS or color PS) and in a variety of sizes.

If you wish to submit your own coverage details, see these instructions.

| Generate plot | Clear form |

Earliest date to include :   2006 May 5 (2006 125)   Latest date to include: 2006 May 12 (2006 132)

Format for plot :   GIF (960 x 720 pixels)   Show   whole sky

Only show search fields that reach below V =   18.0   Color by :

Level of coordinate lines (RA/Decl.):

◉ By default, all reporting observers are included in the plot.

○ **Include the following**   ○ **Exclude the following**
observers: **following observers:**

just big surveys   big surveys
just others   everyone else
LINEAR   LINEAR
CATALINA   CATALINA
LONEOS   LONEOS

| Generate plot | Clear form |

This service utilises the PGPLOT Graphics Subroutine Library and is made possible by Process Software Corporation and their excellent OpenVMS Web server, Purveyor.

SKY COVERAGE
Plot prepared 13 May 2006 12:06:27 by the Minor Planet Center

Morning   Evening
00ʰ   20ʰ   16ʰ   12ʰ   08ʰ   04ʰ
Opposition Point = 15 00.0, −17 06.   Fields reaching fainter than V = 18.0.

Figure 6.7. MPC's "Sky Coverage" plot shows the areas of the sky that have been recently "mined" for asteroids by the professional surveys. (Used with the kind permission of the Minor Planet Center)

6.2.3   Conducting the observations

Creating the images for this project is relatively straightforward, but it can be tedious. With your search area selected, and your search pattern defined, you take one or two images of each FOV in the pattern, and then repeat the search pattern. If you define a search pattern that will take one or two hours to complete, then the timing works out nicely. You finish imaging the search pattern once, and then start over. Covering the pattern twice per night is the bare minimum; three times offers some useful confirmation, if you can fit a third search of the pattern into your imaging schedule.

It is important that the image FOVs in your second pass through the search pattern be pretty accurately aligned with the images from your first pass. Any areas that are covered in one image, but not covered in the second (because the FOVs weren't pointed in exactly the same direction) will not be "blinked" when you compare the images; as a result, any asteroid that happens to be hiding in the "gap" will remain hidden.

When you're establishing the search pattern and estimating how long it will take to complete, be sure to account for the time that you'll need to "fine-tune" the position of each FOV so that it matches the pattern that you've defined on your star chart. It's really aggravating to find that you spend so much time adjusting the FOV positions on the second pass, that you don't get two complete search patterns done before that region sets! If you have a well-behaved "Go-To" mount, you may spend very little time confirming that each FOV in the search pattern is properly positioned. If your mount isn't well synchronized, or suffers from any of a variety of minor woes, then you may well find that you need to spend several minutes fine-tuning the pointing each time you move to a new field in your search pattern.

In general, you'll want to use unfiltered images—you're aiming to detect objects as faint as possible. The exposure duration will of necessity be a bit of a compromise: you want to be able to go as faint as possible (i.e., long exposures), but once the exposure gets so long that an asteroid would be trailing noticeably, you'll get little benefit from taking a longer exposure.

Be sure that you analyze the first-night's images during the next day, before making the second-night's observations. That way, you'll be aware of any candidate new objects before you set up the second-night's search, and can be particularly attentive to the areas of special interest.

6.2.4   Reducing and analyzing your images

You are probably familiar with the technique of "blinking" images to detect moving objects. The idea is simple. Suppose that you take two images of the same FOV, about an hour apart. Align the images so that when you display one, then the other, none of the stars move on your computer screen. Now, rapidly switch back and forth between the two images. The object that bounces back and forth—because it moved during the time interval between the two images—may be an asteroid (or at least *something* other than a normal star). All of the popular CCD imager control and image reduction packages include a version of a blink-comparator; so, doing the alignment between two or three images and "blinking" them is a pretty simple matter.

Blinking images to identify a relatively bright asteroid is a straightforward task—the object is easily visible on your computer screen, and it is almost always easily distinguished from the inevitable background noise fluctuation and cosmic-ray hits. For asteroid discovery, however, you're likely to be searching for a bouncing object that has an SNR <10 and is barely visible on your computer screen. It may be easily confused with random background fluctuations between the images. Summing a few sub-exposures is a good way to increase signal-to-noise ratio, but still you're likely to have a very low signal. Blinking a series of three or four images of the same FOV (instead of just two images) is a good way to protect yourself from false detections. It's not all that rare to have a pair of cosmic-ray "hits" on two images that, when blinked, mimic the stellar-like image and "bouncing" of an asteroid. A third image will almost always give the lie to this type of false detection. Even if there is a third cosmic-ray hit, it isn't likely to be at the right spot on the third image to mimic a true asteroid's constant-velocity, straight-line motion.

The images of your search pattern(s) should be reduced with dark-frames and flat-fields in the usual way. Then, you're ready to perform the blinking process. Adjust your image display so that you are seeing the very faintest possible objects. This usually means setting things so that the "black sky" background is a dark gray, with the background fluctuations clearly visible. Some observers find that using negative images (black stars on white background) provides better sensitivity to faint objects during blinking. Starting with FOV #1, blink the three images (i.e., sequence "A", "B", "C"), looking for moving objects. If you find that two images show a "mover", but the anticipated position of its third incarnation falls outside the boundary of this FOV, check the corresponding image of the adjacent FOV. Continue this process until you've searched all the FOVs in your search pattern.

You will probably find anywhere between zero and a few "movers" in your search pattern. If you find no "movers", then you move on with no further action until the next night's search pattern. Assuming that at least one "mover" is found, your data evaluation becomes more involved. The following steps will be needed:

- Determine the astrometry of the mover, on all images where it appears. For details on performing asteroid astrometry, refer back to Chapter 5.
- Use the Minor Planet Center's "MPChecker" on-line service to see if a known object matches your data. If your "mover" is, indeed, a known object, then you can submit your astrometry report so that MPC can update their database.
- If MPChecker does *not* identify your moving object as a known asteroid, then it may be a new discovery. You need to prepare for a second night of observations, to confirm its reality, and to enable the MPC to calculate a preliminary orbit.

The Minor Planet Center does not give discovery priority to "one-night stands" [10], so if your moving object is a possible discovery, you'll want to schedule observation of it on the very next night to provide confirmation, and then submit the data from both nights in a single report to the MPC.* There are two ways to create a prediction of the object's position on the second night. The simplest is to assume that its motion across the sky is in a straight line at constant velocity (in RA–Dec coordinates). With these assumptions, you can use a spreadsheet or make a graph to extrapolate its position a day or two into the future. These are, to be sure, exceedingly crude assumptions, but for extrapolations of 24 to 48 hours into the future they work acceptably. This extrapolation only needs to be accurate enough that you can find the object on the next night. If you plan on using a 2×2 FOV search pattern, your extrapolation needs to be accurate to about $\pm \frac{1}{2}$ of your imager FOV to give you an excellent probability of recovering your object (assuming that it is, indeed, a real object).

The other way to extrapolate the object's position into the future is to enter your three observations of the object's astrometry into the Minor Planet Center's "New Object Ephemeris Generator". This on-line service accepts your observations (from one night or multiple nights), and generates an ephemeris for future observations.

6.2.5   Reporting your results

Suppose that your candidate object appears on your images from the second night, at approximately its predicted position, and at about the same brightness as it did on the first night. You may very well have discovered something new! First, double-check for known asteroids in the region that you've been observing. You can start by checking with your planetarium program. However, most popular programs (and most amateur astronomers) don't routinely update their asteroid databases with the Daily Orbit Update, so you shouldn't rely on that approach to certify a new object. Here again, the MPC website is the resource that comes to your aid. About halfway down the screen, you'll find a hyperlink to "MPChecker". That utility (shown in Figure 6.8) allows you to enter a time and search window, and it will report all "known" asteroids within the window. In the example illustrated in Figure 6.8, there was a known comet [C/2003 WT42 (LINEAR)] near the location that was entered into the search form. If your confirmed candidate doesn't show up, then run—do not walk—to your computer to prepare the astrometry report and submit it to the Minor Planet Center!

Your report of a new find is identical to the report format used for any asteroid astrometry report (see Section 5.2), except that you'll want to include the words "suspected new object" or equivalent in the comment field.

---

*This, by the way, is another important tidbit of knowledge for the amateur asteroid hunter. The professional surveys are primarily dedicated to discovery of fast-moving near-Earth objects. They typically don't return to make a prompt second-night observation of objects. Since the MPC doesn't give discovery credit to "one-night stands", it is quite possible for the amateur searcher to receive discovery credit by providing two linked nights of astrometry, even though one of the surveys actually saw the object (once) before the amateur's report. The amateur can use his/her diligence and flexibility as a way of outmaneuvering the big surveys.

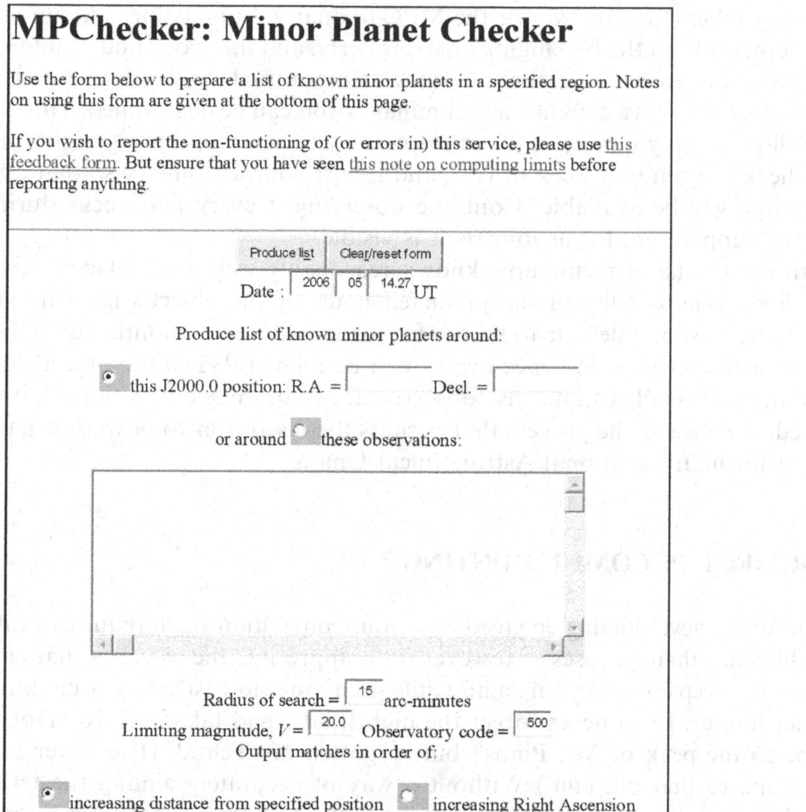

MPChecker: Minor Planet Checker

Use the form below to prepare a list of known minor planets in a specified region. Notes on using this form are given at the bottom of this page.

If you wish to report the non-functioning of (or errors in) this service, please use this feedback form. But ensure that you have seen this note on computing limits before reporting anything.

| Produce list | Clear/reset form |

Date : | 2006 | 05 | 14.27 | UT

Produce list of known minor planets around:

⦿ this J2000.0 position: R.A. = [   ]   Decl. = [   ]

or around ○ these observations:

[   ]

Radius of search = | 15 | arc-minutes

Limiting magnitude, V = | 20.0 | Observatory code = | 500 |

Output matches in order of:

⦿ increasing distance from specified position   ○ increasing Right Ascension

Display motions in arcseconds per ○ minute or ⦿ hour

Display ○ total or ⦿ separate motions

Output:

⦿ all objects

○ just those flagged as needing observations

MPChecker

Here are the results of your search(es) in the requested field(s):

The following objects, brighter than V = 20.0, were found in the 25.0-arcminute region around R.A. = 09 50 45, Decl. = +46 00 00 (J2000.0) on 2006 05 14.15 UT:

```
 Object designation        R.A.       Decl.      V        Offsets
Motion/hr   Orbit   Further observations?
                           h  m  s     °  '  "            R.A.    Decl.
R.A.   Decl.       Comment (Elong/Decl/V at date 1)

    C/2003 WT42 (LINEAR) 09 50 14.1 +46 02 56          5 4W    ? 9N
14+    20-   .0 o  (r =   5.20 AU)
```

Figure 6.8. Use MPChecker to confirm that there are no known objects at the location of your suspected discovery. (Used with the kind permission of the Minor Planet Center)

It usually takes a day or two for the MPC to analyze submissions. Assuming that your astrometry from the two nights links properly, and that your find is indeed new, you will receive a return e-mail that provides a temporary designation for your object. With astrometry from two nights, a preliminary orbit can be determined. This enables you to follow up on your object. It isn't necessary to measure it every night, but you should check it again in a week or two, and report your astrometry so that a longer time baseline will be available. Continue observing it every few weeks during the "discovery" apparition, for as long as it is visible.

With the orbital elements now known reasonably well, your planetarium program will be able to tell you the predicted dates of the object's next apparition. Make a note on your calendar to search for it then (about 18 months in the future), so that its astrometry—and hence orbit—can be refined. When the object's orbit is securely known (usually this means "observed at two or three oppositions"), it will be numbered. That starts the process that permits the discoverer to propose a name for the object to the International Astronomical Union.

6.3   PROJECT P: COMET HUNTING

I will confess to never having engaged in serious comet hunting: it requires a different sort of diligence than I possess. However, I do appreciate the wonders that can pass through your eyepiece and your mind while doing the slow, steady search pattern of the comet hunter. I will never forget the night that I had taken my 16″ Dobsonian telescope to the peak of Mt. Pinos* but forgotten my Telrad. (I've never added a finder 'scope to that big gun.) Without a way of accurately aiming the telescope, I couldn't pursue the deep-sky observing list that I had prepared. Instead, I put in a wide-field eyepiece, and began a reasonably systematic, all-night scan of the sky through its 2-degree field of view. My role model was Herschel, slowly panning and examining everything that passed through his eyepiece. That night was the closest to "comet hunt" scanning that I've ever done, and I remember that night above many other more organized nights of observing. Star clusters, colorful individual stars, tiny nebula, and ghostly galaxies all caressed my retina that night. All were anonymous, I suspect that very few would have found a place on my planned observing list, but I am very pleased to have spent that night with them.

Despite all of the advances in instrumentation and professionally-organized sky surveys, amateurs still discover a few comets each year by keeping their eyes close to the eyepieces of their backyard telescopes. Comet-hunting requires a strong dose of perseverance, and a deep reservoir of patience, but it can yield success in two forms. First, you may indeed discover a new comet. Second, in the course of your search, you are likely to sweep up, and observe, a wide array of interesting, pretty, and unsung deep-sky objects that you would not otherwise have encountered. In doing so, your friendship with the sky is likely to deepen significantly [11].

*Mt. Pinos is a dark-sky mountaintop that is a popular gathering place for southern California amateur astronomers.

6.3.1 Equipment needed

Visual comet hunting requires only the most fundamental equipment in the amateur astronomer's kit:

- Your telescope, with an eyepiece giving you approximately a 1 degree field of view.
- A good star chart.
- (Optional) a planetarium program with a deep database of comets, asteroids, and deep-sky objects.
- Internet access (for use of the Minor Planet Center's databases to check any candidate objects).

And, of course, the visual observing skills that you have learned and exercised over plenty of nights under the stars—the ability to detect faint objects with averted vision, the recognition of galaxies and various types of star clusters, and facility with sketching and star-hopping.

6.3.2 Conducting the observations

The typical amateur comet hunt utilizes time-honored 19th-century techniques. Use a telescope or giant-binocular you're comfortable with and an eyepiece that provides a reasonably wide FOV (a degree or so). Digital setting circles are a nice—but not mandatory—feature. The type of mount isn't important, as long as it permits you to move the FOV smoothly and steadily across your search pattern. Some observers prefer the simplicity and comfort of an alt-azimuth mount, others appreciate the "RA and Dec-oriented" search pattern that an equatorial mount facilitates. Regardless of type, your mount must allow you to make a slow, steady "sweep" to cover your target area of the sky completely and carefully. A good set of star charts (such as Uranometria-2000) is needed to help you filter out the "false alarms". You'll probably use your planetarium program, with its deeper database, to confirm the most promising suspects, but computer screens aren't nearly as protective of your dark-adaptation as is a star chart and a dim red flashlight.

You're searching for a small fuzzy blob. As you know from your visual observing experience, there are plenty of such "faint fuzzies" in the universe. Most are distant galaxies, small nebulae, or unresolved star clusters. On your first few evenings (or mornings) of searching your target region, you'll probably have to check on quite a few of these, using your star charts and/or planetarium program. After a few nights, the population of that region will probably become quite familiar to you and you will more easily recognize them for what they are.

If you find a faint fuzzy that isn't plotted on your star chart, and isn't found in your planetarium program's database, your observing session gets more interesting. First, note the time and the RA–Dec coordinates of the FOV. At this point, you don't need to be too accurate about the coordinates—just sufficient that you can find the field and the object again. Digital setting circles, or the readout of your "Go-To" telescope makes this simple, of course, but in a pinch you can note the pointing direction of your Telrad or finder 'scope. Then, make a sketch of the field of view, taking particular care to get the field stars oriented in the FOV and noting the position of the "target" relative to these field stars. Now, you can continue your "sweep", but reserve time to return to re-examine your candidate in 30 to 60 minutes (sooner if it is going to set, or if morning twilight is rushing toward you). When you return to the candidate, record the time and make another sketch. Did the target move during the time interval between the two sketches? If so, use your planetarium program or star chart to precisely determine the position of the object. Matching your sketch to the star chart of your planetarium program should enable you to estimate the position to within a few arc-minutes.

Those results—the time, location, and detail sketches—are your raw data. Record them directly into your notebook with the care that they deserve (not onto "scratch paper" that can too easily get lost or jumbled). In the course of your searching, you will probably record plenty of false alarms, but those, too, serve an important purpose: your visual skill, and the accuracy of your sketches, will improve with practice. Turn these "false alarms" into learning experiences.

When should you conduct your search, and where in the sky should you be concentrating, in order to improve your odds of success? Since you're searching for an inherently faint, extended object with (probably) low surface brightness, you'll want to select clear nights, free of significant moonlight. Comets brighten as they approach the Sun, so the areas in the western sky after sunset, or the eastern sky before dawn, represent the most favorable regions of the sky. Finally, the full Moon's glare will have impaired everyone's ability to detect low-surface-brightness objects. Hence, as the full moon wanes, you'll want to apply special effort to search the western sky that had been veiled by glare during the evenings surrounding the full moon. The early evening before moonrise is prime comet-hunting time. Similarly, as the Moon waxes toward full, the evening sky is veiled, but the pre-dawn sky (after moonset) offers good conditions for searching.

Aside from the few nights around full moon, either evening or pre-dawn offers potentially fruitful searching time. Use all of those opportunities! Successful comet-discoverers have logged between a couple of hundred to nearly a couple of thousand hours of searching before a successful discovery [12], so you'll need dedication and endurance, diligence and patience if you hope to succeed in this quest. You also have to contend with increasing competition from professional sky surveys. Nevertheless, every year amateur astronomers bag a few new discoveries. It is being done by others, and you can do it, too!

6.3.3   Reducing and analyzing your data

Assuming that you have seen a *moving* faint-fuzzy, the key elements of your data analysis are (1) to determine its position (RA and Dec coordinates), direction and speed of motion, (2) to determine if there are any known Solar-System objects predicted to be close to your target position, and (3) to confirm that it is a real object.

Your planetarium program probably contains a database of comet and asteroid ephemerides. This is the first step in checking your candidate. You may need to update either or both of these databases: most of the popular programs (such as TheSky and SkyMap Pro) contain links to the Minor Planet Center so that they can be updated with the latest catalog information. Assuming that nothing shows up near the candidate's location in this initial sieve, then your next stop is the Minor Planet Center website (*http://cfa-www.harvard.edu/iau/mpc.html*) and their "MPChecker" utility (refer back to Figure 6.8). The MPChecker output includes all asteroids brighter than the magnitude limit entered on the form, plus all known comets regardless of magnitude.

Since your estimated position based on visual observations and sketches is not likely to be particularly accurate, you'll want to be pretty generous in establishing the search radius that you ask for in MPChecker. The default radius is 15 arc-min. I don't recommend using a smaller value. You may want to increase the search radius to 25 arc-min, to be on the safe side.

With a good observation, showing definite motion over an interval of 30 to 60 minutes, and confirmation that there isn't a known comet or asteroid near your candidate's position, you're ready to call the newspapers with a breathless report of your discovery, right? Well, not quite yet. The Central Bureau for Astronomical Telegrams receives far too many erroneous reports. At this point, you should do two or three things. First—if you have a second telescope available, confirm the observation through it. Second—call an experienced amateur astronomer whom you know and trust, and ask him or her to confirm your observation. Third, hold your tongue for 24 hours, while you calculate the predicted position of the suspected comet on the next night. Use the second night to confirm the reality of your discovery. You *must* do at least two of these three checks before making an announcement. Here's why.

The "second-telescope" and "call-a-friend" tests are your protection against ghost images, pointing errors, late-night fuzzy thinking, and other oddities that occasionally happen to astronomers. I've seen a few ghost images that were downright eerie in their ability to simulate a comet. The most spectacular occurred one night at our club's dark-sky site. A friend of mine had been making CCD images of a certain galaxy, when he noticed a tiny, roughly triangular nebular glow in the same field of view. It was bright at the apex and faded as it fanned out—in short, a stereotypical little comet. He checked all the available star charts (this was back in the days before laptop computers with planetarium programs were common at star parties), and no one could find evidence of a neighboring galaxy at the indicated position. He asked me to check the galaxy with my big Dob, but we could find no evidence of his interloper, despite my much larger aperture. We were both able to pinpoint field stars in my (visual) field and his (CCD) image, so we knew exactly where to look, but couldn't find the comet. In the course of an hour, using his CCD images, we saw that the "comet" was moving perceptibly past the nearby field stars. We were all getting pretty excited, but also mystified by the fact that nobody could find the object visually, even in large telescopes. It was bright enough in the CCD images that it should have been easily visible in our big Dobs. I don't recall exactly how we discovered the real cause, but here's what it was: down in the valley below us, one of the neighbors had left a porch light on. The line from that porch light to my friend's telescope was about 60 degrees away from the line of sight to the galaxy he was imaging. Somehow, that porch light's beam had bounced off the front lip of his telescope, entered the corrector plate of his Schmidt–Cassegrain, bounced around inside the optical tube, and by the weirdest chance had made its way to his focal plane, where it appeared as a faint coma of light—his "comet". As the telescope slowly tracked the galaxy (following the sidereal rate), the angles of the internal reflections from that porch light changed and made the smear on his focal plane appear to move relative to the stars. If I hadn't seen it myself, I never would have imagined that such a thing could happen, but it surely did. The bad news is that my friend lost his comet. The good news is that he didn't embarrass himself by reporting the "discovery" and then having to retract it! The moral of the story is: do check things in a second telescope, and do impose on one of your amateur astronomer friends for the favor of privately confirming your suspected discovery. If the comet appears in one telescope, but not in another instrument that should be able to see it, then something is awry, and you need to figure out what that something is, before you make an announcement.

The "second-night" test is virtually mandatory for the same reason: it isn't unheard of for a faint nebula or a small star cluster that is just at the limit of your perception to appear as if it's moving ever so slowly across your field of view. If it's still in the same place after 24 hours, then it was an illusion, not a comet.

6.3.4   Reporting your results

On that happy morning when you've completed your "second-night" test, double-checked the on-line databases, and confirmed that you have, indeed, discovered a new comet, *then* it's time to compose your e-mail report to the Central Bureau for Astronomical Telegrams. The discovery report should be titled "Comet discovery", and must contain all of the following information:

- Your name.
- Address and contact details (e-mail address, telephone/fax number, and "snail-mail" address).
- Date and UT time(s) of observation (see Appendix A if you're not familiar with Universal Time). The preferred format is decimal days, rather than hr-min-sec. For example, 06:15 UT on December 25th is Dec 25.2604. Reported times should be accurate to at least 0.001 day (equivalent to about ±1 minute).
- Coordinates of the object (RA, Dec) to an accuracy of at least 0.1 minute in RA and 1 arc-min in Declination for each observation. Specify the epoch of these coordinates. (See Appendix B for a brief discussion of coordinate epoch. Most current atlases use epoch 2000 coordinates. Some planetarium programs allow you to select the epoch, and at least one that I've used defaults to "epoch of current date". So, do be careful with this step!) Since you've observed the object at least three different times (two on the discovery night, and once more on the

"second-night" check), provide the position information as a table: date, time, RA, Dec, approximate magnitude.

- Brief description of object (size, is it symmetrical or fan-shaped, degree of central concentration, presence or absence of a perceptible tail).
- Observation method used (e.g., naked eye, visual telescopic observation, photographic, or telescopic CCD).
- Telescope description (aperture size, F/ratio, type).
- If the observation was made, or confirmed, with imagery, provide details (type of film or CCD, length of exposure, etc.).
- Location of your observing site (city/town and state/province/country, or other identifying geographical name); longitude, latitude and elevation above sea level.
- Since you are presumably a new contributor to CBAT, provide some background on your observing experience, and describe all of the methods you used to confirm that this isn't a false alarm, including which atlases and databases you referred to for alternate explanations.

This e-mail should be sent to: *cbat@cfa.harvard.edu* Send it in plain ASCII text (not HTML), and do not include any imagery. If for some reason the images are important, you'll receive a follow-up request.

Good hunting, and good luck!

6.4 NOVA AND SUPERNOVA SEARCHES

Stellar explosions are important laboratories for understanding the physics of stars. Before they can be studied, they must be found! Since they are relatively short-lived events, finding them requires continual searching; and amateur astronomers have demonstrated the necessary skill and persistence to succeed in this endeavor. By conducting a nova or supernova search in a consistent way you can contribute to the effort to understand these events, in three ways. By discovering a newly-exploded star you notify astronomers so that they can conduct detailed studies. By a record of your "unsuccessful" observations, you can help astronomers determine the statistics of nova and supernova occurrence. After discovery, you can contribute follow-up observations that determine the lightcurve of the event (using the techniques that were described in Chapter 3 for visual observers, or Chapter 4 for CCD imagers).

First things first: both novae and supernovae are rare events. When you're trying to detect a rare event, you're faced with two challenges: maximizing your "probability of detection", and dealing with the inevitable "false-alarm rate". It seems simplistic to point out that when an event is rare, you need to look long and hard before you're likely to see it. Nevertheless, it's a fundamental truth about these events. Also, when an event is rare, you're quite likely to see things that look like your target, but turn out not to be. Those things that look like a nova, might be a nova, get your heart rate up, and then turn out to be something much more pedestrian, are called "false alarms". The trick is to distinguish them as false before you report them. There are far too many false alarms reported. Although such erroneous reports are certainly not the exclusive domain of amateurs, you should anticipate that the Central Bureau for Astronomical Telegrams will be much more leery of an amateur's discovery report than they would be of a comparable report from a well-known and well-regarded professional astronomer or observatory. They will expect you to take great care to confirm that your discovery is what you say it is before you make a report, and they are likely to query you about how you confirmed your discovery before they accept your report [13].

I'll describe specific advice regarding equipment and search methods in the following sections. Your most essential assets, however, are not equipment or procedures. Considering the challenges of low-probability and high false-alarm rate, amateur searches for novae and supernovae need two things above all. These are: (1) persistence, and (2) dedicated team members.

How rare are these events? Rare enough that you can't expect to make a discovery accidentally. Granted, it has happened [14], but it's sort of like winning the lottery—someone did win the big prize, but it probably wasn't you, and it probably wasn't anyone you know. If you want a secure financial future you need to learn a skill, report to your job, and work hard every day. Similarly, if you want to discover a nova or supernova, you need to learn the techniques that are appropriate for your equipment and your site; and you need to examine a great many star fields, consistently and frequently, for a long time. Bob Evans may be a familiar name to you. He discovered 23 supernovae during his visual search campaign. Over a period of eight years, he logged 74,648 observations of galaxies [15]. That averages out to one supernova for every 3,200 galaxy observations! The Puckett Observatory supernovae search team has discovered 100 supernovae (as of November, 2005), among 850,000 images of galaxies [16]. That's 1 supernova per 8,500 images. Looked at in a somewhat different way, a reasonable estimate is that a typical galaxy will host 1 to 3 supernovae per century [17]. Whichever way you look at it, you'll clearly need to examine a great many galaxies for quite a while before you find a supernova.

Novae are not quite as rare as supernovae, but the amateur nova hunter is restricted to searching a single galaxy—our own Milky Way. How often do novae occur in a single galaxy? Halton Arp once conducted a year-long campaign watching for novae in the Andromeda galaxy (M-31). He found 30 during that year [18]. (The brightest reached mag 15.7, so this is probably not a project that amateurs will want to replicate.) Taking that as a reasonable starting point, and considering that when we're looking at our own galaxy we're restricted to examining only the portion not hidden by the glare of daylight, and that a substantial portion of the galaxy is veiled from us by dense dark clouds of dust, we might expect to be able to detect somewhat fewer than 30 novae per year. Payne–Gaposchkin [19] estimated that the Milky Way should display an average of about 24 novae per year brighter than mag 9.

The fact is that we actually discover only 1 to 5 novae per year (down to magnitude 11). So, one suspects that there is fertile ground for the dedicated amateur astronomer to have a reasonable hope of discovering a new nova. Will the current "era of surveys" make these amateur nova discoveries more difficult? Maybe, but probably not. The surveys are designed to detect faint objects. Even magnitude 11 is a glaringly bright object, too bright for most of them to deal with.

These statistics explain why an above-average level of patience and dedication is needed to be a supernova or nova discoverer. They also are part of the reason that it is useful to have a team, rather than trying to act as the "lone astronomer". A huge amount of data is going to be gathered and examined. It's likely that some team members are more capable of doing one, or the other. It's also likely that the amateur astronomer has other obligations, in addition to the nova/supernova search. By being part of a team, the observational and data-examination duties can be parsed out so that they are fun, rather than a chore, for the team members. The other reason that it is important to be part of a team is that teamwork is the most effective way to deal with the inevitable high rate of "false alarms". If you have other discovery-oriented friends, you can call on them for evaluation of any suspected discoveries. This evaluation is likely to require both independent observation of your target and critical examination of your data. These must be done promptly, since any delay risks losing your discovery to someone else. So, before embarking on either of these projects, do check your level of patience and dedication, and do discuss the project with astronomer friends and members of your astronomy club.

6.4.1   Project Q: nova search

A nova is created in a binary-star system, where one of the stars is a white dwarf that is drawing mass in from its companion. The white dwarf's mass gradually increases. At a certain point the white dwarf responds by undergoing a huge nuclear explosion, blowing away excess mass. The typical nova rises by 10 to 12 magnitudes, and takes only a couple of days to reach peak brightness, followed by a precipitous drop of 2 to 3 magnitudes over the next few days. After that, the nova is described by the nature of its long fade back to "normal" brightness. A "fast" nova fades about 1 magnitude per week. A "slow" nova may fade only 1 magnitude per month.

Your nova search technique will be defined by these characteristics of novae. Because of the rapid rise to peak brightness, very few nova lightcurves cover the "rising" part of the event. Detecting novae early will enable professional astronomers to study these initial hours of the explosion, but doing so requires that the sky be searched every night. If your patrol cadence is "every other night", then you'll probably still find any novae that reach your limiting magnitude, but you may miss the critical early rise. If your cadence is "one night per week", there is a real risk that a "fast" nova may have come and gone without being detected at all. Because novae are generally not bright enough to be seen at intergalactic distances, your search will concentrate on our own Milky Way galaxy. Since white dwarfs are old, highly-evolved stars, your search should concentrate where such stars are most likely to be found: the central portions of the galaxy.

Nova patrols have been successfully conducted visually, using the unaided eye or with binoculars. The two challenges associated with visual search are the need to memorize an enormous number of star patterns (so that you'll recognize a "new" star), and the difficulty of getting rapid independent confirmation that the "new" star is indeed new (and is actually a star). It isn't too great a stretch to imagine that you could learn the binocular-star patterns over a fair swath of the Milky Way. After all, you already probably know the naked-eye constellations quite well, and would easily recognize a magnitude 5 interloper. All it takes is initiative, time, and dedication. Still, having a permanent record is undoubtedly of great value, so consider a photographic patrol.

You'll need:

- A 35-mm camera with "B" (bulb) setting.
- An equatorial-driven mount.
- A finder scope (for accurately pointing the camera).
- Planetarium program.
- A mechanism for "blinking" your images.

Most photographic nova patrols have used film cameras, but if you have a high-pixel-count "electronic 35-mm" camera that can take several-minute exposures with acceptable results, it is certainly worth pressing it into service. If you're using film, it is particularly useful to have a camera with a manual shutter, as opposed to an electrically-actuated shutter. Electronic features require battery power, and batteries are notoriously short-lived in cold weather.

The choice of lens is a matter of tradeoff. A standard 50-mm lens offers a wide field of view (good) and fast aperture (also good), but the stars may be so densely packed in the image that it's difficult to detect individual objects when "blinking" the images (not good). For effective analysis of the images, you may have to restrict yourself to brighter stars (9th or 10th magnitude). A 135-mm or 200-mm lens gives a smaller field of view, so you'll need to take more images to cover the same area of sky that a single 50-mm-lens image would cover. With the longer focal length, you can probably go to 11th or 12th magnitude and still be able to resolve most single stars to that limiting magnitude. Odds are that your 135-mm or 200-mm lens is slower than your "wide-open" 50-mm, so your exposures may be longer (not a significant drawback). Effective patrols have been done using 50-mm lenses, but the consensus seems to be that a reasonably fast 135-mm lens is a better choice if you have one.

The equatorial mount can be one of the commercial camera tracking mounts, a home-made tracking platform, or your main telescope with the camera mounted piggyback. The only requirements are that it be able to handle your camera, that it be aligned and demonstrated to track reasonably well (although at these short focal lengths the tracking accuracy requirements are modest), and that it be equipped with a finder scope that is accurately aligned to the camera.

The longer the camera focal length being used, the larger your finder scope will need to be, in order to accurately point to your selected alignment stars.

In order to prepare for your observations, you'll want to do two experiments. The first is to establish your target fields. The second is to determine the best exposure and film to use at your site. (I'm assuming that the site is likely to be your backyard, since this is pretty much an "every night" project. That means that you'll have to make do with whatever sky conditions you live with.)

Your goal in selecting target fields is to cover as much of the central Milky Way as practical, and be able to accurately return to your selected fields each night.

Checking for good coverage is most easily done with your planetarium program. Make an overlay that matches your camera's FOV, and lay down a pattern that fills the available area of the central Milky Way. (From my southern California site, that means roughly from Sagittarius to Cygnus in the summer, and Canis Major to Cepheus in the winter.) You'll probably want to establish a set of fields for each season, so that you aren't imaging at low elevations (atmospheric conditions will usually be best if you don't go lower than 30 degrees elevation), or during twilight. Examine your selected fields to be sure that each one contains a conspicuous star (easily visible and identifiable in your finder 'scope) that you can use as an "alignment star". You may have to move some of your target field centers to provide yourself with a useful "alignment star". Record the identity of each "alignment star", and where in the FOV it should be placed. Placing your camera FOV accurately, night after night and season after season, is critical to successfully blinking your images.

Your goal in selecting imaging parameters (film type, aperture F-stop, and exposure duration) is to strike a good balance between deep-enough images, and unobjectionable sky glow. If your sky is clear and dark, achieving magnitude 12 is a good goal. Lesser conditions may have to accept a brighter limiting magnitude. Odds are that you'll want to use exposures no longer than about 5 minutes, so that you can cover a good chunk of the Milky Way in a few hours of imaging.

The choice of film is probably going to be driven by easy availability. ASA-400 color print or slide film will offer you a wide range of options, with good sensitivity and fine grain. Higher speed films are likely to be coarser-grained, which may be a problem when it comes to detecting individual star images, but feel free to experiment!

The choice of "print" vs. "slide" film will be driven by your method for "blinking" the images, and the "blinking" is probably going to depend on your skill as a mechanical inventor, and the gadgets that are at your disposal.

With slide film, you can use a once-popular approach that was christened the PROBLICOM ("PROjector BLInk COMparator) by its inventor, Ben Mayer [20]. The idea is to arrange two slide projectors, one above the other, so that their projected images exactly overlap. Last night's slide is loaded into one projector, and the "reference" slide is loaded into the other. Adjust the projectors so that the stars from both images exactly overlap on the projection screen. Then, either with a motorized wheel or a hand-held piece of cardboard, you show first one, then the other, back-and-forth looking for a "flashing" star. A star that flashes on last night's slide, but which is invisible on the reference slide, might be something new, and "something new" might be a nova.

With print film, you'll want to create a stereo viewer. Remember those old gadgets that held "stereo image pairs", and had two eye lenses? Your home-made stereo viewer will be something like that. You'll need two lenses (about 4-inch focal length), a framework to hold them like binocular eyepieces, and a stage where your two prints go: last night's image going to one eye, and the reference image presented to the other eye. Move the prints around so that the star images exactly overlap when you have both eyes open. A star that appears in one, but not in the other image, will "jump out" quite plainly; and you can confirm which print it is on by blinking first one eye and then the other.

A third option is to have your film processor scan your images to digital format. Then, you can use the same PC-based image analysis program that you would have used for blinking your CCD images. Film processors will state the "scanning resolution" in terms of the equivalent print size. Too high a resolution results in enormous file sizes, too low a resolution may cause faint stars to disappear. Compressed images (e.g., JPEG) will be smaller files, but full resolution (usually TIFF) will retain all of the detail of the original negative. If you use this approach, you'll have to do a little experimenting with your film processor to determine the right scanning resolution to use and the right file format. I once made the mistake of asking my film processor to scan 35-mm negatives with the resolution that would be seen on an 8 × 10 inch print, and give me the full-resolution TIFF files. That resulted in a 75-MB file for each image, which was unnecessary overkill! Also, my image-analysis program did not at all appreciate those enormous files when I asked it to do some blinking. Try a few alternatives to find the balance that is "just right" for you and your equipment.

Conducting your observations is relatively straightforward. You point your finder scope and camera at the "alignment star" for your first target field. With your pointing set, you make two images of the target field, one after the other. Then, move on to your second field and repeat the process. And so on, for the next few hours, making two sequential images of each field. The reason you're making two sequential images of each field is to protect yourself from some types of "false alarms". As William Liller has pointed out [21], photographs *can* lie! A flash from a passing satellite, an aircraft's beacon, a point meteor, a film flaw, an electrostatic discharge inside your camera . . . any number of things can make an odd star-like point on your image. Having two sequential images helps to eliminate some of these false alarms.

Along with the images, record in your notebook the relevant information for each image:

- Date and time, preferably in UTC. (If you're not familiar with UTC, refer to Appendix A.)
- Roll #.
- Exposure #.
- Exposure duration.
- Lens (focal length and F/#).
- Identification of field (e.g., RA, Dec of center of field).
- Sky conditions (seeing, transparency).

Analyzing your images is equally straightforward, once you've settled on your method for blinking. Your images from the first night of the season are your "reference" images. Blink your most recent night's images against these reference images, looking for something new. If you see nothing, then carefully archive your images by date and field center. You may need to refer back to them in the future (e.g., to find pre-discovery images of a nova that is discovered at a later date).

Suppose that you do see something new. What then? Then, you begin the process of trying to determine what it is. First, examine both images taken of the field of interest. Does the new star appear on both of them? Is it in the same position? If the answer to either question is "no", then you've found something other than a nova (perhaps a film flaw, perhaps a satellite, perhaps some gremlin that you'll never identify). Did one of your team mates also image that field on the same night, and did he or she see the new object also? If it's not on your team mate's same-night image, it's probably something other than a nova.

If your candidate has passed these initial tests, then you need to check the next-most-likely false alarms. Interrogate the MPChecker at the Minor Planet Center (refer back to Figure 6.8). Are there any asteroids near your point of interest that are within a magnitude or so of your limiting magnitude? Particularly if the "new find" is near the edge of your FOV, there's a good chance that you've found a bright asteroid. Check a couple of the variable-star catalogs for objects near your star of interest (e.g., AAVSO's "VSX" system, and the *General Catalog of Variable Stars*). Your "new object" may be a well-known variable star in outburst. Also use your planetarium program to check for any "too obvious" possibilities: one observer [22] was pretty deflated when his "new star" turned out to be Uranus! Check the IRAS "point source catalog", just in case.

Finally, suppose that your new star has passed all these tests. As with comets, it's a good idea to have a trusted, experienced astronomer whom you know double-check the reference catalogs for the possible identity of your find, while you arrange for confirmation with a second night of imaging. This confirmation will ideally include a repeat of your "patrol" image, plus either visual or CCD image confirmation with another telescope (your team mates may be able to help here, too). The CCD image can be compared with the Palomar Sky Survey.

Odds are that most of your candidates will turn out to be some sort of false alarm. That is not cause for dismay. If you're finding asteroids, then you know that your data analysis is effective at detecting objects that weren't in your reference images. A regular nova patrol is likely to find quite a few variable stars. Most will be known, but a few may be new discoveries in their own rights. If one of your finds is a cataclysmic variable in outburst, that's an observation worthy of note, and deserves a report to AAVSO (see Section 3.1).

If your "new" discovery has passed all of these tests for false alarms, then—and only then—it's time to skip ahead to Section 6.4.3, "Reporting your discovery".

6.4.2   Project R: supernova search

The peak brightness of a supernova is about 10 magnitudes brighter than a nova. They are bright enough to be visible to amateur observers at least as far away as the Virgo galaxy cluster, and farther if you have a large telescope. As mentioned above, they are also rare. The last supernova observed in the Milky Way Galaxy was seen 400 years ago (Tycho's star of 1604). Hence, your supernova search will monitor external galaxies, rather than the Milky Way.

The typical supernova displays a rapid rise in brightness (of as much as 20 magnitudes) over the first couple of days, with a slower rise (of a few magnitudes) over the next month or two. It then fades gradually (at about a half-magnitude per month). Hence, an observing cadence of "check each galaxy in your program once every week" will probably assure that you'll see any detectable supernovae. It might not get you the discovery credit, however. Several amateur and professional searches have a more rapid observing cadence.

This is a CCD-imaging project, for which your standard set-up will be used:

- Telescope (equatorially mounted, polar aligned, and tracking).
- CCD imager.
- PC, with CCD-control software.
- Planetarium program (for locating target galaxies).

You'll want to monitor relatively nearby galaxies (so that any supernovae that actually occur will be bright enough for you to detect). You'll also need to monitor a great many galaxies to have a reasonable probability of detecting a supernova in a reasonable amount of time. The most successful amateur supernova search patrol is the one led by Tim Puckett (see *http://www.cometwatch.com*). They image over 1,000 galaxies every clear night, and strive to examine the images with a one-day turn-around time. That pace has earned them 100 discoveries in an 8-year period. In order to keep up anything approaching that pace, you'll need a well-automated imaging system, and a dedicated team of imagers and data-examiners.

Making your observations is pretty straightforward, in theory. Using your established list of target galaxies, you make images of each galaxy, acquiring as many galaxies as possible each night. As with the nova search project, you'll want to establish a target list for each season, and arrange them to minimize the amount of time you spend slewing and acquiring, so that you maximize the time you spend actually gathering photons. Take two or three images of each galaxy, to protect against the myriad things that can go wrong in any single CCD image (cosmic-ray hits, hot pixels, passing aircraft or satellites, etc.).

Whatever "standard" exposure duration you use should be kept constant from night to night and season to season. That way, you can confidently compare each night's images with the "reference" images that may have been taken months (or years) previously. During your first cycle through your list of target galaxies, you should also take substantially deeper images of each galaxy (i.e., use long enough exposure to reach 1 or 2 magnitudes fainter than in your "standard" exposure). The reason for taking these "deep images" is that it isn't uncommon to be misled when a foreground star happens to be a long-period or cataclysmic variable. When a "new" star appears in your standard exposure image, you can refer to your "deep image" of the same galaxy to see if there's any evidence of a faint foreground star at the same location.

Reducing your observations is equally straightforward, in theory: blink last night's images against the reference images of each galaxy, looking for "something new". As with nova and comet hunting, the real chore is to investigate all of the myriad things that the "something new" could be, aside from a supernova. This step is *absolutely* essential! Follow all of the same guidance that was given in the previous sections [23]:

- Use a second telescope, and/or a trusted and capable friend, to confirm your observation.
- Search the asteroid and comet ephemerides, to see if one of these is near your galaxy.
- Check your "deep images" of the galaxy for evidence that your "discovery" might be a foreground variable star.
- Hold your tongue until you can observe the galaxy again, on the next night, to confirm that your suspect is still there, and hasn't moved.

6.4.3 Reporting your discovery

On that exciting morning when you have confirmed your "suspected discovery", and ruled out any other explanations for your observation, it's time to make a report to the Central Bureau for Astronomical Telegrams. The most convenient approach for communicating with them is via e-mail, at: *cbat@cfa.harvard.edu* Your e-mail should be titled "Possible nova (or supernova)". In the case of supernova your title should identify the galaxy by Messier or NGC number. This e-mail should be sent in plain ASCII text (not HTML or other encoding), and must include the following information:

- Your name.
- Your address and contact details (preferably e-mail address, otherwise telephone/fax number).
- Date and UT time of observation.
- Observation method (e.g., naked eye, visual telescopic observation, photographic, or telescopic CCD).
- Specific details on instrumentation (aperture size, F/ratio, etc.) and exposures (type of film or CCD, length of exposure, etc.).
- Observation site (name of location, giving either city/town and state/province/country, or some other geographical name nearby); longitude and latitude and elevation above sea level can be useful.
- Position of the object. In the case of novae, RA and Dec, with an indication of the accuracy of the position. In the case of supernova, identify the galaxy, and give both the RA/Dec coordinates of the suspect supernova as well as its location relative to the galaxy center (i.e., separation and position angle from the galaxy center).
- Times of observation, in UTC. The preferred format is decimal days.
- Give full information of the sources that you checked, in order to rule out possible "false alarms" (atlases, catalogs, etc.). Describe the ways in which you confirmed your observations (e.g., names and locations of other observers, dates of observations, limiting magnitude of the "deep field" you checked), and

any other information that will bolster your evidence. It is absolutely essential that you be able to report that the object has been seen on multiple images, to eliminate the possibility of a cosmic-ray hit or other image flaw.

Do not include any images in this message. If images are required, the CBAT will request them separately.

6.5   PROJECT S: SERENDIPITOUS DISCOVERIES

Many amateur astronomers observe, or photograph, the popular celestial objects on a regular basis. I remember one cynic (whose name I've forgotten) who asked, "Does the world really need another image of M-78?" Anyway, why spend big bucks for astro-imaging equipment when you can download better images from the NASA website? Well, it turns out that "one more image" of M-78 made quite a stir in early 2004.

In January of that year, an amateur astronomer named J.W. (Jay) McNeil had made some images of this small, bright nebula in Orion. You've probably seen this nebula yourself, and may have imaged it more than once. It's in a pretty busy region—the bright nebula, a dense star field, and several smaller and fainter knots of nebulosity (both light and dark). Mr. McNeil is an experienced amateur astronomer and quite an expert astrophotographer. He is also a diligent observer, who takes meticulous care with his images. While he was processing these particular images, he noticed an area of bright nebulosity that had not been apparent on previous images he had seen of this area. He searched several professional databases, and confirmed his initial suspicion—this was something new. He reported it to the IAU [24], and within 48 hours several professional observatories were altering their telescope schedules to collect data on "McNeil's Nebula". What he had discovered was an outburst of a not-yet-fully-born star, still enshrouded in the cloud of gas and dust in which it was gestating.

That's amazing enough. Here's a second dimension of amazement: the telescope that he used to make the discovery images was a little 3-inch refractor!

This sort of thing doesn't happen by accident. Fortune favors the prepared (and experienced) mind. Mr. McNeil had 20 years of experience as an amateur astronomer. He was well familiar with most popular deep-sky objects (including M-78)—so that they were, in a sense, his home neighborhood. He had studied images (both his own, and those made by others) with care, so that even before he understood what he had found, he could tell that there was something different about that night's image. He had, over the years, prepared himself with a more-than-casual knowledge of the information that is available to sophisticated amateurs (such as the on-line Palomar Optical Sky Survey, to which he referred for confirmation of his suspicion that he'd found something new). Finally, he knew who to call for a discussion of his tentative discovery.

So, I encourage you to take advantage of the many opportunities for continuing education that we amateur astronomers have available; to not be shy about repeating observations (or images) of "familiar" objects; to make all of your observations carefully; to keep a notebook of both your observations and your evaluation of them; and to be ever aware of the possibility that there's a peculiarly valuable pearl hidden in your most routine observation.

If you've made a series of images of an object, in addition to all of the routine image reduction that you do, add the effort to "blink" the series, just on the off-chance that there's a moving object or a variable star in that field. Compare tonight's images with ones that you—or other astronomers—have made of the same region, to see if you're seeing the same things or if something has changed. Things do change!

On the night of August 17, 2005, German amateur astronomer Jeorg Hanisch took a very nice image of the Dumbell Nebula. The next night, Swedish amateur astronomer Hans-Goeran Lindberg happened to be imaging the same object. By chance, both men noticed something odd in their images: if they blinked that night's image against one taken previously, a tiny blue star jumped out [25]. They had, serendipitously and independently, discovered a nova.

Like McNeil, Hanisch and Lindberg were experienced and capable amateur astronomers. Like him, they took the critical extra steps: they paid special attention to how each night's observation fit into their prior knowledge of the object; they had learned how to effectively search for, and recognize, something new; and they took the trouble to report what they had found.

Even if you didn't set out to do a "science project" on a particular night, I encourage you to keep in mind the possibility that there's some science hidden in your observations. Or, if you are doing one sort of science project, take advantage of any anomalous data that you may run across. Sometimes that "anomalous" data is valuable in its own right. More than one amateur photometrist has been doing an asteroid lightcurve project (see Section 4.4), and found that one of his comp stars was fading or brightening. If that happens to you, check the databases at AAVSO and IBVS: you may have just discovered a previously-unknown variable star. It's not as rare as you may think—and even "accidental" discoveries accrue to your credit.

6.6   REFERENCES

[1] Sagan, Carl, *Cosmos*, Ballantine Books (1985).
[2] Haas, W.H., "That Stuart Brilliant Flare and the search for a new lunar craterlet", *The Strolling Astronomer*, Journal of the Association of Lunar & Planetary Observers, vol. 47, no. 1, pp. 46–55 (Winter, 2005).
[3] "Lunik II Impact on the Moon", *Sky & Telescope*, November, 1960.
[4] ALPO, "Making and Submitting Lunar Meteor Observations" at: *http://www.lpl.arizona. edu/~rhill/alpo/lunarstuff/lunimpacts.html*
[5] NASA Press Release: "An Explosion on the Moon" (December 23, 2005).
[6] NASA Press Release, "A Meteor Hits the Moon" (June 13, 2006).
[7] Latham, G.V., Dorman, H.J., Horvath, P., Ibrahim, A.K., Koyama, J., and Nakamura, Y., "Passive Seismic Experiment: A summary of current status", *Proceedings of the 9th Lunar and Planetary Science Conference*, p. 3609 (1978).

[8] Dorman, J., Evans, S., Nakamura, Y., and Latham, G., "On the time-varying properties of the lunar seismic meteoroid population", *Proceedings of the 9th Lunar and Planetary Science Conference*, p. 3615 (1978).

[9] Lowe, Andrew: Webcast (March, 2006).

[10] Kowalski, Richard, in a note posted on the Minor Planet Mailing List.

[11] Levy, David H., "Introducing the Levy List", *Sky & Telescope*, vol. 110, no. 5, p. 80 (December, 2005).

[12] Levy, David H., *The Quest for Comets*, Plenum Press, New York (1994).

[13] Liller, W., *The Cambridge Guide to Astronomical Discovery*, Cambridge University Press, Cambridge UK (1992).

[14] Evans, R.O., *AAVSO Supernova Search Manual*, American Association of Variable Star Observers (1993).

[15] Van den Bergh, S. and McClure, R., "Rediscussion of Extragalactic Supernova rates derived from Evans's 1980–1988 observations", *The Astrophysical Journal*, vol. 425, p. 205 (April 10, 1994).

[16] Aguirre, E.L. "Amateur Team Finds 100 Supernova", *Sky & Telescope*, vol. 110, no. 5, p. 101 (November, 2005).

[17] Maza, J. and van den Bergh, S., "Statistics of Extragalactic Supernova", *The Astrophysical Journal*, vol. 204, p. 519 (March, 1976).

[18] Arp, H.C., "Novae in the Andromeda Nebula", *The Astronomical Journal*, vol. 61, no. 1235, p. 15 (February 1956).

[19] Liller, B. and Mayer, B., *The Cambridge Astronomy Guide*, Cambridge University Press (1990).

[20] Mayer, B., "Steblicom/Problicom/Viblicom—An international search for comets", in *Proceedings of the 20th ESLAB Symposium on the Exploration of Halley's Comet, Heidelberg, October, 1986* (ESA SP-250, December, 1986).

[21] Liller, B. and Mayer, B., *The Cambridge Astronomy Guide*, Cambridge University Press (1990).

[22] Liller, B. and Mayer, B. *The Cambridge Astronomy Guide*, Cambridge University Press (1990).

[23] Evans, R.O., *AAVSO: Supernova Search Manual, 1993*, American Association of Variable Star Observers (1993).

[24] IAU Circular 8284: "IRAS 05436-0007" (February 9, 2004).

[25] AAVSO Alert notice #325 "1955+22C VAR VUL 05—new variable near M27" (August 23, 2005).

7

Further avenues

In the previous chapters I described observational research projects that can be done using equipment that is commonly available in the amateur astronomer's toolkit, or which can be added at modest expense. I avoided projects that require the use of math beyond standard high-school algebra. There are, of course, valuable projects that go outside these arbitrary boundaries. The purpose of this chapter is to identify additional projects that may be of interest to the amateur researcher. If you are willing to invest in some specialized equipment, or do some more complex math, then these projects can be brought within the boundaries of your universe. The equipment needed for some of them is likely to cost you a couple of thousand dollars, and may also require that you do some custom design, construction, and de-bugging. Depending on your budget and your enthusiasm for a particular area of research, these can be extremely rewarding investments in your hobby.

The following sections will offer only the barest outline of projects that you may want to investigate. They will identify the project—what it is, and why it's important—and offer a short description of the special equipment that is required. For detailed information on how to prepare the equipment, conduct the observations, reduce the data, and prepare results for publication or professional use, you will need to do some in-depth research of your own. I'll offer a few pointers to get you started in your education in each of these project areas.

7.1 METEOR AND FIREBALL PHOTOGRAPHIC/VIDEO NETWORKS

In Section 1.3, I described the technique for plotting meteor paths onto a star chart, to determine their radiant positions (and possibly identify new radiants). Suppose that two observers, separated by 20 miles or so, were to observe and plot the same meteor. Theoretically, their plots would display the parallax of the meteor, which would enable them to determine its trajectory. Careful measurement of the starting and ending points of the trail would show the meteor's height at its start and end. Some straightforward geometry could then be used to calculate the path of the meteor, in three dimensions. Theoretically!

As a practical matter, visual observations and manual plots aren't accurate enough to determine the meteor's path. However, if you and a partner are willing to invest in a special-purpose imaging set-up, you can indeed determine an individual meteor's path through the atmosphere, measure its altitude, make quite precise measurements of the radiant positions of meteor showers, and contribute data that enables astronomers to calculate the pre-entry orbital paths of meteors. Several amateur organizations have made very valuable contributions to meteor science in this way, notably the Nippon Meteor Society in Japan, and the Dutch Meteor Society in the Netherlands. Despite the ongoing work by professional astronomers using radar to detect and track meteors, there are two reasons that amateur meteor-imaging patrols are still needed: (1) the photographic method is the most accurate method for measuring a meteor's path [1]; and (2) catching a meteor photograph at two separate stations is a bit of a chancy affair, so the more teams that attempt it, the more likely it is that some will succeed on any given night.

How valuable is a two-station meteor observation? The IAU Meteor Data Center's catalog contains only a few thousand meteor orbits [2] (going back to 1936). This database of meteor orbits has been used to determine precise radiant positions for major streams, estimate the motion of radiants over decades, identify fine structure (filaments) within meteor streams, and set bounds on the flux of interstellar particles into the inner Solar System. Every year, new papers are published that are based (at least in part) on the Meteor Data Center catalog—testimony to the ongoing need for more high-accuracy meteor observations. This is important work, and it depends to a great extent on diligent data gathering by amateur groups.

There are three technologies that amateurs are using for imaging and measuring meteor paths: photography, video, and CCD. The photographic method is in some ways simplest, but the video method offers the potential for more detections and more highly automated data gathering and data reduction. The CCD method is a relatively new concept that shows great promise.

The requirements imposed on the imaging equipment are driven by the nature of the data that must be gathered. Therefore, before describing the data-gathering equipment that can be used, let's consider the data that is needed in order to completely describe a meteor's trajectory.

When a meteor is sailing through interplanetary space, it follows an elliptical path—its orbit around the Sun. It can follow this orbital path, largely unperturbed, for millennia. (We'll neglect the minor gravitational perturbations from other planets and moons, since we're most interested in its orbit just before it encounters the Earth; and we'll neglect small forces such as radiation pressure, because we are primarily interested in particles that are large enough to create fairly bright meteors.) When a meteor encounters the Earth's atmosphere, several things happen in very rapid succession. Atmospheric drag decelerates the particle, so it moves ever more slowly as it penetrates downward. It ablates (getting smaller as it travels), it might break up under the stress of atmospheric entry, and on rare occasions a meteor might display aerodynamic forces ("lift") that change its direction as it flies. Finally, the meteor either completely evaporates (often with a terminal flash) or, more rarely, it reaches the ground. The complete description of the meteor's trajectory through the atmosphere encompasses both its path (in three dimensions) and its speed (which is *not* constant along the path). With accurate data on the path and speed of the meteor during its atmospheric flight, it is possible to mathematically "backtrack" and determine the velocity vector that it had *before* it entered the Earth's atmosphere. Knowing the exoatmospheric velocity and the precise time of the meteor's entry, it is possible to calculate its pre-entry orbital elements. They tell us what path it was on before it was swept up by the Earth, and enables it to be correlated with other meteors, and with comet orbits. For shower members, this information can illuminate fine detail in the meteor stream [3], and give tantalizing hints about the stream's orbital evolution [4, 5]. For sporadic meteors, this information enables multiple meteors to be correlated, and may identify hitherto unrecognized minor showers [6, 7].

The fundamental information contained in an image of a meteor trail is the path of the meteor across the sky, as viewed from the camera location. The path only shows the meteor's motion perpendicular to the camera's line of sight. If the meteor is coming directly toward or away from the camera, there will be a point of light, but no trail.

Two images of the same meteor, taken from different locations, enable you to determine the three-dimensional path of the meteor, relative to the surface of the Earth. That gives you the meteor's trajectory, but not its velocity vector, because the trailed image of a meteor on a photograph doesn't tell you anything about the meteor's speed. Determining the speed is important because a meteor that enters our atmosphere at 60 km/hr is on a different orbit than one that enters at only 30 km/hr, even if they follow the same path. Therefore, you need a way of determining the meteor's speed. This speed determination must be taken at several points along the meteor's path, because the meteor is decelerating as it dives deeper into the atmosphere. With several speed measurements, it is possible to extrapolate backwards and make a good estimate of the meteor's pre-entry speed through space. Since the typical meteor flash lasts only a second or so, you need to be able to measure its speed several times per second.

The spacing between stations in a two-station patrol should be a sizable fraction of the height of the meteors, in order to generate a large parallax angle. This means that you must take care in determining the pointing directions of the cameras at the two stations. As illustrated in Figure 7.1, the two stations' pointing directions should be set so that they overlap at the height where meteors typically appear—about 60 miles altitude (100 km).

The path, speed, and deceleration of the meteor completely define its trajectory through the atmosphere, and its state of motion before it encountered the Earth. In order to translate that trajectory into orbital coordinates, you need to also know exactly *when* the meteor entered our atmosphere. Hence, accurate time-keeping is also valuable for meteor imaging.

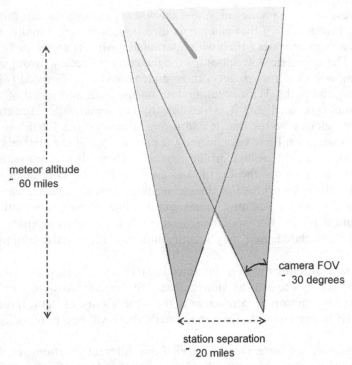

meteor altitude
˜ 60 miles

camera FOV
˜ 30 degrees

station separation
˜ 20 miles

Figure 7.1. Two-station meteor patrols must be aimed so that their FOVs overlap at approximately 60 miles (100 km) altitude. This sketch is approximately to scale, for vertically-aimed fields.

To summarize, ideally you would like your meteor-imaging system to provide simultaneous images, from different locations, covering the entire meteor path; determine the meteor's speed at several points along the path; and record the precise time of the meteor's appearance. This is a tall order. Happily, valuable information can be obtained even if any one of the desired parameters is absent; and happier still, there are ways to gather all of this desired information if you set up your equipment and your observing program with these goals in mind.

7.1.1 Photographic meteor patrol

Many meteor patrols use networks of relatively simple film cameras. In designing your set-up, you must choose the lens, film, and exposure duration, plus include a way of timing the meteor's passage.

If you've ever tried to capture a meteor on film, you know that it is a frustrating enterprise. I read somewhere that if you set up on the peak night of Perseids, using a standard 50-mm lens on a 35-mm film camera, and take images all night, you're probably going to capture between zero and three meteors. That matches my experience, and explains why small-camera meteor patrols are so challenging. There are only three ways to increase the number of meteors captured on film: use a wider field of regard, use a faster lens, and use a faster film. A standard 50-mm lens will offer you a speed of about F/1.5 at its widest opening. Demanding a significantly faster lens is likely to be technically and financially prohibitive. Film speed of ASA 400 to ASA 1000 is the fastest that you're likely to be able to buy, and there isn't much you can do to change that (meteor photographers don't dominate the market for film!). You can consider using a shorter focal length lens (e.g., a 28-mm or 20-mm "fisheye") to get a wider field of view. Unfortunately, those wide-angle lenses tend to have smaller apertures (e.g., F/4 or slower), so what you gain in field of view is lost in terms of limiting magnitude. On balance, most amateur photographic patrols find that a standard 50-mm lens and 35-mm film of ASA 400 is the most convenient and best balance of cost and performance.

The FOV of a 50-mm lens on a 35-mm camera is about 20 by 30 degrees—a tiny fraction of the entire sky. One way to increase the field of regard is to use multiple cameras, pointing in different directions, and mounted on a single platform, as illustrated in Figure 7.2. I have seen arrays ranging from two to eight cameras used in this way. The platform may be stationary, or equatorially mounted. It isn't mandatory that the two (or more) stations have identical set-ups in this regard, but the wider their field of regard, the more of the sky is being monitored and the more likely they are to have mutual observations.

The challenge of measuring the speed and deceleration of the meteor is usually handled by adding a motorized chopper wheel to the camera platform. This may challenge your inventive and handyman skills. You'll want a multi-bladed chopper that can cover/uncover each camera's aperture at least 5 times per second, and preferably as rapidly as 25 times per second. Each blade should be wide enough to completely block or pass the entire clear aperture of the camera. This criterion is likely to drive you to a relatively large chopper wheel.

The speed of the chopper wheel must be well-characterized, and well-controlled. Controlling the chopper wheel's speed is most easily done by driving the wheel with an AC synchronous motor. The rotational speed of this type of motor is established by the frequency of the commercial power source (60 Hz in North America, 50 Hz in Europe), which is quite accurately controlled by the power utility. Alternatively (much more accurate, but also more complex), you can design and build a crystal-controlled oscillator circuit to drive the motor. In either case, you are striving for chopper accuracy and stability of a few percent, so that your calculations of the asteroid's angular speed and its deceleration will not be thrown off by variations in the chopper wheel's rotational frequency.

The final data-collection challenge is to record the exact time when the meteor occurred. If you simply record the beginning and ending time of each exposure, your timing of the meteor's appearance is likely to be uncertain by several minutes. For example, suppose that you're using standard 24 exposure roll film, and you want the roll to last through a 6-hour observation period. Twenty-four exposures divided into 6 hours means that you'll need to use exposures that are at least 15 minutes long. (Fifteen to 30 minute exposures are typical for photographic meteor patrols.) If you simply assume that the meteor appeared at the midpoint of the exposure, then your

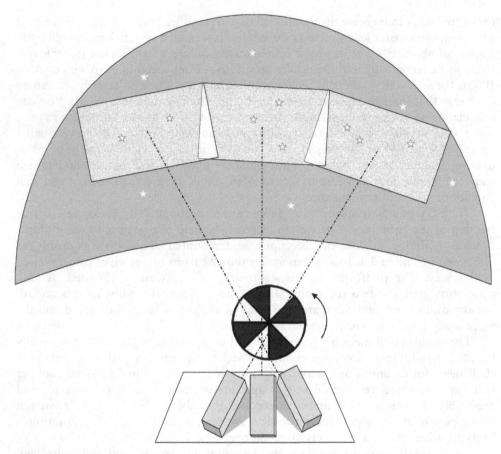

Figure 7.2. Using multiple cameras will increase the field of regard of a photographic meteor-patrol system.

timing error is ±7.5 minutes, which isn't very good. The very successful Dutch photographic patrol solved this problem by combining photographic and visual monitoring [8]. The visual observers recorded the time of any meteor that passed through the region of sky being photographically monitored. In that way, meteor timing to within a few seconds was achieved.

With a photographic set-up like this, you can reasonably expect to capture meteors down to about magnitude 0, or possibly a bit fainter, depending on the speed of the meteor. That translates into—at best—a few meteors per night. With so few meteor trails, this isn't an attractive way of determining the radiants of low-rate showers, but it is a useful way to gather the data needed for meteor trajectory and orbit determination of relatively bright meteors. It has the advantage of being relatively inexpensive compared with the intensified-video, WFOV-video, or CCD set-ups.

7.1.2   Video patrols

The second technology that has been used for gathering meteor images and trajectory data is video. Video is a particularly attractive method for determining the trajectory, speed, and deceleration of the meteor. With two-station arrangements this information can be used to determine the meteor's orbit.

There are two variants of video technology in use for meteor observations: intensified video systems for monitoring of all meteors within a relatively small field of view, and wide field of view (WFOV) video systems for fireball patrols.

The development of a video meteor-observing system is not a trivial undertaking. The most complete system will require you to integrate video equipment (possibly intensified), wide-field-of-view optics, time-code insertion, and meteor-detection software. However, construction of such a system is well within the capability of a group of electronically-savvy amateurs, and will make a fine project for your local astronomy club. With a pair of stations separated by 10 or more miles, you can make interesting observations and contribute valuable information on meteor trajectories.

7.1.2.1   Intensified video

Image intensifiers amplify the light from a faint source by a thousand-fold, making even faint meteors detectable by a video camera. They are becoming affordable (a couple of thousand dollars for a new unit, or several hundred dollars for a used one), and can be coupled to video cameras and fast lenses. The combination of a Generation-II image intensifier, a fast lens providing a 10 to 40-degree FOV, and a video camera creates a powerful meteor-observing system. It can be capable of recording meteors as faint as magnitude 7 (possibly fainter). That is nearly twice as faint as a visual observer can expect to detect. The video image offers position/trajectory accuracy of a few arc-minutes. That is vastly superior to a visual observer's plot on a star chart. The video system's accurately-controlled frame rate is ideal for meteor timing and speed determination, with an accuracy of a fraction of a second. With such systems, amateur observers are generating professional-quality meteor data in a variety of ways.

A single-station intensified video set-up can accurately determine the radiants of minor meteor showers, provide objective data for determination of a shower's population index, record the statistics of fainter-than-visual meteors (a relatively unexplored territory), and record unusual effects such as trails, trains, and sonic booms associated with meteors. A two-station intensified video network can provide the trajectory, speed, and deceleration data that are needed to determine a meteor's orbit.

Intensified video has several advantages over a comparable photographic set-up for meteor orbit determination. First, since it operates at standard video rates (30 frames per second with North American NTSC video, or 25 frames per second with European standard PAL video), there is no need for a chopper wheel. The video frame rate acts as the "chopper". Second, accurate video time-insertion devices are available at modest cost, to place accurate time display directly onto the video record (e.g., the KIWI OSD, available from PFD Systems at *http://www.pfdsystems.com/kiwiosd.html*). This eliminates the need for visual observers to record meteor times. Finally, software is becoming available that will search the video record and reliably detect meteors (discriminating against the most common types of false alarms such as satellite trails and aircraft). This eliminates the need to visually examine the entire videotape from each night's patrol. An example of automatic meteor-detection software is MetRec,* developed by Sirko Molau. It is a "shareware" program, available for download at *http://www.metrec.org/#1* This hardware/software combination is being used by many members of the International Meteor Organization's IMO Video Meteor Network (described at *http://www.serve. com/wh6ef/imo-mirror/video/*). If you are an Apple Macintosh computer user, a similar piece of software is MeteorScan, developed by Peter Gural.

7.1.2.2   WFOV video

A video set-up without image intensification can provide a very wide field of regard (up to "all sky") but will detect only the brightest meteors and fireballs. The enormous field of regard simplifies the challenge of aligning multi-station systems (in an "all sky" set-up, no alignment between the stations is required). The combination of accurate astrometry (based on background star images), fast time resolution, and multi-station coordination makes it possible to determine the orbits of the fireballs that are detected, and to estimate the ground-impact location of any debris that survives entry. Several such fireball networks are active around the world, and more are needed.

7.1.2.3   CCD imaging

A third approach that has been used to monitor meteor trajectories is to combine a CCD imager with a short-focal-length lens to create a very wide FOV CCD system. You can find commercial lenses with extremely short focal length (a few millimeters) and reasonably wide apertures (F/2 or so). These lenses can be combined with sensitive CCD imagers. Such a system is capable of recording meteors that are significantly fainter than the photographic limit, while providing a large field of view. As with video, software is available that can take the drudgery out of examining the images. It will automatically scan through the sequence of images, and identify those that probably have a meteor trail in them, so that you can critically examine and measure only those images that have "hits". The software package "Meteor Scan"— a plug-in for use with MaximDL—is freely available from the developer (John Winfield at *http://winfij.homeip.net/maximdl/meteorscan.html*). Such a system is an excellent tool for determining meteor counts, determining radiants, and conducting photometric (and possibly spectral) studies of bright meteors.

---

*Beware that MetRec requires a fairly high-end PC running DOS or the "DOS mode" of Win95/98—it will not run under newer versions of Windows such as Win XP. It is compatible with only one specific type of frame-grabber card, the Matrox *"Meteor II"* frame-grabber family.

The recent commercial introduction of the SBIG "Meteor Camera" will, hopefully, generate increasing interest in meteor patrols by dedicated amateur groups.

7.1.3 Resources

In both photographic and video patrols, the amateur's job is to collect accurate observational data. The data reduction is sufficiently complex that I won't describe it here. You'll want to collaborate with a specialist who can either do the data analysis, or instruct members of your team in the proper use of the relevant computer codes.

The most active organization conducting meteor-imaging patrols is the International Meteor Organization (IMO). An array of useful information and references is available at their website, *http://www.imo.net/*

7.2 SPECTROSCOPY

The application of spectroscopy to celestial objects was without question the most important development in astronomy after the telescope. It was by using spectroscopy that astronomers learned what the Sun was made of, gathered compelling evidence that stars were distant suns, measured the masses of binary stars, discovered extra-solar planets, and identified the expansion of the universe (among other things). Therefore, as a research-oriented amateur astronomer you're likely to find yourself wondering about the feasibility and value of amateur spectroscopy.

There is indeed a small, but active and growing, community of amateur spectroscopists around the world. Some have designed and built their own spectrographs, but many are taking advantage of the excellent commercial units available from SBIG.

The nature of research that can be done by an amateur spectroscopist will be limited by the resolution of your spectrograph and the aperture of your telescope. Even if you are restricted to relatively bright targets, you can take advantage of the amateur's ability to devote great amounts of telescope time to the search for, and investigation of, transient spectral events. With a typical amateur spectroscopy set-up, you probably won't solve the problem of the expansion of the universe, but you may very well contribute to our understanding of other phenomena.

Amateur spectroscopy equipment comes in three classes. The least expensive and simplest is the class characterized by the Rainbow Optics "star spectroscope". With a 200-l/mm grating, it is a modest-resolution instrument, capable of displaying prominent spectral features in bright stars. For research purposes you will probably need to couple it to a CCD imager (to quantify the spectral data), and will most likely want to craft a way of injecting standard calibration wavelengths into the spectrum.

On the next step up are the commercial spectrographs available from SBIG. These units feature excellent resolution (a fraction of an angstrom), a calibrated wavelength source, and straightforward integration with the SBIG imagers and software. These instruments are being successfully used in amateur–professional collaborations, such as examination of transient spectral events in variable stars [9].

The final step is designing and crafting your own high-resolution instrument. A few ambitious groups are moving amateur spectrograph design into new frontiers. The Astronomical Ring for Amateur Spectroscopy has created a wonderful set of very capable, high-resolution spectrographs, and coordinated studies with amateur and professional astronomers [10]. One particularly remarkable effort is the project by the Spectrashift group (*http://www.spectrashift.com*) to create an amateur instrument that is capable of detecting the radial velocity shift of stars with systems of extra-solar planets [11]—a measurement that is challenging even for professional observatories. This team has succeeded in detecting the radial velocity shift caused by a planet orbiting the star tau Bootis [12].

In addition to acquiring a spectrograph and learning the intricacies of its use, you will be challenged to identify useful projects that are within the range of your capabilities. The trick is to find projects that can be conducted with your equipment and your observing schedule, and which are valuable to the professional astronomical community. By far the best way to do this is to introduce yourself to a professional astronomer, and discuss collaborating, along the following lines. He or she identifies the project of interest, and advises you on data collection methods. You collect the data, and provide it to your professional collaborator for detailed reduction and evaluation.

Spectroscopy is a new frontier for amateur astronomy. If it sounds like your cup of tea, the standard reference is:

Tonkin, Stephen F. (ed.), *Practical Amateur Spectroscopy*, Springer-Verlag (2002).

A second book that's a good read, with plenty of mathematical background information, is:

Kitchen, Chris, *Optical Astronomical Spectroscopy*, Institute of Physics Publishing (1995).

7.3   FILAR MICROMETER BINARY-STAR MEASUREMENTS

In Section 5.3, I described the use of modern CCD imagers to measure the separation and position angle of binary pairs. That is a valuable project that uses widely-available equipment and software. However, as I noted there, the CCD presents some difficult limits: if one of the target stars is bright (limiting your exposure time), you may not be able to gather a sufficient number of astrometric reference stars to develop accurate plate constants; if the pair's separation is too small, the CCD may not have the resolution to adequately separate the pair; if the pair presents a significant brightness contrast, the CCD may have difficulty adequately imaging both components. You can overcome these modern limitations by taking a careful step into the past.

In the pre-electronic era of astronomy, the telescope's clock drive was crafted from spring-and-weight watch works, and images were created by viewing and sketching at the eyepiece. Astronomers also wore jackets and ties while they manned the telescopes (and they *were* virtually all men, back then), but that's another story. For accurate measurements of the separation and angles of close double stars, they used an opto-mechanical device called a "filar micrometer". There are several types of filar micrometer that differ in their details, but the essence of the device is that it places one or two extremely fine wires at the focal plane of a high-power eyepiece. One of the reticle wires can be moved, by turning a precise micrometer screw. The entire assembly can be rotated by another precision gimbal. Despite the low-tech perspective, those astronomers made remarkably accurate measurements of binary-star orbits, and even today these instruments are competitive with CCD binary-star measurements. In cases of faint pairs, pairs with significant magnitude difference, or sparse fields, the filar micrometer may do a better job of precisely measuring the separation and position angle than a CCD set-up will. A long focal-length refractor equipped with a filar micrometer, and an observer who is skilled in its use, can still provide valuable double-star measurements.

The primary issues to be faced when dealing with the filar micrometer are its rarity and expense. As of this writing, there is only one commercially available filar micrometer that I am aware of (the "Needle Eye" bi-filar micrometer, by Van Slyke Engineering). It will cost you between $2,000 and $3,000 depending on the options and extras that you select. Alternatively, if you are an amateur telescope maker with sufficient skill and a well-equipped precision machine shop, you may choose to design and build your own unit.

A venerable reference on the construction and use of the filar micrometer is:

Jones, Kenneth Glyn, *Webb Society Deep-Sky Observer's Handbook: Volume 1 Double Stars*, Enslow Publishers (1979).

A more modern and extraordinarily valuable reference is:

Argyle, Robert (ed.), *Observing and Measuring Visual Double Stars*, Springer-Verlag (2004).

7.4 "CLOUDY NIGHT SCIENCE": DATA MINING

Since most professional astronomical research is funded by government grants, the data are ultimately the rightful property of the taxpayers who paid for it. The world wide web offers researchers a simple and convenient way to make their raw data readily available to interested users, professional or amateur. Large professional projects are increasingly doing so. This treasure trove of professionally-gathered astronomical data is potentially a valuable resource for the amateur investigator. Many modern professional projects generate an enormous flood of data. One estimate [13] is that 1% of that data is carefully evaluated by the principal investigators, or delegated to their grad students. The other 99% may contain all sorts of interesting information that the knowledgeable amateur astronomer with time on his or her hands can profitably utilize. For example, the Large Angle and Spectrometric Coronagraph (on board the SOHO spacecraft) has been used to discover over 1,000 comets. Roughly half of these were identified by amateur "data miners" who followed the imagery as it was posted on the web for public use.

Another interesting example of data mining used raw data from the Rossi X-ray Timing Explorer spacecraft to (probably) detect meter-size objects in the Kuiper belt of our own Solar System [14]. The spacecraft was designed to monitor rapid bursts of X-rays from galactic and extra-galactic sources such as variable stars, neutron stars, and active galactic nuclei—certainly it wasn't designed with small, cold bodies at the edge of our Solar System in mind. However, Chang and his collaborators realized that the brightest X-ray source in the sky, Scorpius X-1, is not too far from the ecliptic; and it seemed reasonable to assume that small Solar System bodies would also be concentrated near the ecliptic. So the investigators retrieved all of the RXTE data on Scorpius X-1, and did a frame-by-frame examination of the flux data, looking for short, deep fades—the signature of occultations. They searched through an enormous amount of data—nearly 90 hours of flux measurement, examined in 2-msec-long frames (whew!)—and found 58 occultation signals. This particular "data-mining" project was conceived and conducted by professional astronomers, but it is a fascinating example of what can be done with readily available data that was collected for completely different reasons, if you have the insight to think of it, and the energy to do it.

Some potentially useful resources for the amateur data miner are:

- SOHO (Solar and Heliospheric Observatory): The SOHO spacecraft has kept the Sun under continuous observation, 24 hours per day, since 1995. The Coronal Diagnostic Spectrometer has returned over 10 million images, and it is still going strong. Virtually all of that data is readily available to anyone who will take the time to learn about the format and contents of the data set. Point your browser to *http://sohowww.nascom.nasa.gov/*
- OGLE (Optical Gravitational Lensing Experiment): When you monitor 130 million stars for five years, who knows what may be hidden in the data? A description of the data and access to the archive is available through *http://bulge.princeton.edu/~ogle/*
- MACHO (Massive Compact Halo Object): This project has gathered 27,000 images, and 18 million two-color lightcurves of stars near the galactic core. A description of the project and an explanation of how to access the database is available at *http://wwwmacho.mcmaster.ca/*
- TASS (The Amateur Sky Survey): This amateur enterprise watches the sky with three telescopic cameras every clear night. It has gathered 77 million two-color measurements of 8 million stars, down to about magnitude 15, and more data is coming in every night, at about 1 Gbyte/hour. For an explanation of the TASS project and the data archive, refer to *http://www.tass-survey.org/*

Finally, do not overlook the long-term data in the AAVSO archive, which may yield useful new insights into variable stars.

Probably the greatest challenges in "data-mining" projects are imagining how these databases can be applied beyond their original purposes, and learning enough about your chosen topic that you can figure out how to make use of the existing data. There are virtually no limits to the uses that you might make of these stockpiles of data, if you put your mind to it!

7.5 "RAINY DAY SCIENCE": PERUSING THE
PROFESSIONAL LITERATURE

Any amateur astronomer who is pursuing research activities will want to gain a bit of exposure to the professional studies that are being done in your area of interest. You will find it quite valuable to see the types of data that the professionals are using, the ways in which they display their results, and the conclusions that they draw from the data. There are two fine resources available to help you in this regard: the internet, and the library.

On the internet, the NASA Astrophysics Data System, located at *http://adswww.harvard.edu/index.html*, is a wide-ranging, searchable archive of most of the astronomically-oriented literature. This is a handy tool if you're searching for published papers about a specific topic. If you're reporting an asteroid lightcurve, for example, a search of the ADS will identify virtually all published data on that asteroid. If you're curious about a particular star or a specific phenomenon (e.g., gamma-ray bursts), the ADS will bring to your attention the literature on that topic.

There is a pitfall in the ADS, however, when it comes to recent references (within the last few years): while it can provide abstracts of current articles, the full text of current articles is available only by paying a fee to the publishers.

In order to avoid copy charges, and to have a more leisurely opportunity to scan the latest goings-on in astronomical research, your best resource is the local college or university library. I try to spend an afternoon haunting the current periodicals in the science library at the nearby university, about once a month. I don't read all of the articles in the major journals—in fact, I barely read one article per journal per month. My method is quite random: I pick the current issue of, say, *Icarus*, and I scan down the table of contents. If an article sounds interesting to me (e.g., it relates to asteroid photometry or meteors), then I read the abstract. If I don't understand the abstract (or in some cases, I don't understand the title), then I move on. If I can follow most of what is written in the abstract, then I'll read the article. Often, I have no idea what I'll do with what I've read, but occasionally I run across a real gem (such as that happy afternoon when I found M. Kaasalainen's original results from his inversion of asteroid lightcurves to determine the 3-D shapes of the asteroids).

It is fascinating to compare "professional" data with similar data gathered by amateurs. The professional often has to make do with a very sparse data set, whereas the amateur can make profligate use of his telescope time, gathering dense, complete, overlapping data sets. It also isn't unusual for the amateur's data to display features that were indiscernible, or difficult to confirm, in the professional's data.

What journals should you browse through? The following list is far from exhaustive, and some journals may not be available at your local university library. See what's available, spend a few days skimming over them, and after a while you'll have a good feel for the journals that are most likely to carry articles that are aligned with your interests.

The two most prestigious general science journals are *Nature* (Nature Publishing Group) and *Science* (the Journal of the American Association for the Advancement of Science). Both are weekly publications, and both endeavor to present the most significant research results across all of the sciences. In any given issue, you are likely to find only one or two astronomical papers, but those are likely to be very heavy-hitting news. For example, the first evidence for gravitational lensing was reported in *Nature* [15], and the first detection of a brown dwarf was reported in *Science* [16]. It is a worthy use of time, and a very educational practice, to at least skim through both of these journals on a regular basis.

One of the ways that scientific journals are ranked is the "impact factor" of their articles. This is roughly defined as "the average number of times that articles published in this journal are referenced by subsequent articles published anywhere in the scientific literature." The theory is that the most significant results will be frequently referenced by later researchers. Not everyone is happy with the use of impact factors as a way of judging the significance of scientific results, but they do give at least a rough guide to the relative prestige of the journals. (The most prestigious journals are likely to be the most widely-read, and therefore the most frequently referenced.) Within the astronomical universe, the three highest impact journals are *The Astrophysical Journal* (including *ApJ Letters* and *ApJ Supplement Series*, published by the University of Chicago Press for the American Astronomical Society), *The Astronomical Journal* (published by the American Astronomical Society) and the *Monthly Notices of the Royal Astronomical Society* (published by Blackwell Publishing for the Royal Astronomical Society). These are almost certainly available in your local university library.

Despite the prestige of the "high-impact" journals, the vast majority of really important astronomical research results are published in journals that have smaller circulation, and/or more focused audience. Journals that are of particular interest to the amateur researcher, because they are likely to include articles related to the projects described in this book, include:

- *Planetary and Space Science* (Elsevier).
- *Earth Moon and Planets* (Springer).
- *Icarus*—International journal of solar system studies (American Astronomical Society).
- *Proceedings of the Astronomical Society of the Pacific* (published by the University of Chicago Press for the Astronomical Society of the Pacific).
- *Astronomy & Astrophysics Supplement Series* (EDP Sciences).
- *The Minor Planet Bulletin* (Association of Lunar and Planetary Observers).

- *Journal of the American Association of Variable Star Observers* (AAVSO).
- *Journal of the British Astronomical Association* (BAA).
- *Journal of the Royal Astronomical Society of Canada* (RASC).
- *Journal of the Royal Astronomical Society of New Zealand* (RANZ).
- *Journal of Double Star Observations* (University of South Alabama).
- *Meteoritics and Planetary Science* (Meteoritical Society).
- *The Observatory.*
- *WGN*—The Journal of the International Meteor Organization (IMO).

Your local college or university library may not carry all of them, but you will probably find a goodly number in the "current periodicals" room. Spend some rainy afternoons with them!

7.6   REFERENCES

[1] Betlem, H., Ter Kuile, C.R., de Lignie, M., van't Leven, J., Jobse, K., Miskotte, K., and Jenniskens, P., "Precision meteor orbits obtained by the Dutch Meteor Society—Photographic Meteor Survey (1981–1993)", *Astronomy & Astrophysics Supplement Series*, vol. 128, p. 179 (February, 1997).

[2] Lindblad, B.A., Neslušan, L., Svoreň, J., and Porubčan, V., "The Updated Version of the IAU MDC Database of Photographic Meteor Orbits", *Proceedings of the Meteoroids 2001 Conference*, Swedish Institute of Space Physics, Kiruna, Sweden.

[3] Yano, H., Abe, S., Ebizuka, N., Fujino, N., and Watanabe, J., "Fine Structure within the Leonid Dust Trail: Resonant Filament Model Examined by HDTV Video Observations", *Meteoroids 2001 Conference, Kiruna, Sweden, August 6–10, 2001.*

[4] Porubcan,V., Kornos, L., and Williams, I.P., "The Taurid complex meteor showers and asteroids", *Contributions of the Astronomical Observatory Skalnaté Pleso*, vol. 36, no. 2, pp. 103–117 (June, 2006).

[5] Jones, D.C., Williams, I.P., and Porubčan, V., "The Kappa Cygnid meteoroid complex", *Monthly Notices of the Royal Astronomical Society*, Online Early (August, 2006).

[6] Campbell-Brown, M.D. and Jones, J., "Annual variation of sporadic radar meteor rates", *Monthly Notices of the Royal Astronomical Society*, vol. 367, issue 2, pp. 709–716 (April, 2006).

[7] Galligan, D.P, "A direct search for significant meteoroid stream presence within an orbital data set", *Monthly Notices of the Royal Astronomical Society*, vol. 340, pp. 893–898 (April, 2003).

[8] Betlem, H., Ter Kuile, C.R., de Lignie, M., van't Leven, J., Jobse, K., Miskotte, K., and Jenniskens, P., "Precision meteor orbits obtained by the Dutch Meteor Society—Photographic Meteor Survey (1981–1993)", *Astronomy & Astrophysics Supplement Series*, vol. 128, p. 179 (February, 1997).

[9] Mais, D., Stencel, R., and Richards, D., "Mira Variable Stars: Spectroscopic and Photometric Monitoring of This Broad Class of Long Term Variable and Highly Evolved Stars—II", *Proceedings of the Symposium on Telescope Science*, Society for Astronomical Sciences (May, 2004).

[10] Desnoux, V. and Buil, C., "Spectroscopic Monitoring of Be-Type Stars", *Proceedings of the Symposium on Telescope Science*, Society for Astronomical Sciences (May, 2005).

[11] Kaye, T.G., "Radial Velocity Detection of Extrasolar Planets", *Proceedings of the Symposium on Telescope Science*, Society for Astronomical Sciences (May, 2004).

[12] Kaye, T., Vanaberbeke, S., and Innis, J., "High-precision radial-velocity measurement with a small telescope: Detection of the tau Bootis exoplanet", *Journal of the British Astronomical Society*, vol. 116, no. 2, pp. 78–83 (April, 2006).

[13] Feinberg, R., during informal remarks at the 2006 Symposium of the Society for Astronomical Sciences.

[14] Chang, Hsiang-Kuang, King, Sun-Kun, Liang, Jau-Shian, Wu, Ping-Shien, Lin, Lupin Chun-Che, and Chiu, Jeng-Lun, "Occultation of X-rays from Scorpius X-1 by small transneptunian objects", *Nature*, vol. 442, p. 660 (August 10, 2006).

[15] Walsh, D., Carswell, R.F., and Weymann, R.J., "0957+561A,B—Twin quasistellar objects or gravitational lens", *Nature*, vol. 279, p. 381 (May 31, 1979).

[16] Glanz, J., "Found: A star too small to shine", *Science*, vol. 270, no. 5241, p. 1435 (December 1, 1995).

Appendix A

Some notes on time

Several of the projects described in this book require that your observations be accurately timed. For some projects, the concern is "time duration"—how long the event took, or the time interval between two events. For others, it is "absolute time"—time referenced to a widely-known time base that enables the analyst to combine observations from many observers in different locations.

Time-duration and time-interval measurements are the sort that are made with a stopwatch. You determine the duration of an event, or the time interval between two events, but do not attempt to identify the exact starting or ending time. For example, suppose that you're the judge of a 100-meter dash. You will be meticulous about determining the interval between the starting gun and the first runner crossing the tape, but you are not at all concerned with the exact time that the starting gun went off. Were it 11 a.m. or 11:02 a.m., the absolute time of the starting gun doesn't really matter.

The standard time interval is the "second", defined by International Atomic Time (TAI) as the interval in which an atom of cesium-133 makes 9,192,631,770 oscillations. This is sometimes colloquially referred to as "Atomic Clock Time".

Measuring the duration of an asteroid occultation is an example of a project where a "time-duration" measurement has some scientific value, even if it isn't linked to a standard time base. However, linking the observation to a standard time base greatly increases the value of the observation, because then it can be combined with other observers' data of the same event. For observations that are to be linked to a standard time base, there are four time bases that the amateur astronomer/researcher will encounter: Coordinated Universal Time, Civil Time, GPS Time, and PC Time.

A.1 COORDINATED UNIVERSAL TIME

Coordinated Universal Time (UTC) is the fundamental time base used by scientists worldwide. Its precise definition is a little tricky. Start with a measurement of the rotation of the Earth, made at your observatory location. You could (theoretically) use observations of solar transits to synchronize your clock to exactly noon, each day. But you know that the Earth's orbit around the Sun is slightly elliptical, so that the Earth travels faster when it is closer to the Sun, and more slowly when it is farther from the Sun. As a result, the time interval between successive solar transits is a bit shorter or longer, depending on the season. This seasonal drift is referred to as the "equation of time", which describes how the Sun "runs slow" during the northern-hemisphere winter, and "runs fast" during the northern-hemisphere summer. If you were to average the Sun's motion over the year, you could define a concept called "the Mean Sun". This is a theoretical Sun, whose motion across our sky doesn't have this seasonal drift. With this done, 1 day = 24 hours = 86,400 seconds is the time it takes for the Earth to rotate exactly once relative to the Mean Sun (i.e., the time interval from transit to transit of the "Mean Sun"). The result is called "UT0". The key thing about UT0 is that it is anchored directly to the actual rotation of the Earth.

We also know that the Earth's rotational axis wanders (on a time scale of decades). From the standpoint of a single observatory, making the measurements for UT0, this polar wander is equivalent to a gradual shift in the longitude and latitude of the observatory. Thus, each observatory's UT0 will gradually drift, relative to other observatories. Taking UT0 and applying the corrections for polar wander, to relate all time to the location of Greenwich, England, results in a time base called "UT1". UT1 is therefore independent of your location on Earth, and it is unaffected by polar wander.

Alas, the rotation rate of the Earth isn't exactly constant, even in the frame of reference of UT1. Even if there weren't any polar wander, careful measurements would show that the Earth's rotation is slowing down. This is mostly a result of the effect of the Moon: the tidal drag and exchange of angular momentum between the Moon's orbital motion and the Earth's rotational motion. For a variety of reasons, it is desirable to have the universal time base stay in synchronization with the Earth's rotation. This is accomplished by "UTC" (Coordinated Universal Time). UTC is defined as having a "second" equal in length to the atomic standard (TAI), and a "zero point" that keeps it within 0.9 sec of UT1. As the Earth's rotation slows down, UTC occasionally adds a "leap second", to keep it synchronized with UT1 (the actual rotation of the Earth). Historically, we've needed to add a "leap second" every one or two years. They have all been added just before the turn of a new year. The most recent leap second was added at the end of 2005. The last minute of December 31st, 2005 had 61 seconds. Whereas at the "normal" turn of a minute the time tick would go $23:59:59 \ldots 00:00:00$, on the last minute of 2005 the time ticks went $23:59:59 \ldots 23:59:60 \ldots$ then $00:00:00$.

Coordinated Universal Time, UTC, is the fundamental time base that you should use for your astronomical observations. The most reliable sources of UTC are the standard time broadcasts from shortwave radio stations WWV or WWVH (in the

USA), or the standard time services provided by other nations (e.g., Canada's CHU or Great Britain's MSF). Most shortwave receivers can pick up these broadcasts. Broadcast frequencies for several national services around the world are listed in Table A.1.

Table A.1. National standard time broadcast services

Station	Transmit frequency (kHz)
CHU (Canada)	3,330, 7,335, 14,670
DCF77 (Germany)	77.5
JJY (Japan)	40, 60
LOL, Buenos Aires, Argentina	5,000, 10,000, 15,000
MSF, Rugby, United Kingdom	60
RWM, Moscow, Russia	4,996, 9,996, 14,996
WWVH, Kekaha, Hawaii	2,500, 5,000, 10,000, 15,000
WWV, Ft. Collins, Colorado	2,500, 5,000, 10,000, 15,000, 20,000
YVTO, Caracas, Venezuela	5,000

In addition to the general-purpose shortwave receiver, there are a couple of special-purpose alternatives. If you search the used-equipment markets or websites, you may find special-made WWV receivers (e.g., the Radio Shack "TimeCube") that are permanently tuned to WWV. There are also retail manufacturers of "atomic time clocks". These are quartz clocks with built-in WWV radio receivers that keep them synchronized to WWV. If you're an electronics technician, there have also been designs published that enable you to make your own portable WWV receiver.

In older books, you may find references to GMT (Greenwich Mean Time). For most practical purposes, you can assume that GMT meant the same thing that we now call UTC. However, if for some reason you need to make a very precise calculation of time intervals that reach into the past, then you'll need to determine what the reference to GMT meant to the original author (usually it's what we now call UTC, sometimes it meant what we now call UT1); and you'll need to account for the "leap seconds" that were inserted into UTC since the time of the historical observation.

A.2 CIVIL TIME

Civil Time, or local time, is the time that you read on your watch. It is based on UTC, with an offset that is determined by the longitude of your time zone. For example, in the western USA we are on "Pacific Standard Time", which is UTC minus 8 hours. You are generally discouraged from recording your observations in "Civil Time". If you do use Civil Time as your time base, then you must take care to record all of the information that will be needed later to convert your observations to UTC. If you're using Civil Time in your observing notebook or computer clock, then the following are absolutely essential:

- Specify the time zone (e.g., Pacific, Central, etc.). Note that there may be situations in which your physical location does not match the time zone of your recorded time, so be careful to separately record in your notebook both your location and the time zone being used.
- Specify if the recorded time is "standard" (e.g., PST) or "Daylight Saving" Time (PDT).
- Specify the date of observation. This is particularly tricky for astronomical observations because the night will bridge two dates: you may begin your observations on the evening of Saturday August 12, but at midnight the date becomes Sunday August 13. Beware of the potential for confusion, and make sure that your observing notes can be deciphered!
- Be sure that your "Civil Time" is synchronized to UTC. For accuracy of ±a few seconds, that's not a problem. Many astronomical projects require fraction-of-a-second accuracy, and most of the common methods of setting your watch (or computer clock) are unsatisfactory in this regard. Many local phone companies offer "Current Time" phone numbers, but their accuracy is not guaranteed to meet astronomical requirements. Commercial radio-station time markers are generally not reliable to the fraction-of-a-second requirements of astronomical observations. As a practical matter, the only sure way to synchronize your Civil-Time watch is to set it to WWV's UTC time tick; and as long as you're going to do that, you may as well set it to UTC and thereby avoid time conversions from Civil Time.

A.3   GPS TIME

Your GPS receiver probably displays both your location, and the time. GPS Time is derived from the atomic clocks aboard the GPS satellites, which are in turn synchronized to UTC [1]. The signal that your GPS receiver gets from the satellites contains an amazing array of information: the identity of each satellite it hears and that satellite's orbital details, plus the time as measured by the satellite's atomic clock, plus a correction factor that describes the drift between the satellite's clock vs. UTC. So, theoretically, your GPS receiver has all of the information it needs to give you accurate UTC.

There are, however, a couple of details that need to be considered in order to bridge from the time that your GPS receiver reports, to UTC. These are the problem of "leap seconds", and the problem of delays in the display.

When the GPS system became operational (in 1980), their clocks were synchronized to UTC. Since then, additional leap seconds have been inserted into UTC, but not into the GPS system time. As of January 2006, therefore, GPS satellite-clock time was 14 seconds ahead of UTC. Some GPS receivers attempt to compensate for the leap seconds in the time that they display; others don't. If you have a GPS receiver that compensates for leap seconds, so that it displays UTC, you should be aware of how it accomplishes this adjustment.

Even though their clocks aren't adjusted for leap seconds, the GPS satellites do know about them. The system that checks the drift of their atomic clocks against UTC tells each satellite about the drift—including the leap seconds. The satellite, in turn, sends a signal that tells your GPS receiver about the "leap second". However, this "leap second" signal is transmitted only once every 15 minutes. Your GPS receiver may save the last-known leap-second corrections. However, if you need to "re-initialize" the receiver after a long trip, it might not apply the leap-second correction until after freshly receiving it from the satellites—which means that you might need to let the receiver listen to the satellites for 15 minutes before you can be sure that it is displaying UTC instead of GPS Time.

As you can imagine from the array of information that your GPS receiver is manipulating, it is a very busy little device. Some of its calculations are more critical than others, and the designers have had to give some functions priority over others. In most units, the task of refreshing the display is a relatively low-priority function. As a result, even though the GPS receiver's internal knowledge of time is extraordinarily accurate, the time display that you see on its screen may be noticeably delayed (by as much as a second or so), compared with UTC, because of the display refresh rate.

These effects—leap-second compensation and display delay—also apply to the "Go-To" telescope mounts with built-in GPS receivers. Do not blindly assume that the time display of the telescope's hand control is accurate UTC (or GPS Time, for that matter).

You'll want to do some experimentation to find out exactly what the "time" display of your GPS receiver means. A shortwave receiver tuned to WWV will enable you to do this. For example, when I checked my Magellan hand-held GPS receiver, I discovered that it does, indeed, display UTC. However, its display is just about exactly 0.5 second late (compared with the WWV time ticks). This most likely represents the time delay in the refresh of the LCD screen. On the other hand, my Celestron NexStar-11-GPS reports a time that is exactly 3 seconds behind UTC. I have no idea why.

Your mileage may vary ... what's important is that you confirm the meaning of your GPS receiver's time display before you rely on it.

A.4 PC TIME

There are many projects where a PC is involved in the data-acquisition process. For example, if you are taking CCD images (for lightcurves, or astrometry), the computer that controls the imager is almost certainly writing the time of the image to the header of the image FITS file. This "PC Time" is read from the computer's internal clock. There are several issues to beware of in this regard: accuracy, stability, the meaning of the time written to the FITS header, the "midnight turnover" problem, and the "Daylight Saving Time" problem.

In order for the PC clock to be accurate, it must be set to a precise time reference. As you may have guessed by now, WWV is the gold standard if you're setting the PC clock manually. Listening to WWV and hitting the "apply" button on the PC clock's window can, with care, get you accurate to within ±0.5 second. Watching the PC clock display while listening to WWV for a minute or so can establish the correction offset between PC Time and UTC to within a fraction of a second. If you have an internet connection to the PC, there are software/hardware packages that enable the PC to periodically re-sync its clock to WWV throughout the night. This is a very attractive method for ensuring both accuracy (to within better than ±0.01 sec) and night-long stability.

Most PC clocks are inherently quite stable, over modest time intervals, if they are left alone to do their own thing. Alas, the PC is busy doing a lot of things, many of which entail some sort of interaction with the internal clock, so you cannot blindly assume that the clock's reported time is stable. In fact, as certain software applications and events take control of priorities in the PC, the clock display may actually stop for a few seconds, and it may or may not "catch up" afterwards. I saw a really shocking example of this when I was using a fairly old laptop (an IBM ThinkPad) to run a parallel-port imager, on a very cold night. The PC clock was set to WWV at the beginning of the evening. It turned out that the imager-control software I was using had been designed to give absolute priority to image downloads at the completion of each exposure, stopping virtually all other PC functions for the ~10 seconds that a download required. This apparently included even the interrupts that synchronized the quartz oscillator in the PC to the "clock" function. So, the clock would appear to stop during image downloads. Usually it would "catch up" after the image download was done. Occasionally, randomly, however, it would not catch up. As a result, as the night wore on the PC clock lagged farther and farther behind UTC. By morning, the image time that was being written to the FITS header was more than a minute behind UTC—a seriously unacceptable situation!

Therefore—especially when your project requires accurate time in the image header—you should carefully check the night-long stability of your PC's clock, and the time that it is reporting to the image FITS header.

You will also want to run a few experiments to determine exactly what the "time" reported in the FITS header means on your system and software. Different CCD imager-control software records different times. For example, the version of CCDSoft that I use to control my SBIG ST-8XE camera writes the *starting* time of the exposure to the FITS header. Other software/imager combinations that I've used wrote the *midpoint* of the exposure to the FITS header. I suspect that there is other software that writes the *ending* time of the exposure to the FITS header. It doesn't particularly matter which time event is selected by your software. What's critically important is that you *know* the meaning of the time in the header, so that your data can be properly reduced and reported.

The phenomenon of PC clock interrupts being inhibited during certain functions can also give rise to an effect known as the "midnight turnover" problem. Suppose that some high-priority task (e.g., image download) is underway just at the stroke of midnight. The PC clock display may pause just before midnight while the high-priority task is performed, and recover just after midnight. After the clock display is re-started, it may show the correct time ("00:mm:ss"; i.e., just after midnight) but still be showing the previous date. There is a long, complex explanation of why the brains inside the PC can correctly roll the time, but "forget" to roll the date. I don't know enough about computers to understand the explanation. But it is clear that the research-oriented amateur astronomer will want to determine if his images are affected by the "midnight turnover" problem, and take measures to prevent it. Determining if you've been bitten is simple: check the FITS header of an image you took before midnight, and one from sometime after midnight, and confirm that the date did, in fact, roll correctly. If the midnight turnover bug has bitten you, it's not hard to edit the FITS headers of the affected images to correct the date. But once you've had to manually make that correction to every one of a sequence of 50 images (for an asteroid lightcurve project), you'll want to be sure that you never have to do it again! (Trust me; this is the voice of experience speaking.)

For astronomers in the USA, the simplest thing that you can do to prevent the "midnight turnover" problem is to set your PC's clock to UTC. Midnight UTC happens when it is daytime in the USA, so that your astronomical observations will not be occurring at midnight UTC (apologies to you solar observers). Setting the time zone of your PC clock is easy (see "Control Panel" in Microsoft Windows, or your operating system instructions if you're using a different operating system). You won't find "UTC" as an option in the time zones within Windows, but you will find "Greenwich England". You'd think that would give you GMT, which is equivalent to UTC, and you'd be almost correct. The "almost" relates to the way that Daylight Saving Time is treated. Thereon hangs a tale.

I live in southern California, and after experiencing the "midnight turnover" problem in late 2004, I determined to avoid it by adjusting my PC clock and selecting the time zone for "GMT Greenwich England" from the drop-down menu. I also unchecked the box for "automatically adjust for daylight saving time". In CCDSoft, I checked the button for "Save images with coordinated universal time (UTC)" in the Camera/Setup window. And all was well ... until late March 2005.

I was aggravated when I discovered that the FITS headers for images taken during the last few days of March were all in error by exactly one hour. Here's what had happened:

In Windows-98 when you uncheck the "automatically adjust for daylight saving time" box, the computer apparently interprets that to mean: "do not change the clock, but *do* set the internal flag that says daylight time is in effect". Then, CCDSoft saw the "daylight time" flag, and assumed that it should subtract one hour from the PC clock time to adjust for Daylight Saving Time. Hence, when it was 05:00 UT, my PC clock read "05:00", but CCDSoft wrote "04:00" to the FITS header (because it was misled by the "Daylight Saving Time" flag that must be hidden somewhere in the Windows code).

In the USA we don't change to daylight time until the first Sunday in April (= April 3, 2005). So why did I see this problem on images taken in the last few nights of March, when we were still on regular standard time? Because in Europe, they change to daylight time on the last Sunday in March (= March 27, 2005). Since my

PC was using GMT, it thought that it was in Greenwich England, so it set the "daylight time" flag on March 27th.

How to avoid this problem? There seem to be two ways; either one will work. The first is to check the "automatically adjust for daylight time" box in Windows. Then, the PC clock will display Greenwich *Daylight* Time, and CCDSoft will recognize the "daylight time" flag, hence subtract one hour, putting the FITS header back onto GMT = UT. I'm not in favor of that approach, because I'll still have to remember whether the displayed PC time is "regular" or "daylight" time. My preferred approach is to select "GMT Casablanca" as the time zone in Windows, and leave the "automatically adjust for daylight time" box *un*checked. In Casablanca (Northwest Africa), they don't do daylight time, so the flag never gets set. In that time zone, the PC clock will display UT all year, and CCDSoft will correctly write UT to the FITS header.

There are a couple of morals to this story. First—as I've learned before—no matter how many mistakes you've made, there is at least one more mistake waiting for you. Therefore, double-check everything in your observational set-up, and again in your data reduction. Second—as I've often told students—astronomy forces you to learn about many other subjects. Who would have thought that an astronomer in southern California would need to learn about European habits regarding daylight time? Or that it would be useful to know that western Morocco doesn't use Daylight Saving Time?

In any case, most amateur researchers will find it most convenient to simply set their PC clocks to the "GMT Casablanca" time zone, and synchronize them to WWV, so that they read UTC all year.

A.5 HELIOCENTRIC TIME AND LIGHT-TIME CORRECTIONS

The speed of light is finite. This means that over astronomical distances, the question of "what time did an event occur?" gets wrapped up with the questions "where did the event occur?" and "where was the clock located?"

You may recall that one of the first accurate measurements of the speed of light was based on observations of the orbits of Jupiter's Galilean moons. This observation and its analysis provides an illustrative example of the importance of the effects of finite speed of light, location, and time determination. Referring to Figure A.1, we have the Sun at the center of the Solar System, Earth orbiting at 93 million miles from the Sun, Jupiter orbiting at 483 million miles from the Sun, and one of Jupiter's Galilean moons orbiting the giant planet. Occasionally, the moon's orbit around Jupiter will carry it into the shadow of the planet. You may have witnessed such an event yourself, since it's a popular and interesting visual observation.

When the moon passes into the shadow of the planet it disappears from view. Suppose that you carefully measure the time of disappearance. At the time of this measurement, you (and your clock) are, of course, located on Earth. So what you have measured is not the actual time that Jupiter's moon entered its shadow: you have measured the time that the signal of the event reached the Earth. That light signal

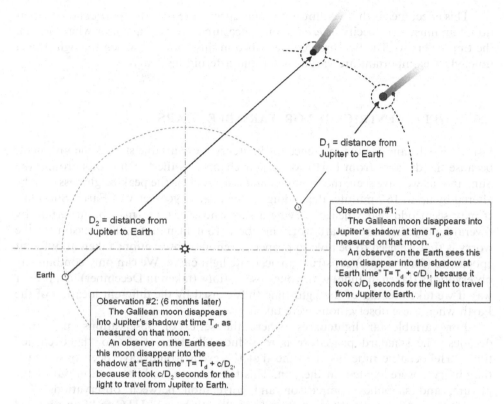

D_1 = distance from Jupiter to Earth

D_2 = distance from Jupiter to Earth

Earth

Observation #1:
  The Galilean moon disappears into Jupiter's shadow at time T_d, as measured on that moon.
  An observer on the Earth sees this moon disappear into the shadow at "Earth time" $T = T_d + c/D_1$, because it took c/D_1 seconds for the light to travel from Jupiter to Earth.

Observation #2: (6 months later)
  The Galilean moon disappears into Jupiter's shadow at time T_d, as measured on that moon.
  An observer on the Earth sees this moon disappear into the shadow at "Earth time" $T = T_d + c/D_2$, because it took c/D_2 seconds for the light to travel from Jupiter to Earth.

Figure A.1. Because of the finite speed of light, you must specify both "where" the clock was, and "what time" the clock read.

takes about 35 minutes to travel from Jupiter to Earth, so what you've measured is the "Earth time" that the signal arrived, which is about 35 minutes later than the "Jupiter time" at which the event actually occurred.

To finish the story of determination of the speed of light, imagine that using Newton's laws and some careful measurements of the orbits of Jupiter's moons, you make accurate predictions of these "eclipses", for a year or so in advance. Then, you measure the time that you observe the eclipses to occur. You'll find that the eclipses will happen several minutes "early" (compared with the prediction) when Jupiter is near opposition (i.e., opposite to the Sun as seen in the sky), and several minutes "late" when Jupiter approaches conjunction (i.e., close to the Sun in the sky). Figure A.1 explains why this is so. When Jupiter is near conjunction, the "signal" that the eclipse has occurred has to travel nearly $2 \times 93 = 186$ million miles further, before we receive it. This is (approximately) the method that Danish astronomer Ole Roemer used to determine the speed of light, in 1675. Using modern values, the time difference between predicted (neglecting Earth's motion) vs. observed signal arrival can amount to $2 \times 93 \times 10^6$ miles/186,000 miles/sec $= 1,000$ sec $= 16.7$ minutes.

This effect means that you must not only specify the *time* of a particular observation, you must also specify *where* the time measurement was taken, or what *location* the time refers to. The two projects described in this book for which the light travel time effect is important are variable-star and asteroid lightcurves.

A.5.1 HELIOCENTRIC JD FOR VARIABLE STARS

Using "Earth time" as the reference for lightcurves of variable stars is inconvenient, because the distance from Earth to the star changes as the Earth orbits around the Sun, and, as we have seen, the timing of an event (such as the peak brightness) can be altered by up to 16.7 minutes depending on the relative geometry of Earth–Sun–star. It may seem odd that when observing a star whose distance may be uncertain by several light years, here we are worrying about 16 light-minutes. The reason for the worry is that we need to be able to properly combine observations taken at different epochs, to (for example) construct an accurate light curve. We can only combine my observations (taken in June) with your observations (taken in December) in a proper way if we take account of the light-time shift caused by the different location of the Earth when these observations were taken.

For variable-star lightcurves—where the series of observations may span decades—the standard procedure is to reduce all observations to "heliocentric" time. "Heliocentric time" is the time that the event would have been observed if the observer were located on the Sun. This takes Earth's orbital motion out of the picture, and simplifies comparison and correlation between observations. The geometry is tricky, since the correction from "Earth-based" UTC to "Sun-centered heliocentric time depends on the three-dimensional geometric locations of the Sun, Earth, and star. For those who are interested, the equation for the heliocentric time $(T_\star)$ is:

$$T_\star = T + 5.787 \cdot 10^{-3} [\sin \delta \sin \delta_\star - \cos \delta \cos \delta_\star \cos(\alpha_\star - \alpha)]$$

where $T = $ UTC of observation, as measured on Earth (days + fraction);

$(\alpha, \delta) = $ RA, Declination of star at time of observation (expressed in radians);

$(\alpha_\star, \delta_\star) = $ RA, Declination of Sun at time of observation (expressed in radians).

In this equation, T and $T_\star$ are expressed in days.

Because the observational records for a variable star may span many decades, the astronomers who analyze them are often faced with the problem of determining the time interval between two dates. Suppose you had to figure out the number of days between August 5, 1915, and June 5, 1945. You'd have to account for the differing number of days in each month, and the leap years that make some years longer than others. Even with a computer, that calculation would be frustrating and error-prone. In order to simplify this problem, astronomers invented the concept of "Julian Day Numbers". The Julian Day calendar/time begins at JD $0.0 =$ UTC $12:00:00$ on January 1, 4713 BC. The Julian Days are numbered consecutively from that point. Time is handled as a fraction of a day (e.g., 3 hours after the start of JD 1, the "date" is $1 + \frac{3}{24} = \text{JD } 1.125$). Calendars are available that show JD as well as civil dates (e.g., from the AAVSO), and there are UT-to-JD conversion programs easily accessible on the internet (see the AAVSO website). Using one of these, the calculation of time intervals over many years is easy. For example:

$$\text{UT } 05:00 \text{ on August 5, 1915} \rightarrow \text{JD } 2,420,724.7083$$

$$\text{UT } 12:15 \text{ on June 5, 1945} \rightarrow \text{JD } 2,431,612.0104$$

and the time interval between these two observations is 10,887.3021 days.

For variable-star reports, you won't need to actually use the equation for heliocentric time conversion, nor convert any observations to JD. Your variable-star reports to AAVSO are to be made in (Earth-centered) UTC. The analysts at AAVSO will translate them into heliocentric time, and calculate the "heliocentric JD" when they enter your observations into the data archive. Further, many of the commercial photometry programs contain the necessary equations, and will give you the option of plotting your results in UTC or heliocentric time when you're analyzing your own data.

A.5.2 Light-time correction for asteroid lightcurves

The other project where light-time corrections are important is asteroid lightcurves. The situation is similar to that shown in Figure A.1, if you take away Jupiter and put your asteroid in its place. For main-belt asteroid (those between Mars and Jupiter), the distance from Earth to the asteroid changes significantly over a surprisingly short time interval. The "light travel time" between Earth and the asteroid can change by several minutes over the course of a month or so. Since typical asteroid lightcurve periods are about 8 hours, a shift of several minutes between observations made only a month apart is a noticeable offset of the lightcurve's phase. If it isn't taken into account, your lightcurves will be blurred by the scatter in observation times.

For asteroid lightcurves, the convention is to report observation times as if you were located on the asteroid—that is, apply the "light-time correction" to each observation before submitting your results for publication. This convention is less consistently followed than that for variable stars, so all asteroid observations and lightcurves that you prepare should be labeled either "light-time corrected" or "no light-time corrections applied".

A.6 REFERENCE

[1] Allan, D.W., Ashby, N., and Hodge, C., *The Science of Timekeeping*, Hewlett Packard Application Note 1289 (June, 1997).

Appendix B

Some notes on astrometry

This appendix will provide some background information on the International Celestial Reference Frame, astrometric reference catalogs, and on the purpose of "plate constants". These topics aren't required reading for the astrometric projects described in Chapter 5, but they may be interesting, if you're conducting astrometry.

The novice star-gazer knows that the stars rise in the east, and set in the west—aside from those near the "North Star", that make small circles around the pole—and that each star rises just a bit later each night than it did the night before. Combine that observation with a desire to record the positions of the "fixed" stars, and it's a small step to imagine the heavens as a giant hollow sphere, with the Earth at the center, and the stars marking tiny embers on its inner surface. This "celestial sphere" spins around in just about 24 hours, its axis of spin passing through a point quite near to the North Star. The Sun, Moon, and planets move across the sky, relative to the "fixed stars".

Since the heavens are pictured as a sphere, it is reasonable to denote the positions of the "fixed" stars by a spherical grid that is analogous to the grid of latitude and longitude that we use to define the position of a point on the spherical Earth. You've seen such a grid described in elementary astronomy texts, with a picture such as Figure B.1. The North Pole and the Equator of the Earth can be projected into the sky, where they are known as the north celestial pole and the celestial equator. The distance of an object north or south of the celestial equator is called its Declination. The celestial equivalent of longitude is called Right Ascension. On the Earth, geographers had to pick a particular line to define the zero-point of longitude (any great circle line of longitude would do, as long as it could be defined). For complex historical reasons, the meridian of Greenwich, England, marks the zero-point of longitude on Earth. Astronomers at least didn't have to fight wars and build empires in order to reach a common agreement on the location of the zero-point of Right Ascension. There is a convenient happenstance that the path of the Sun's motion—the ecliptic—crosses the celestial equator at two points, and these are in some sense

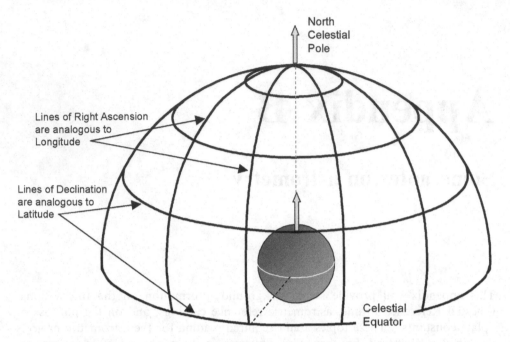

Figure B.1. Simplified description of the celestial coordinate system.

"special enough". The point where the ecliptic crosses the celestial equator in the northward direction is defined as the zero-point of Right Ascension.

That simple view, and the resulting coordinate frame of RA–Dec, is quite satisfactory for many purposes, as long as you don't measure positions and motions with too great a precision. With it, you can locate the stars, and point your "Go-To" telescope. You can measure the positions of the planets relative to the "fixed" stars, and try to figure out the rules that govern their motion. Astronomer-astrologers as long ago as 1000 BC probably made this sort of measurement, in some places with astounding precision.

Alas, when you make sufficiently precise measurements over long-enough periods of time, all sorts of complications are discovered. Sometime before 127 BC, Hipparchus recognized that there was compelling evidence that the sky's rotation axis wasn't exactly constant. That is, the orientation of the north celestial pole moves, slowly but surely, and its motion is measurable. We call that motion "precession" today.

B.1   THE INTERNATIONAL CELESTIAL REFERENCE FRAME

Modern astrometry has moved far beyond the simple picture of RA–Dec that is described in Figure B.1. When measurements are accurate to a fraction of an arc-second, then all sorts of effects must be considered; and amateur astronomers are quite capable of making measurements at that level of accuracy. Professional astrometry is now at the milli-arc-second level, thanks to very-long-baseline interferometry (VLBI) radio telescopes, and the Hipparcos astrometry satellite. The ICRF provides a celestial reference frame that meets the accuracy needs of modern astrometry. It is consistent with the simple picture of Figure B.1 but adds a variety of detailed prescriptions.

For example, it is one thing to glibly say that "the celestial equator is the projection of Earth's equator onto the sky", and quite another to deal with the complexities of the real, moving Earth. The orientation of the Earth's equator wanders due to precession, by a few milli-arc-seconds per year. In order to be precise, you need to specify the date on which you set the orientation of the Equator when you projected it into the sky. Similarly, the position of the origin of Right Ascension isn't a fixed position in space—the Earth's orbit around the Sun changes slightly over time, so the precise position of the "equinox" isn't a fixed point. You have to specify the epoch that you assumed. These two effects are the primary reason that you see celestial positions specified as "the RA and Dec of epoch J2000". That means that they are based on the Equator and equinox that (theoretically) existed at the instant of 12:00:00 January 1, 2000 UT.

The gridlines of RA and Dec aren't traced onto the sky. The situation is similar to that of Earth-bound surveying. The concepts of latitude and longitude are clear, and their definitions can be precisely formulated, but you still don't really have direct access to the lines of latitude and longitude. They just aren't drawn onto the ground in most parts of the world! What we do have access to are landmarks and benchmarks—real, physical, tangible things—that we can use as "defined" locations. Then, we determine the precise location of our position by measuring the distance from us to the nearest benchmark.

A similar method is used for astrometry. The concepts of "north celestial pole", "celestial equator", and "zero-point of Right Ascension" are well-defined, and the rules by which you determine them are carefully specified. These definitions form the International Celestial Reference *System* (ICRS). The System is a theoretical prescription—it doesn't have a physical existence in the real world. Since there aren't gridlines anywhere in the sky, we need to establish fixed benchmarks—real, physical objects that can be observed. These benchmarks form the Celestial Reference *Frame*—the physical realization of the ICRS. Stars would be troublesome benchmarks because they're floating around at the milli-arc-second level. They exhibit both real ("proper") motion relative to an inertial reference frame, and "apparent" motion due to parallax and/or the aberration of starlight. Galaxies are far enough away that they should be free of parallax, and with careful measurement we can subtract out the effect of the aberration of starlight. But galaxies are extended objects, so using them as reference points would generate arguments about what point in the galaxy is the benchmark.

The fundamental defining objects for the ICRF are 212 extra-galactic radio sources. They are scattered across the whole sky, and the position of each one (RA, Dec for epoch J2000) has been precisely established. Together with the mathematical methods for interpolating between these points, these objects are the physically-observable realization of the ICRF. Extra-galactic point sources (mostly quasars) were selected as the basis of the ICRF because their great distance implies that they should have no observable parallax or proper motion, and because the methods of VLBI radio telescopy could define their positions with extreme precision (less than a milli-arc-second). If you've read Chapter 2 (Occultations), it's sort of fun to know that the ICRF's "zero-point" of Right Ascension was precisely defined with the help of lunar occultations of the quasar 3C-273 that were observed by radio telescopes.

The optical realization of the ICRF comes from the measurements made by the Hipparcos satellite. It generated precise positions and proper motions for 117,955 stars (epoch J1991.25). This network of stars, and specific mathematical methods for using them, provides the basis for interpolating the positions of other objects to very fine accuracy.

B.2   ASTROMETRIC CATALOGS [1]

The odds are that your astrometry projects will not use either the quasar reference objects or the Hipparcos stars directly. The situation is analogous to that of photometry: it is interesting to know that the zero-point of the magnitude system is defined by Vega, but you will probably never actually use Vega as a comparison star. Similarly, it is interesting to know how the ICRF is defined, and where it came from, but for practical projects your astrometry will be based on interpolation from secondary catalogs that provide very dense networks of well-measured stars. Examples of these astrometric reference catalogs are the *Tycho Catalog*, the *Guide Star Catalog*, the USNO catalogs, and the UCAC catalogs. All of the popular astrometric data-analysis programs that are available to amateur astronomers can utilize one or more of these catalogs as the basis for astrometry.

The *Tycho Catalog* is quite complete down to between magnitude 10.5 to 11.5. It has excellent position accuracy. Most positions are accurate to about 0.025 arc-sec, and proper motion is included. Because of this relatively bright magnitude limit, Tycho stars are fairly widely spaced. Recall that for accurate astrometry it is important to match your imager's pixel scale to the focal length of your telescope, to achieve well-sampled star images. Unfortunately, this leads to a relatively small CCD field of view with typical amateur equipment. You are not likely to have very many Tycho stars in your CCD FOV. For the calculation of plate constants, a relatively dense mesh of astrometric reference stars is desired. The simplest formula for plate constants requires that you have three astrometric reference stars in the image, but it is preferable to have many more (at least six are typically desired, and up to a dozen or more can be used to improve the accuracy of your calculations). In order to provide a denser mesh of reference stars, most planetarium programs use the *Guide Star Catalog*.

The *Guide Star Catalog* (GSC) prepared for the Hubble Space Telescope has become the *de-facto* faint-star database in almost all of the popular planetarium programs. It is quite complete to about magnitude 15, which means that you are virtually guaranteed to have plenty of GSC stars in your CCD image. The stellar positions in this catalog are accurate to about 0.2 arc-sec. For many purposes, it is quite acceptable to use the GSC stars as your astrometric references. However, it is not the best choice for the most critical work, because its database does not include proper motion.

The USNO SA-2 is an astrometric catalog containing 54 million stars covering the entire celestial sphere. It is convenient because it provides a dense network of stars, and it fits onto a single CD-ROM. It is not normally included in planetarium programs because it is not "complete". There are many stars in the sky that fall well within its magnitude limits, but which are not included in its database. The reason for this is that it was never intended to contain every star within its magnitude range. Its stars were selected to provide a relatively dense mesh of accurately-measured stars distributed uniformly across the entire celestial sphere. With so many stars in this catalog, you can be confident that virtually any CCD image you take will have several USNO SA-2 reference stars available. The stars are generally faint, so that they are appropriate for CCD astrometry. The USNO SA-2 stellar positions were measured to an accuracy of about 0.25 arc-sec, but they do not include proper motion, so this catalog will provide a slightly lower level of accuracy than the UCAC2 catalog.

The USNO B-10 catalog contains over 1 billion stars, with position accuracy of about 0.3 arc-sec. It includes proper motion, so it is more accurate than the USNO SA-2 catalog. It is about 80 GB in size, making it a beast in your computer. As of the date of this writing I don't know of any widely-used amateur astrometry programs that can handle this enormous catalog.

The UCAC2 (*Second US Naval Observatory CCD Astrograph Catalog*) was first released in 2000. The purpose of the UCAC2 project was to provide an all-sky, dense network of stars with positions and proper motion measurements accurate to about 0.02 arc-sec (for the brighter stars) or 0.07 arc-sec (for the fainter stars). It contains stars in the R-magnitude range of about 7.5 to 16, which makes it very useful for astrometric measurement of CCD images taken from amateur-size telescopes. The version released in 2000 contains over 48 million stars, but it does not cover the entire celestial sphere. The regions north of 40 degrees Declination are poorly covered, and there are no UCAC2 stars above 52 degrees Declination. As of this writing, the all-sky version has not yet been released, but it will probably be available soon. It is the best current choice as an astrometric reference catalog. MPO Canopus will accept its database for astrometric purposes.

As I mentioned in Chapter 4, even though these catalogs include magnitude data for their stars, the accuracy and consistency of their magnitudes is not acceptable for the demands of photometric projects. These catalogs were developed specifically for astrometry, and that is how they should be used.

D.3 PLATE CONSTANTS

The illustrative example presented in Figure 5.1 was described as if the pixel array in your CCD image formed a perfect representation of the RA–Dec grid of the celestial coordinate frame. That made the explanation simple to follow (I hope), and isn't too misleading since the more realistic assumptions—and corresponding detailed calculations—that are actually involved are done by your astrometry-reduction software, not by you. So, in most cases you can treat the astrometry software as a "black box", into which you put CCD images and an astrometric catalog, and out of which you get the coordinates of your target object. Nevertheless, it may be of interest to get a deeper feel for what's going on inside that "black box".

You can be pretty sure that your CCD's grid of pixels isn't oriented perfectly to RA and Dec, for two reasons. First, there's a limit to how carefully you can adjust the rotation of the imager in the focus drawtube. Second, there's a good chance that you intentionally rotated the imager, in order to attractively compose the image, or to put a guide star into a convenient position.

If you've studied optics, you know that all optical designs render a somewhat distorted replica of the world onto the plane of the image. These distortions can be minimized by careful (and sometimes complex) optical design, but at the accuracy levels demanded by modern astrometry, there always remains some residual amount of distortion. If there were perfect RA–Dec gridlines in the sky, their image on your focal plane would not be perfectly square, nor perfectly mensurated. The RA–Dec grid as focused onto your image plane isn't perfectly "square"—the RA and Dec axes may not be exactly at right angles—due to slight misalignments in your optical train. If your field of view is large, it isn't unusual to have detectable amounts of "barrel" or "pin-cushion" distortion, which make the angle between RA and Dec axes vary across the image. A little coma or astigmatism isn't unheard of either! The RA–Dec grid on your focal plane isn't perfectly mensurated because such things as field curvature create a varying magnification across your FOV.

The purpose of "plate constants" is to provide a transformation from your measured pixel location (x, y) to the true RA and Dec (X, Y) coordinates. The most simple case uses a linear transformation [2]:

$$X = Ax + By + C$$
$$Y = Dx + Ey + F$$

where A, B, C, D, E, and F are the "plate constants".

This formulation accounts for the displacement of the origin (i.e., the [0, 0] coordinate of your pixel grid probably isn't RA = 0 hr, Dec = 0 deg) and recognizes your image scale (the number of pixels per hour of RA and degree of Dec). It also accounts for the fact that your pixel grid is rotated from the RA, Dec grid, and that the RA, Dec grid may not be perfectly orthogonal ("square") on your image plane.

The way that the plate constants are calculated—deep in the bowels of your astrometry software—starts with the selection of an array of stars in the image that have accurately-known positions, from one of the astrometric catalogs such as UCAC2. The software measures the (x, y) coordinates of each star in terms of pixel position, and it knows the accurate RA and Dec (X, Y) of the star from the catalog.

If there are a good handful of such stars, scattered across the image, then the software can use a least-squares calculation to determine the best-fit plate constants for your image. That is, it uses the grid of "known" stars to calculate the constants A through F. Since there are six unknown plate constants to be found, you need at least three astrometric stars in the image (each star provides two equations, so you have six equations and six unknowns). It is far better to have at least twice this many astrometric stars so that a true least-squares calculation can be made.

The quality of the calculation of the plate constants can be checked by using the "best-fit" constants to calculate each star's RA, Dec from its position in the image plane (x, y). The difference between "calculated" and "catalog" position of a star is its "residual". Most astrometric software allows you to view the residuals for the catalog stars that were used in the calculation of plate constants. The residuals provide two useful pieces of information. First, they provide a useful estimate of the accuracy that you can expect in your measurements of the position of an "unknown" target. The residuals should (for most projects) be no more than a few tenths of an arc-second. The second value of the residuals is that they give you a way to judge the quality of your reference star images.

It isn't unusual to find that most of the residuals are small (a few tenths of an arc-second), but that a few are much worse. There are several reasons that any particular star may show an unacceptably large residual. If you are using one of the catalogs that does not include proper motion data for the reference stars, you may see large residuals for a few stars with relatively high proper motion. If a star's image is affected by a cosmic-ray hit, or a saturated "hot" pixel or a dead pixel, the image defect may have corrupted the software's centroid calculation for that star. Assuming that you have a sufficient number of reference stars in your image, the best strategy is to simply tell your software to ignore the "high residual" stars from the solution, and re-calculate the plate constants. Most astrometric software gives you this option.

For most amateur projects, using small-field-of-view imagers and modern telescopes, the linear formulation of plate constants will be sufficient. As the FOV of the image gets larger, higher-order terms may be necessary in order to reduce the residuals. One common formula that accounts for all of the "first-order" effects, plus field tilt and distortion is:

$$X = Ax + By + C + Gx^2 + Hy^2 + Ixy$$
$$Y = Dx + Ey + F + Jx^2 + Ky^2 + Lxy$$

If you are using an unusually wide-field imaging system, you may need even higher-order plate constant formulae [3]. If that happens to you (as evidenced by a characteristic "trend" in the residuals, betraying a higher-order aberration), then you'll have to move to software that can calculate the higher-order plate constants. The most easily available is IRAF. The higher-order aberration corrections are almost never needed for most amateur CCD set-ups (but beware of focal reducers).

B.4   REFERENCES

[1] Zacharias, N., Gaume, R., Dorland, B., and Urban, S.E., *Catalog Information and Recommendations*, US Naval Observatory Astrometry Dept. (November 7, 2004).

[2] Dept. of Physics, Gettysburg College, Astrometry of Asteroids, Document SUG 9, version 0.70.

[3] Rizvanov, N., Dautov, I., and Shaimukhametov, R., "The comparative accuracy of photographic observations of radio stars observed at the Engelhardt Astronomical Observatory", *Astronomy & Astrophysics*, vol. 375, p. 670 (2001).

Appendix C

The scientist's record book

The projects in this book are scientific investigations. As an investigator, you may be gathering data that is unique—you are seeing something that no one has ever seen before—and which will therefore be subject to verification by other researchers who follow you. Or you may be adding to a series of observations begun by other researchers. In both cases, your records must provide the basis for combining and verifying the observations. It is not unusual for questions to arise after publication, and there will be cases where the raw data and observational records are needed for analyses that were never envisioned on the night that the data was gathered.

Therefore, every scientist—an astronomer in the observatory, a chemist at the lab bench, or an industrial engineer in the factory—follows the discipline of recording his or her observations and activities in a notebook. The value of these original records cannot always be known at the time they are made. Hidden in a routine observation may be the first record of a new discovery. The apparently-discordant data point may be the clue to a new scientific understanding. Therefore, records should be carefully, clearly, completely, and—most importantly—legibly made during the evening's observations.

The exact format of the notebook is not too important, and it will probably be different according to the particular project being undertaken and the tastes of the researcher. What is critical is that there be a notebook, chronologically ordered, and containing a complete, objective record of the circumstances and observations of each night's effort. Spiral-bound or book-bound record books are preferred. Individual sheets, tipped into a three-ring binder, are not a good choice. They present too great a temptation to put off the recording until a later time, they will call into question the exact order in which things were done, and they present a risk that some records will either fall out, or never be entered into, the binder. Your observations are too valuable to handle in such a haphazard way!

Each night's records should include everything that may be needed to reconstruct the circumstances of the observations and the data that was gathered. For visual observations, this will include such things as the telescope, eyepiece and/or magnification, sky conditions ("seeing" and transparency), Moon or light-pollution effects. If observations are not made at a fixed observing site (as, for example, in the case of occultation projects), each night's record should include the location and the way that it was determined. For CCD projects, the notebook should contain all of the information needed to reconstruct the circumstances of the images: telescope, filters, camera used, camera orientation (i.e., which direction is "north" in the image), time of each image, exposure of each image, imaging chip temperature, sky conditions, any "odd" events that occurred during the course of the observations (e.g., computer malfunctions, aborted images, etc.). Both visual and CCD records might include descriptions, sketches or tabulated data. For example, a night of meteor observing may result in sketches of meteor paths, sketches of the trails left by bolides, and tabulated data for the time, brightness, and identity of each meteor observed. These records should all be made at the time the observation is made—that is, right at the eyepiece in the case of sketches or written descriptions—so that you can compare what you wrote/sketched with what you see in the eyepiece, and ensure that it is as accurate as you can make it.

Do not rely on next-day recollections to create these data records. Your goal is for the notebook's records to be as objective as humanly possible. Next-day memories are notoriously affected by second-guessing and wishful thinking. They are too-easily corrupted by considerations of "what I should have seen." The objective record is a description of what you *actually* saw, regardless of what you may have expected!

There will be situations where mistakes must be corrected, or additional information added to your notes. When you make a clerical error in entering information during the course of the night, do not erase or obliterate it. Make a single line-out so that the "erroneous" data can still be read, if necessary. (Brains sometimes get addled when you're tired, cold, and in the dark: it may turn out that analysis will show that your "corrected" record is in fact the mistake, and that your original "erroneous" record is indeed the correct one!)

There will also be cases where some additional remarks will add to the value of the notebook's records. Such after-the-fact notes should be clearly distinguishable from the original on-the-spot records. Some authors recommend using different color ink for later comments. I prefer a system wherein the right sheet of my notebook contains only "on the night" notes, and the facing (left) sheet is available for later comments regarding data reduction, anomalous or discordant observations, or corrections made during data reduction (e.g., in the case where the computer's clock was found to be in error, and FITS header records were adjusted during data analysis).

In the case of electronically-collected observations (e.g., CCD or spectrograph data), the pen-and-paper notebook is still needed. There are some things (such as sky conditions, and equipment malfunctions) that are not adequately written to the electronic data. More critically, considering the complexity and quantity of data that the CCD or spectrograph will be writing to the data record, it is a good idea to have the ability to spot-check that this all got done correctly. Maybe your experience is better, but I have seen plenty of situations where the sophisticated equipment I'm using didn't quite work the way I thought it would (or maybe what I told it to do wasn't quite what I meant!) A comprehensive written record will help you unravel those situations.

CCD images will be reduced and analyzed in the days following the observation. My preference is to retain an archive of the raw image data—exactly as it was collected, without any flat- or dark-frame reductions—and also archive the darks and flats to be used with each night's images. That way, if there are any problems with data reduction, or any need to re-analyze the night's data, there is no question about starting with the purest raw image.

As a goad to any amateur scientists who are tempted to be lax in regard to their observing notebooks, here's a cautionary tale [1]. You remember from your general astronomy education that William Herschel discovered Uranus in 1781. He first noticed its slightly non-stellar disk, and then over several nights confirmed that it was a planet by monitoring its motion past the "fixed" stars. He carefully recorded these observations in his notebook, so that they could be accurately reported, correlated from night to night, and analyzed to determine the orbit of his new discovery.

Herschel's name is writ large in any history of astronomy, thanks both to good luck and to careful discipline with his notebook. One of his contemporaries, the French astronomer Pierre Lemonnier, rarely gets even a footnote in most histories. Yet, Lemonnier observed Uranus in January 1776—five years before Hershel's announcement of his discovery of the planet! And this was not a chance single-night glance (such as Galileo's sketch of a star field that is now known to include Uranus). In fact, Lemonnier recorded Uranus six times over a period of nine days, including on four consecutive nights. He was, alas, a particularly sloppy record keeper. As a result, these particular observations were scattered haphazardly among his papers, and he never did succeed in putting them together in a coherent way. He had made the discovery of a lifetime, but didn't know it, didn't report it, and as a result he didn't get credit for it.

Do not let this happen to you! Be disciplined, accurate, complete, objective, and orderly in recording your observations in your notebook!

REFERENCE

[1]   Sidgwick, J.B., *The Amateur Astronomer's Handbook*, Faber & Faber (1961).

Appendix D

Measurement and uncertainty

In your everyday activities when you measure something, you're concerned only with the *value* of the measurement. If a recipe calls for one cup of flour, then that's what you use. If your scale reads 165 lb, then that's what you weigh. However, in science and engineering, an additional parameter is required: every measured *value* is accompanied by an estimate of the *uncertainty* of the measurement (colloquially called the "accuracy" or "quality" of the measurement). These two features of a measurement—its value and its uncertainty—always go hand-in-hand in a scientific report. They can be displayed in different ways, depending on the purpose and format of the report. You may not have noticed, but most of the examples in this book followed the rule of reporting both the *value* and the *uncertainty* of a measurement. Look back at Figure 2.5: when you report your observation of a lunar occultation, you are expected to include the estimated accuracy of your location measurement and the accuracy of your measured times of disappearance and reappearance. Look back at Figure 4.26: the asteroid's lightcurve period is reported as 4.55 ± 0.01 hours. Look back at Figure 4.36: on the curve of brightness vs. phase angle, each data point is accompanied by an "error bar" that describes the estimated uncertainty in the measurement. All of these formats are quite acceptable. The key point is that any measurement that you report should be accompanied by the estimated uncertainty in the measurement.

D.1   ESTIMATING THE UNCERTAINTY OF A MEASUREMENT

Why do scientists and engineers demand an estimate of uncertainty? For two reasons. First, at a fine enough level of examination, no two measurements will be exactly identical. Second, there is a tricky distinction between what "my measurement" says, and what the underlying "truth" is. In order to illustrate these reasons, consider the simple problem of measuring the length of a bar. Let Figure D.1 represent the bar we are measuring.

Figure D.1. A simple measurement problem: determine the length of this bar.

If you put your ruler against the page, you'll say "it's 2.5 (6.35 cm) inches long". Now, measure it again, more carefully. Is it exactly 2.500 inches long, or is it perhaps actually 2.501 inches long? In a precision assembly, that difference of 0.001 inch might make the difference between a smooth fit vs. a part that can't be assembled into the slot that was designed for it. How accurately can you measure this bar? If you're using a common desk ruler, then the finest lines on your scale are probably separated by $\frac{1}{16}$th of an inch, so you can be quite confident that you've measured the length of the rectangle to within $\pm\frac{1}{16}$th inch ($= 0.0625$ inch). You might be able to interpolate the length to a fraction of the smallest interval on your scale. By interpolating carefully, your measurement is probably good to $\pm\frac{1}{32}$nd inch (i.e., half of a division, or 0.03125 inch). So, you report that the length of the bar is 2.5 ± 0.03 inch. Put another way, you are pretty confident that the length of the bar is somewhere between 2.47 and 2.53 inches.

Could you improve the quality of the measurement by using a magnifying glass to get a closer view of the rectangle and the ruler? Maybe, but eventually you'll face a quandary, illustrated in Figure D.2.

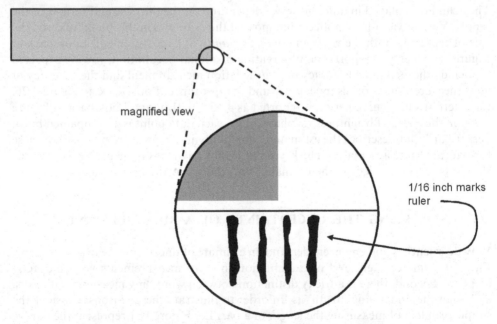

Figure D.2. Magnified view of ruler and the bar being measured.

The ink marks on the ruler aren't perfect—so where exactly is $\frac{1}{16}$th of an inch? Is it the distance between the "center" of the marks? Where, exactly, is the "center" of each mark? Odds are, if you measure the length of the bar several times you'll get a somewhat different result each time because of the vagaries of your "eyeball" guess as to where the center of each mark is, and of where the edge of the bar falls between the marks. As you can see, if you want to measure more accurately than about $\frac{1}{32}$nd of an inch, you need a different sort of measuring device.

A micrometer can easily read a half of a thousandth of an inch, and units are available that can reliably read to a tenth of a thousandth of an inch. So, suppose you use one of those instruments. At a sufficient level of fineness, you'll discover that there is some randomness in the measurements you make because no surface in the "real world" is mathematically perfect. Figure D.3 shows a highly magnified view of the corner of our bar, to illustrate one way of looking at this problem. The length that you measure will depend on the position where you place your micrometer's anvils, because at the tenth-of-a-thousandth of an inch level, the bar's surface has detectable roughness. The uncertainty in your measurement of the bar's length depends not only on the quality of the measuring instrument that you use, but also on the roughness of the bar's surface, and on the position where you make the measurement.

In most projects, there are many possible sources of measurement error. It is theoretically possible to calculate the total effect of multiple sources of error, thereby determining the total uncertainty of the estimate. (Any advanced statistics book will explain the procedure.) For example, if you knew the statistics of the bar's surface

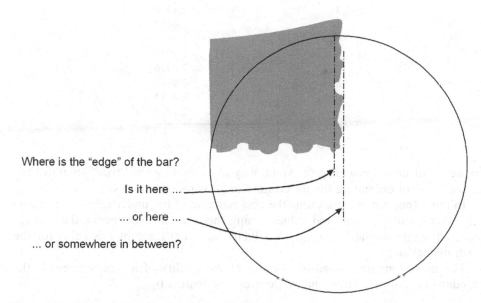

Where is the "edge" of the bar?

Is it here ...

... or here ...

... or somewhere in between?

Figure D.3. When making very meticulous measurements, the quality of the result can depend on the definition of the parameter being measured ("what do we mean by the 'length' of the bar?") and by detailed characteristics of the object being measured.

roughness, you could calculate what effect the roughness has on the expected uncertainty in your length measurement. However, there is a tendency for such calculations to be optimistic, under-estimating the errors in the measurements. The reason for this optimism is that the calculations can only account for error sources that you know about (and in the real world there are often many sources of error that you simply didn't expect), plus the calculations assume that you have a valid characterization of the statistics of each source of error (which you often don't have). Therefore, in most cases, the preferred approach is to make multiple measurements, and use the internal consistency of the measurements as a way of estimating the quality of your result.

Let's continue with our example of measuring the length of a bar. You probably don't have a good way of characterizing the roughness of the surface, but you can get a handle on the magnitude of the effect by making several independent measurements of the bar's length. Suppose that with your "tenths" micrometer, you set up and make a measurement of the bar, then remove the micrometer and zero it, then make another measurement of the bar, etc. You'll get a slightly different reading each time you make a measurement—one reading might be 2.4984, the next 2.4977, and the third might be 2.4978 inch, etc. The results of making ten measurements of the bar are shown in this table

Measurement #	Value
1	2.4984
2	2.4977
3	2.4978
4	2.4983
5	2.4982
6	2.4982
7	2.4976
8	2.4981
9	2.4983
10	2.4977

This series of measurements gives you a way of estimating the "true" length of the bar, and also of estimating the uncertainty in your measurements.

In most (but not all) situations the best estimate of the underlying "true" value is the average of the measured values. Using these ten measurements, the average measured length of the bar is $L_{avg} = 2.49803$ inch. Let's accept this value for the length of the bar.

The most commonly-used description of the quality of a measurement is the "standard deviation". The standard deviation is defined by:

$$\sigma = \sqrt{\frac{\sum(x_i - \mu)^2}{N}}$$

where x_i = the ith measurement;
        μ = the average measurement;
        N = the number of measurements made;

and the summation extends from $i = 1$ to $i = N$.

In this example, the standard deviation of the ten measurements is $\sigma = 0.0003$ inch. Therefore, you would report your measurement of the length of the bar as $L = 2.4980 \pm 0.0003$ inch. That is, you are pretty confident that the length of the bar is somewhere between 2.4977 and 2.4983 inches.

The concept of "standard deviation" gives us a way of quantifying what we mean when we say we are "pretty confident" that the length of the bar falls within a certain interval. Most (but not all) measurement situations present several sources of random error, and the overall error has a normal ("bell curve") distribution, as illustrated in Figure D.4.

In a normal distribution, 68% of the measurements will fall within one standard deviation of the average measurement. That is, for our micrometer measurement of the bar, we can say that there is a 68% probability that the "true" length of the bar is between 2.4977 and 2.4983 inches. If a 68% probability doesn't make you "pretty confident", then you might extend your interval to 2σ. For a normal distribution, 95.5% of the measurements will fall within two standard deviations of the average. That means that you can be 95.5% confident that the length of the bar is between 2.4974 and 2.4986 inches.

Note that the ruler's low-quality measurement of the length of the bar (2.5 ± 0.03) is consistent with the micrometer's high-quality measurement (2.498 ± 0.0003), in the sense that each one's uncertainty interval overlaps the uncertainty interval of the other.

Consider the relationship between measurement uncertainty and the quality of the measuring instrument for this example. When you used a ruler to measure the

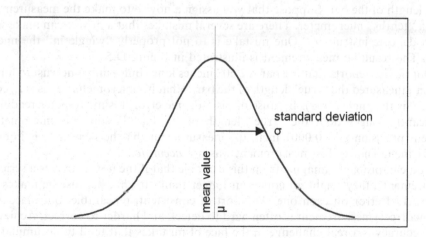

Figure D.4. Many random processes can be described by the "normal distribution".

length of the bar, the relatively coarse ($\frac{1}{16}$th-inch) marks on the ruler established the quality of your result, and you reported the uncertainty of ±0.03 inch based on the ruler's scale marks—the ruler's inherent accuracy. In the case of the "tenths" micrometer, your instrument is assumed to have an inherent accuracy of 0.0001 inch, but your reported measurement quality (based on the standard deviation of 10 measurements) was reported to be ±0.0003 inch. That is quite a bit worse than the inherent capability of the micrometer. This is not an unusual situation. It reflects the fact that there were additional sources of uncertainty in the measurement, beyond the inherent accuracy of the measuring instrument.

So, we've seen that the quality of a measurement can never be much better than the inherent uncertainty of the measuring instrument. It can, however, turn out that the quality of our measurements may be much worse than the accuracy of our instrument, if other sources of error are present (such as the surface roughness, in our example).

D.2   UNCERTAINTY: ACCURACY VS. PRECISION

There is an important distinction to be made in evaluating the quality of a measurement: the difference between *accuracy* and *precision*. These two terms can be roughly explained by considering the ruler vs. the micrometer. *Precision* relates to the number of significant decimal places in the measurement, or the instrument's ability to reliably and repeatedly read out a measurement. As we have seen, with the ruler you are capable of making a reading to about 0.03 inch, whereas with the micrometer you can make a reliable, repeatable reading to 0.0001 inch. These are different levels of *precision*.

Accuracy relates to how closely the measurement matches the underlying "truth". Unfortunately, a very precise reading may be far off the mark—it may not be a very good match to the underlying "truth". Here's an example, based on our measurement of the length of the bar. Suppose that you assign a novice to make the measurement, using a "tenths" micrometer. There are several mistakes that a novice can make with such a delicate instrument. One mistake is to not properly "wiggle in" the micrometer. The resulting measurement is illustrated in Figure D.5.

Our novice reports that the bar is 2.508 inches long. But—unbeknownst to him—he hasn't measured the "true" length of the bar. What his micrometer reads is $L/\cos\theta$ (where θ is the angle error). Because of his 5-degree error, his micrometer reading is significantly longer than the "true" length of the bar. Despite his micrometer's excellent precision of ±0.0001 inch, the measurement that he has made is in error by 0.01 inch. That is, his measurement was not *accurate*.

You can probably imagine from this example that if the novice makes a series of measurements, they might all come out longer than "truth", because he makes the same sort of error on each one. This sort of consistent, repeatable, but inaccurate (i.e., incorrect) measurement is often hard to detect, and harder to correct. Achieving good accuracy is a real challenge in the face of mistakes that tend to accumulate in only one direction. When our novice inspector used the micrometer to measure the

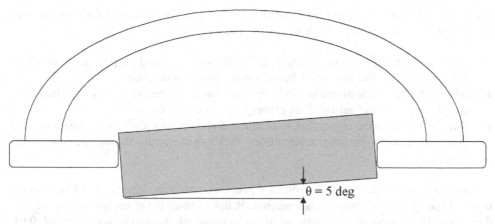

Figure D.5. A measurement procedure that results in repeatable, but incorrect, answers has poor *accuracy*, even though the measuring device is capable of excellent *precision*.

length of the bar, his mistake made the bar appear longer than "truth". He might make a series of measurements in an attempt to "average out" the mistake, but this type of mistake has a one-sidedness to it. The angle error always makes the bar appear longer than it really is. There is no way for the angle error to make the bar appear shorter than it really is. The best you can do is to position your micrometer correctly, so that there is zero angle error, and hence your measurement is accurate. Any angle error makes the bar appear longer—your inaccuracy is always in the same direction.

D.3 BACK TO ASTRONOMY

What does this all have to do with astronomy? Most of the projects described in this book entail some sort of measurement. You measure the time and duration of an occultation, the position of an asteroid, or the brightness of a star. Therefore, you have to be aware of the possible sources of error and the different types of error, so that you can report an estimate of the uncertainty in your measurements.* With this information, other researchers can fairly compare your results with other investigators' measurements. Even though the parameters that you'll measure in your observational research projects are unique to astronomy, the principles for evaluating the quality of your measurements, and estimating your measurement uncertainty, are

---

*A few projects entail counting, rather than measurement. For example, you don't measure the number of meteors per hour, you count them. In general, when you're actually counting something, then you don't report an error. If you counted 12 meteors, then 12 is the number, period.

the same as those used in the machine shop to evaluate the quality of a dimensional measurement, such as the length of a bar. The principles are:

- Take prudent steps to minimize your random errors, and (if possible) calculate your expected random errors based on an appropriate theory.
- Make multiple measurements and use their "internal consistency" as the basis for estimating your actual random errors (i.e., your *precision*).
- And beware of, and search for, systematic errors in your data, your instruments, your procedures, or your assumptions that might impair the *accuracy* of your measurements.

In many photometry and astrometry projects, it may be impractical to calculate *a-priori* the quality of your measurements. Rules of thumb are useful, of course. As described in Section 4.5, if you want to achieve photometric accuracy of 0.01 magnitude, then you need to have a signal-to-noise ratio of at least 100:1. But getting a good SNR is not by itself a guarantee of good photometry! Hence, just as in the case of our measurement of the length of a bar, in your photometric and astrometric projects it is valuable—really, mandatory—that you make multiple measurements and examine their average and standard deviation. The standard deviation is probably a better estimate of your measurement quality than is any *a-priori* calculation or "rule of thumb".

Recall back to the discussion of timing accuracy for occultations, in Chapter 2. If you're using a stopwatch, you are faced with both a *precision* and an *accuracy* challenge. There will be a certain randomness in your measurements, because sometimes you'll be a little more "primed" than at other times. Most likely, these errors will be random, and normally distributed, and might be as small as 0.1 sec (standard deviation). These random errors represent the *precision* of your occultation timing. However, your "reaction time" in an occultation is always a one-sided (systematic) error. You'll always click the stopwatch *after* the event. You'll never have situations where you click the watch now, in response to an event that will happen three seconds in the future! Your "reaction time" is also likely to be consistently larger than the randomness in your individual measurements. Suppose that it takes about 0.2 second for your brain to sense an event (such as a star blinking out) and send a signal to your finger to squeeze the button on the stopwatch. Then, even if you could magically eliminate the randomness in your response, you'd still be left with a consistent, repeatable error, in which you always click the button 0.2 seconds after the event occurs. That is your *accuracy* challenge. If you know in advance that there is likely to be a systematic error in your measurement, then you can do some special experiments to determine the magnitude of the offset, and adjust your reported results to compensate for it, thereby improving your accuracy.

The problem of systematic errors is particularly frustrating, because it can be so hard to imagine how they might creep into your data. Sometimes they are the astronomical equivalent of the butcher who leaves his thumb on the scale while weighing your purchase—the scale gives a precise reading, but it's precisely *wrong*! Hopefully, you can recognize—and prevent—that sort of problem in your data, or in your instruments. Other times, a systematic error can be generated by an inadequate procedure. For example, consider the accuracy of your CCD photometry. Recall the discussion of dust donuts in Section 4.5.4.4. If you fail to properly flat-field your photometric images, the dust donut that sits over your target, but not over your comp star, will create a systematic error in your brightness calculations. If you make multiple images, they will all show essentially the same systematic error, so you won't recognize the problem just by looking at your data—you need to be sure that your *procedure* is correct (i.e., that you do, in fact, flat-field your images), and that you use a flat-frame that is taken through the same filter, and at the same camera orientation, as your science images.

The trickiest type of systematic error is the one that is hidden in your *assumptions*. This sort of error is most often encountered during data analysis and reduction. For example, early estimates of the distance to galaxies were based on the brightness of Cepheid variable stars—basically the same approach that is taken today—but the initial calibration of the Cepheid brightness scale was way off the mark. Why? Because hidden in the data analysis was the assumption that interstellar space was empty, and hence that there was no need to correct for extinction [1]. So, beware of assumptions that seem so reasonable, and so self-evident, that you may not even be conscious of them. They might cause trouble!

In any case, all of your reported measurements should include a realistic, quantitative estimate of the quality of the measurement. In most cases, this will be the standard deviation. In general, the number of decimal places that you use when reporting your results should be appropriate to the precision of your measurement. If you used a ruler to measure the bar, you should report that the bar is 2.50 ± 0.03 inch, not 2.50000 ± 0.03 inch—all those extra zeros are inappropriate!

D.4   REFERENCE

[1]   Trimble, V., "H_0: The Incredible Shrinking Constant 1925–1975", Publications of the Astronomical Society of the Pacific, vol. 108, p. 1073 (December, 1996).

Index